Characterization and Modelling of Composites

Characterization and Modelling of Composites

Editor

Stelios K. Georgantzinos

MDPI • Basel • Beijing • Wuhan • Barcelona • Belgrade • Manchester • Tokyo • Cluj • Tianjin

Editor
Stelios K. Georgantzinos
Department of Aerospace
Science and Technology,
National and Kapodistrian
University of Athens
Greece

Editorial Office
MDPI
St. Alban-Anlage 66
4052 Basel, Switzerland

This is a reprint of articles from the Special Issue published online in the open access journal *Journal of Composites Science* (ISSN 2504-477X) (available at: https://www.mdpi.com/journal/jcs/special_issues/Characterization_Modelling_Composites).

For citation purposes, cite each article independently as indicated on the article page online and as indicated below:

LastName, A.A.; LastName, B.B.; LastName, C.C. Article Title. *Journal Name* **Year**, *Volume Number*, Page Range.

ISBN 978-3-0365-0714-9 (Hbk)
ISBN 978-3-0365-0715-6 (PDF)

Contents

About the Editor

Stelios K. Georgantzinos is currently Assistant Professor in the Department of Aerospace Science and Technology at the National and Kapodistrian University of Athens. Stelios conducts research in the fields of mechanical design, structural analysis, and additive manufacturing, with applications in composite structures, machine elements, and nanotechnology. He has more than 100 publications in refereed journals, conference papers, and book chapters, and he has edited technical reports, textbooks, and laboratory manuals. His research work has been cited more than 1100 times in open literature. He is a member of the Editorial Board of the journal *Mathematical Problems in Engineering* (Hidawi) and a Topic Board member of the journal *Molecules* (MDPI). He has also served as Guest Editor in *Materials* (MDPI) and *Journal of Composites Science* (MDPI). He is a member of the Greek Association of Computational Mechanics (GR.A.C.M.) and the Technical Chamber of Greece. He is also a registered mechanical and electrical engineer in Greece.

Editorial

Editorial for the Special Issue on Characterization and Modelling of Composites

Stelios K. Georgantzinos

Department of Aerospace Science and Technology, National and Kapodistrian University of Athens, 34400 Psachna, Greece; sgeor@uoa.gr

Received: 29 January 2021; Accepted: 31 January 2021; Published: 3 February 2021

The papers published in this Special Issue of the *Journal of Composites Science* will give the composite engineer and scientist insight into what the existing challenges are in the characterization and modelling for the composites field, and how these challenges are being addressed by the research community. The papers present a balance between academic and industrial research, and clearly reflect the collaborative work that exists between the two communities, in a joint effort to solve the existing problems.

Developing an advanced monitoring system for strain measurements on structural components represents a significant task, both in relation to the testing of in-service parameters and in the early identification of structural problems. Arena and Viscardi [1] provide a state-of-the-art review on strain detection techniques in composite structures. The review compares different novel strain measurement techniques. The challenges for the research community are discussed by opening the current scenario to new objectives and industrial applications.

Composites are susceptible to unnoticeable damage as they experience various loading conditions in-service such as fatigue, bird impacts, lightning strikes, etc., which can alter their dynamic characteristics ultimately leading to failure. The formation of cracks in a structure may lead to catastrophic events. Govindasamy et al. [2] present a novel technique called the node-releasing technique in Finite Element Analysis (FEA), whichis used to model the perpendicular cracks as well as slant cracks of various depths and lengths for unidirectional laminate composite layered configurations simulating the actual damage scenario. Furthermore, Saadati et al. [3] studied the interlaminar fracture toughness and delamination behavior of unidirectional flax/epoxy composite under Mode I, Mode II, and Mixed-mode I/II loading.

González and Fernández-León [4] have developed a supervised machine learning model to detect flow disturbances caused by the presence of a dissimilar material region in the liquid molding manufacturing of composites. The machine learning model can predict the position, size and relative permeability of an embedded rectangular dissimilar material region through the use of only the signals corresponding to an array of pressure sensors evenly distributed on the mold surface.

Additive manufacturing (AM) has continued to grow exponentially since its inception for its extensive benefits. Landes and Letcher [5] investigated an additive manufactured composite material that is a greener alternative to other composites that are not reinforced by natural fibers. A bamboo filled polylactic acid (PLA) composite manufactured by fused filament fabrication was evaluated to gather mechanical strength characteristics. Moreover, lattice cell structures can be easily manufacturing via 3D printing and have many scientific and engineering applications. Alwattar and Mian [6] studied the equivalent quasi-isotropic properties required to describe the material behavior of the body-centered cubic (BCC) lattice unit cell. Finite element analysis was used to simulate and calculate the mechanical responses of the BCC unit cell, which were the mechanical responses of the equivalent solid. In addition, cell specimens were fabricated on a fused deposition modeling 3D printer using acrylonitrile butadiene styrene (ABS) material and tested experimentally under quasi-static compression load demonstrating the validity of the proposed method.

Rouhana and Stommel [7] investigated a highly ordered, hexagonal, nacre-like composite stiffness using experiments, simulations, and analytical models. Polystyrene and polyurethane were selected as materials for the manufactured specimens using laser cutting and hand lamination. A simulation was conducted using material data based on component material characterization. Available analytical models were compared to the experimental results, and a more accurate model was derived specifically for highly ordered hexagonal tablets with relatively large in-plane gaps. Additionally, Kriwet and Stommel [8] have used the new developed Arbitrary-Reconsidered-Double-Inclusion (ARDI) model to describe stiffness and damping. A homogenization equation was used to derive the transversal-isotropic stiffness and damping tensors. By rotating and weighting these tensors using orientation distribution functions, it was possible to create a material database.

Singh et al. [9] present a model for the fiber-matrix interface in polymer matrix composites. Finite element models were developed to study the interfacial behavior during the pull-out of a single fiber in continuous fiber-reinforced polymer composites. It was determined that the force required to debond a single fiber from the matrix was three times higher if there was adequate distribution of the sizing on the fiber. It was observed that the interface debonded first from the matrix and remained in contact with the fiber even when the fiber was completely pulled out.

Geopolymer concrete (GPC), due to its capability to minimize the consumption of natural resources, has attracted the attention of researchers. Azarsa and Gupta [10] studied fly-ash based GPC and bottom-ash based GPC, which were exposed to harsh freeze-thaw conditions. The dynamic elastic modulus of both types of GPC was determined by resonant frequency testing. The results showed that bottom-ash based GPC had better freeze-thaw resistance than fly-ash based GPC. Moreover, the leachability of bottom-ash based GPC was also investigated to trace the heavy metals. The results showed that all the heavy metals could be effectively immobilized into the geopolymer paste.

The collection of papers in this issue may help advance technology and bring industry closer to understanding these approaches of the characterization and modelling of composites, and thus being able to confidently implement them into a variety of applications.

Conflicts of Interest: The author declares no conflict of interest.

References

1. Govindasamy, M.; Kamalakannan, G.; Kesavan, C.; Meenashisundaram, G.K. Damage Detection in Glass/Epoxy Laminated Composite Plates Using Modal Curvature for Structural Health Monitoring Applications. *J. Compos. Sci.* **2020**, *4*, 185. [CrossRef]
2. Arena, M.; Viscardi, M. Strain state detection in composite structures: Review and new challenges. *J. Compos. Sci.* **2020**, *4*, 60. [CrossRef]
3. Saadati, Y.; Chatelain, J.F.; Lebrun, G.; Beauchamp, Y.; Bocher, P.; Vanderesse, N. A study of the interlaminar fracture toughness of unidirectional flax/epoxy composites. *J. Compos. Sci.* **2020**, *4*, 66. [CrossRef]
4. González, C.; Fernández-León, J. A Machine Learning Model to Detect Flow Disturbances during Manufacturing of Composites by Liquid Moulding. *J. Compos. Sci.* **2020**, *4*, 71. [CrossRef]
5. Landes, S.; Letcher, T. Mechanical Strength of Bamboo Filled PLA Composite Material in Fused Filament Fabrication. *J. Compos. Sci.* **2020**, *4*, 159. [CrossRef]
6. Alwattar, T.A.; Mian, A. Developing an Equivalent Solid Material Model for BCC Lattice Cell Structures Involving Vertical and Horizontal Struts. *J. Compos. Sci.* **2020**, *4*, 74. [CrossRef]
7. Rouhana, R.; Stommel, M. Modelling and Experimental Investigation of Hexagonal Nacre-Like Structure Stiffness. *J. Compos. Sci.* **2020**, *4*, 91. [CrossRef]
8. Kriwet, A.; Stommel, M. Arbitrary-Reconsidered-Double-Inclusion (ARDI) Model to Describe the Anisotropic, Viscoelastic Stiffness and Damping of Short Fiber-Reinforced Thermoplastics. *J. Compos. Sci.* **2020**, *4*, 37. [CrossRef]

9. Singh, D.K.; Vaidya, A.; Thomas, V.; Theodore, M.; Kore, S.; Vaidya, U. Finite Element Modeling of the Fiber-Matrix Interface in Polymer Composites. *J. Compos. Sci.* **2020**, *4*, 58. [CrossRef]
10. Azarsa, P.; Gupta, R. Freeze-thaw performance characterization and leachability of potassium-based geopolymer concrete. *J. Compos. Sci.* **2020**, *4*, 45. [CrossRef]

Publisher's Note: MDPI stays neutral with regard to jurisdictional claims in published maps and institutional affiliations.

Journal of
Composites Science

Review

Strain State Detection in Composite Structures: Review and New Challenges

Maurizio Arena * and Massimo Viscardi

Department of Industrial Engineering, Aerospace Section, University of Naples "Federico II", Via Claudio 21, 80125 Naples, Italy; massimo.viscardi@unina.it
* Correspondence: maurizio.arena@unina.it

Received: 26 April 2020; Accepted: 17 May 2020; Published: 25 May 2020

Abstract: Developing an advanced monitoring system for strain measurements on structural components represents a significant task, both in relation to testing of in-service parameters and early identification of structural problems. This paper aims to provide a state-of-the-art review on strain detection techniques in composite structures. The review represented a good opportunity for direct comparison of different novel strain measurement techniques. Fibers Bragg grating (FBG) was discussed as well as non-contact techniques together with semiconductor strain gauges (SGs), specifically infrared (IR) thermography and the digital image correlation (DIC) applied in order to detect strain and failure growth during the tests. The challenges of the research community are finally discussed by opening the current scenario to new objectives and industrial applications.

Keywords: characterization; composite; measurements; testing; structural monitoring

1. Preface and Motivation

For the development of lightweight structures, in recent decades, the use of advanced composite materials and modern manufacturing methods has intensified the need to provide effective means of performing experimental investigations to support analytical and numerical analyses. The complexity and high associated costs of full-scale experiments indicate enhancing the reliability of the experimental research to improve precise and effective techniques capable of reducing full-scale testing to subcomponent testing. In the aerospace industry, for instance, several application areas have garnered significant interest. In effect, structural monitoring can improve the characterization and prediction of effects associated with failure that affect the structural safety. In addition, non-destructive techniques can offer fault tolerance to component/subsystem/system-level, thus offering the possibility to reduce costs associated with ordinary/extraordinary maintenance [1]. Chia et al. illustrated an ambitious plan towards developing smart hangar technology, where noncontact measurements would be used to inspect aircraft structures [2]. Rapid inspection also regarded space structures for pre-launch verification using health monitoring technologies [3,4]. Advanced composite materials have encouraged considerable interest in the research community owing primarily to their increased application in both military and general aviation vehicles. Next-generation marine vessels are also adopting orthotropic materials and, for similar reasons, the need for capable monitoring systems is crucial [5]. In this large framework, many different sensing methodologies have thus been deployed for these structural monitoring applications. Many of these methods are based upon strain measurements, which can be detected by conventional gauges or other techniques [6]. Since 1940, the resistance strain gauges (SGs) exemplified the most powerful tool concerning the experimental stress evaluation, even today representing a common choice for monitoring material deformations and damage on in-service composite vehicles. The physical environment of the strain gauge is, in any case, a crucial parameter that has to be considered in gauge selection and protective coating. Owing to their relatively high surface area and the need for protective coatings, electrical resistance strain gauges have found a lot of difficulties in acceptance of their filling into laminated composite materials. A main drawback to such an approach is given by their surface limitation, so a large quantity of them would be necessary to monitor an entire vehicle, yielding a complex network with a lot of wires and cables. For this reason, the strain gauges' location is often optimized for the most critical areas. Optical fibers have been introduced recently just to overcome the weaknesses given by conventional SGs. In many approaches, fiber optic sensors are used to achieve static or quasi-dynamic strain, providing better resolution of up to two to three orders of magnitude compared with conventional SGs [7–9]. Structural monitoring requirements have increased rapidly in the last few decades, and these requirements have prompted many new developments in various sensing technologies. The present article deals with a review of the strain characterization techniques; the technological approach and introduction to experimental techniques are topics illustrated within the following chapters. In this perspective, Fibers Bragg grating (FBG)-based sensors represent of course the most advantageous application to monitor in situ strains over the life of a component, providing more reliable decisions regarding maintenance and replacement of the system. Additionally, semiconductor-based SGs represent another contact transducers category for embedded structural monitoring applications, revealing a high fatigue life, which makes them very attractive for long-term installations. Infrared (IR) thermography and digital image correlation (DIC) represent attractive non-contact techniques offering an interesting full-field investigation of the material response. Their combination could allow a coupled analysis of different specimen aspects from the early stages of testing. A further, very appealing possible integration of the hybrid monitoring system is that which provides the use of non-destructive and non-contact controls such as online monitoring with infrared thermography or DIC in order to highlight possible hot-spots during the tests. The inspection is accomplished in a remote way, avoiding any direct contact, thus preserving the controlled surface from any contamination. Moreover, it can be performed far away from any dangerous environment, safeguarding the safety of the operator. All these techniques are analyzed here and numerous examples are provided for different damage scenarios and aerospace components in order to identify the strength and limitations of each approach. The development of an efficient monitoring system inevitably has repercussions on the life-cycle of the product. Several scientific studies [10] have shown how the proper implementation of structural monitoring may have a positive effect on a structure's life-cycle cost, thus achieving a positive cost/benefit ratio. Commercial insights and industry perceptions pointed out many scenarios where immediate, near-term, and long-term cost savings outweigh the cost of the measurement system, confirming the benefits of its implementation.

2. Contact-Based Techniques

2.1. FBG Sensing Network

2.1.1. Theory Background

One of the alternatives to strain gauge-based sensors for structural monitoring is given by FBG optical sensors. The theory of fiber Bragg gratings may be developed by considering the propagation of modes in an optical fiber. Although guided wave optics is well established, the relationship between the mode and the refractive index perturbation in a Bragg grating plays an important role in the overall efficiency and type of scattering allowed by the symmetry of the problem [11]. Being de facto small and flexible, FBG sensors can be embedded discretely into composites at locations of interest, thereby allowing for the detection of local strain distribution and progression without compromising the structural integrity of the host material. Often, it is mandatory for getting the strain state in many stations of the structure; the need for a dense network of sensors, that is, strain gauges, could lead to a considerable size increase of the test set-up, comprising a large amount of electric wirings and so on. From this perspective, optical FBG sensors are lightweight and versatile, they can be non-intrusively inserted into adhesive materials as well as multiplexed into an array of single FBG sensors. Taking into account the advantages of electromagnetic noise tolerance, high sensitivity, and multiplexing capability, the fibre optic sensing technique has been recognised as a reliable means of measuring structural strain response (static and transient), and FBGs have been successfully developed and employed in structural health monitoring (SHM) for strain, temperature, refractive index, or loads measurements. An FBG is a delimited and systemic discontinuity, imprinted on an optical fibre wire, which reflects different wavelengths. Narrowband reflections can be obtained by deploying discontinuities to allow this system to filter those frequencies [12–17]. FBGs are formed by exposing an optical fiber to an ultraviolet interference pattern, producing a periodic change in the refraction core index. These periodic changes cause a reflection when the light in the waveguide is of a particular wavelength (owing to the constructive interference of grating plane reflections), while other wavelengths are transmitted in the fiber, as shown in Figure 1 [18].

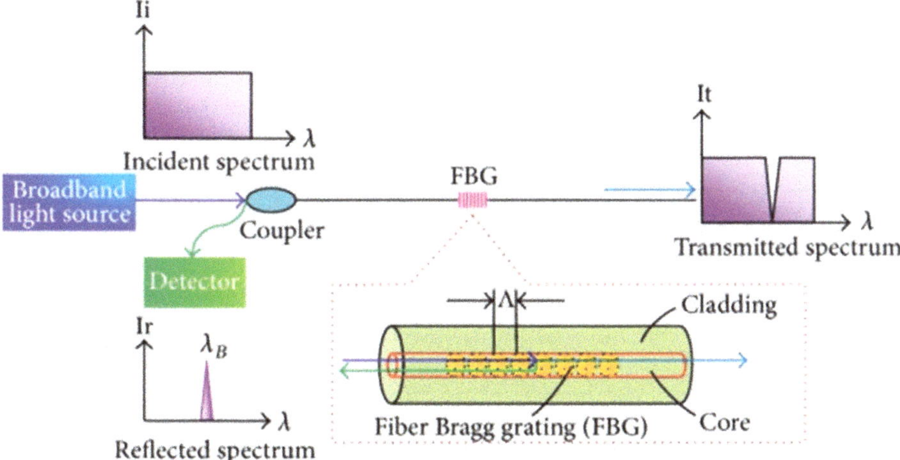

Figure 1. Reflected and transmitted light through a Bragg grating [18].

FBG operation is based on the Fresnel reflection: a broadband light band is diffused into the optical fiber. Once the signal goes through grating, a specific wavelength (i.e., a narrow frequency band, centred at the nominal frequency) is reflected and detected by a spectrometer as well. When fiber is stretched, the gratings time is contracted or expanded and the wavelengths reflected are modulated in turn. The most common way to print gratings inside a glass core fibre is to perform a periodic modulation of the effective refractive index using lasers or UV sources and a suitable method to realize the spatial pattern, such as phase mask. The reflection of light signal by the grating occurs at the Bragg wavelength λ_B. At each plane, light waves are scattered, interfering constructively. Bragg's law defines the requirement for constructive interference from several grating planes, so this narrow energy component is reflected back and missed in the spectrum of transmission. The particular reflected wavelength λ_B is dependent upon the FBG's period of core index modulation Λ and the effective core index of refraction n_0, according to Equation (1) [19].

$$\lambda_B = 2n_0\Lambda \tag{1}$$

The grating's period length varies and the reflected wavelength changes when an FBG is subjected to a local deformation. The typical period of a Bragg grating is about 0.5 μm. For silica core fibres ($\rho_e = 0.22$), the typical strain sensitivity is nearly 1.2 pm/με. A crucial issue is related to embedding optical fiber sensors into laminates; it represents a mandatory step towards the industrialization of the

SHM. In the case of fibre-optics strain sensors, the engineering strain ε could be treated as the ratio between the total wavelength shift $\Delta\lambda$ and the initial wavelength λ, in particular, by Equation (2):

$$\varepsilon = \frac{\Delta\lambda}{\lambda} = \frac{\Delta L}{L_0} \tag{2}$$

where $\Delta L = L - L_0$ is the length change between final L and initial value L_0. Different gage factors hwere estimated, that is, 0.79 for a conventional single mode fibre [20]. When an FBG is strained, it stretches or contracts, causing Λ and n_0 to increase or decrease. This, in turn, produces a differential increase or decrease in λ_B directly related to Λ and n_0, as represented in Equation (3).

$$\delta\lambda_B = \delta(2n_0\Lambda) = 2\Lambda(\delta n_0) + 2n_0(\delta\Lambda)\delta\lambda_B = \delta(2n_0\Lambda) = 2\Lambda(\delta n_0) + 2n_0\delta\Lambda \tag{3}$$

Because λ_B is scalar, the multi-dimensional differential operator, δ, could be expressed by a 1D discrete symbol, Δ, representing changes or shifts. The same for changes in Λ, which, under the assumption of constant strain along the single FBG length (and not necessarily the entire fiber), relates to the 1D linear mechanical strain in the fiber direction, as per Equation (4).

$$\frac{\Delta\lambda_B}{\lambda_B} = \frac{2\Lambda(\delta n_0)}{2n_0\Lambda} + \frac{2n_0(\Delta\Lambda)}{2n_0\Lambda} = \frac{\delta n_0}{n_0} + \frac{\Delta\Lambda}{\Lambda} = \frac{\delta n_0}{n_0} + \varepsilon \tag{4}$$

Equation (4) can be further simplified into a constitutive relationship between strain and index of refraction change using photoelasticity laws. This will make the left side of Equation (4) exclusively a function of measured wavelength and the right side exclusively a function of constant material properties and strain in the fiber direction. A strain-optic tensor is used to relate the index of refraction changes to strain analogous to a stiffness tensor relating stress to strain [21]. The strain-optic tensor components encompassed material properties, that is, photoelastic constants, which correlate index to strain changes. The set of equations collapses to a single expression (analogous to a 1D stress–strain relationship), shown in Equation (5), when the strain field is mainly uniaxial along the fiber grating axis.

$$\frac{\Delta\lambda_B}{(\lambda_B)_0} = (1 - p_e)\Delta\varepsilon \tag{5}$$

where λ_{B0} is the unstrained Bragg wavelength and p_e is a reduced, first order strain-optic coefficient [22]. While the equations determining λ_B are not influenced by additional FBGs written onto the same fiber, the signal processing of multiple gratings must provide each individual FBG's contribution to a combined reflection signal. The wavelength shift changes linearly with both strain and temperature. When the grating part is subjected to external disturbance, the period of the grating will be changed and the Bragg wavelength is varied accordingly. The variation of the Bragg wavelength can be given by Equation (6):

$$\frac{\Delta\lambda_B}{(\lambda_B)_0} = (\alpha + \xi)\Delta T + (1 - p_e)\Delta\varepsilon \tag{6}$$

where ΔT is the temperature change, α is the coefficient of the thermal expansion, and ξ is the thermo-optic coefficient. The authors of [23] reviewed FBG strain sensors with a high focus on the description of physical principles, the interrogation, and the read-out techniques. Their operative performance are reported and compared with the conventional architectures, promoting advanced applications in technological key sectors. The following subparagraph deals with collecting FBGs applications in the field of composite materials characterization.

2.1.2. FBGs Application

Small FBG sensors inserted in a composite layup can provide in situ data on polymer curing (strain, temperature, refractive index). Composite manufacturing processes such as resin transfer molding (RTM) and resin film infusion (RFI) could be really improved by implementing these sensors. Moreover, their application is very attractive in monitoring the composite "health" and impact detection [24]. Among applications in this field, French aeronautics equipment manufacturer Ratier-Figeac, in collaboration with the CEA-LIST (the Atomic Energy Commission, Laboratory for Systems and Technology Integration), developed an advanced measurement system, based on embedded FBG sensor technology in order to get insights into the following: the process sequence, map resin flow by detecting air-to-resin transitions during the injection, checking for dry zones or voids in the structure, and consequently improving the manufacturing quality and reducing development costs. The use of optical-fiber sensors led to supervising the manufacturing airplane propeller blades, as shown in Figure 2 [25].

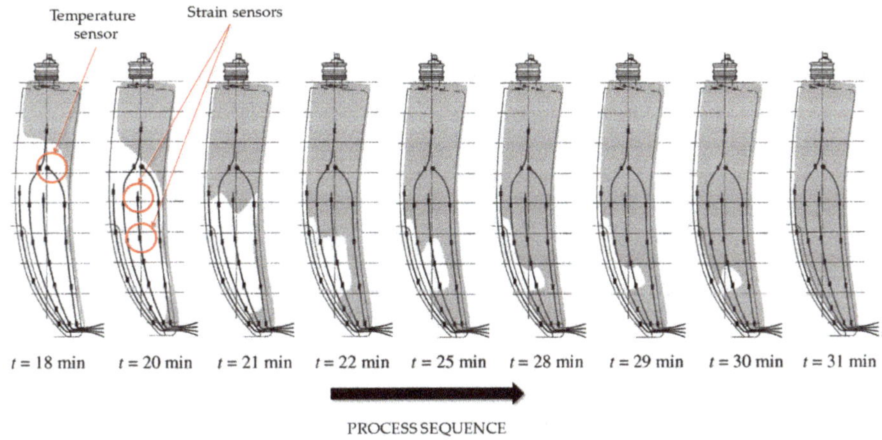

Figure 2. Progression of the resin flow in an airplane propeller blade mold [24,25].

Green and Shafir overviewed the main approaches to overcome this limitation owing to their high fragility, especially when working with a molding process [26]. It is a practical method comprising the addition of a loose tube shield around the optical fiber at the edge of the composite materials specimen, Figure 3. The tube may be realized in polyvinylidene fluoride (PVDF) or polytetrafluoroethylene (PTFE, well-known as Teflon) [27–42]. The study of [43] investigates the influence of strain state distribution on the accuracy of embedded FBGs used as strain sensors with structural loading coaxial to the fiber optic direction. Finite element analysis (FEA) helped for evaluating the fiber optic sensors output, both at far field and in near field areas of the constraining grips. A direct comparison among testing fiber optic strains, strain gauges, and FEA outcomes provided good correlation in the far field, with error of less than 1%. However, in the near field location, some differences found are owing to birefringence arising from complex strain distributions. Some insights are schematized in Figure 4; sensors' installation and the main results are reported. In the work of [44], both analytical and testing investigations on the strain transfer mechanism of surface-attached FBG sensors on composite structures under thermal loading have been performed. Sensitivity analysis indicated that the strain path is mainly affected by the bonded length; on the other side, the strain transfer efficiency is relatively less sensitive to the bonding thicknesses and to the thermal expansion coefficient of the host material. FBGs could be applied to monitor structures exposed to repetitive high strain amplitude dynamic loads: common strain gauges have low fatigue resistance, which causes unreliable strain monitoring in cyclic loading conditions. Aluminum and polypropylene (PP) plates with different thermal expansion coefficients were surface-bonded with FBG sensors and then exposed to various temperatures field (30–70 °C). The proposed error-modification formula can be used to effectively improve the strain measurement accuracy and instruct the optimum design of FBG-based sensors. The work of [45] illustrates the strain monitoring of biaxial glass-fiber reinforced epoxy matrix composites subjected to a constant, high strain uniaxial cyclic loading. The results pointed out that such damage mechanisms observed in fatigue causes FBG strains follow the stiffness degradation trend where damage mechanism occurred; such a strain rate allowed for detecting the specimen onset failure.

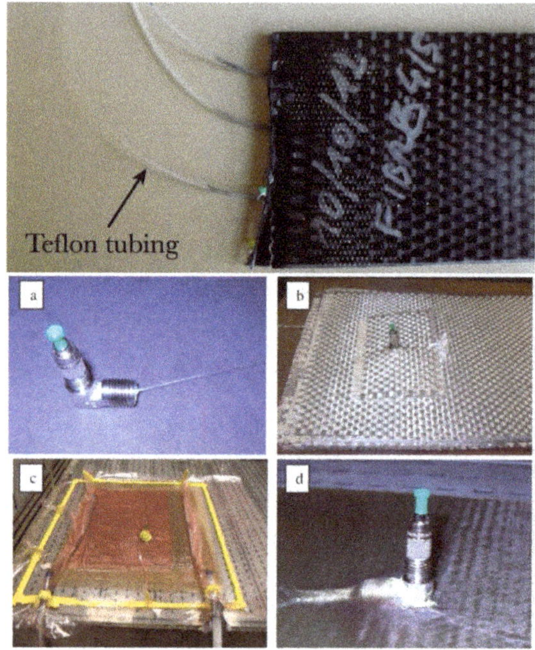

Figure 3. Teflon protective loose tube of optical fiber integration of surface-mounted connector [42].

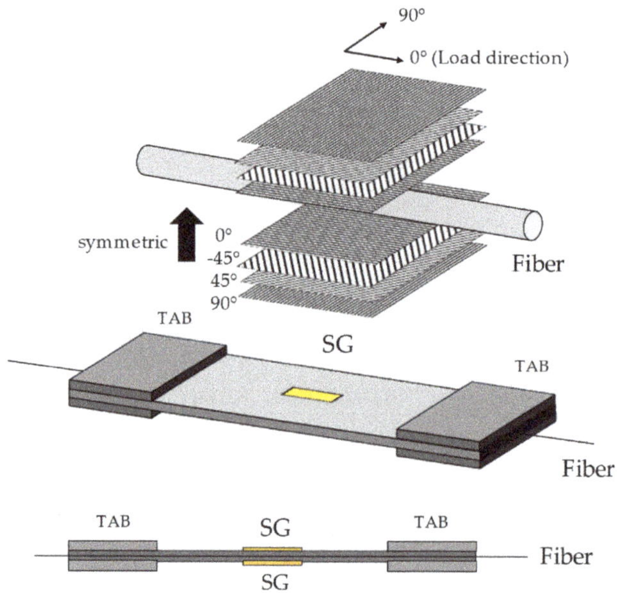

Figure 4. Composite specimen layup and strain trend. FEA, finite element analysis [43].

Another paper [46] presents a cross-correlation function-based method applied for a skin-stringer debonding detection and length estimation. The proposed methodology is applied to investigate the debonding line extension caused by low/medium (45 J) energy impact on 24-ply carbon fiber-reinforced plastic (CFRP) stiffened panels. The results showed a good coherence with respect to the NDI (nondestructive inspection) performed by ultrasonic C-scan flaw detector. Application studies consisted of two different impact events, two different sensor layouts, and two different load conditions after impact (unloaded and quasi-static load, respectively). The main details concerning sensor layout and spatial strain distribution after impact are represented in Figure 5. The aim of the works of [47,48] is to evaluate the feasibility of an efficient hybrid SHM system on a CFRP panel demonstrator, inserting an FBG network within the stack of the preform according to the design requirements. The proposed hybrid monitoring system was successfully applied to different CFRP panels, achieving good results in both low velocity impact tests and guided wave propagation experiments. The signals acquired through the FBG sensors are compared with those gathered from piezo-electrics (PZTs), demonstrating a very satisfactory matching. An experimental configuration is sketched in Figure 6.

PANEL 1

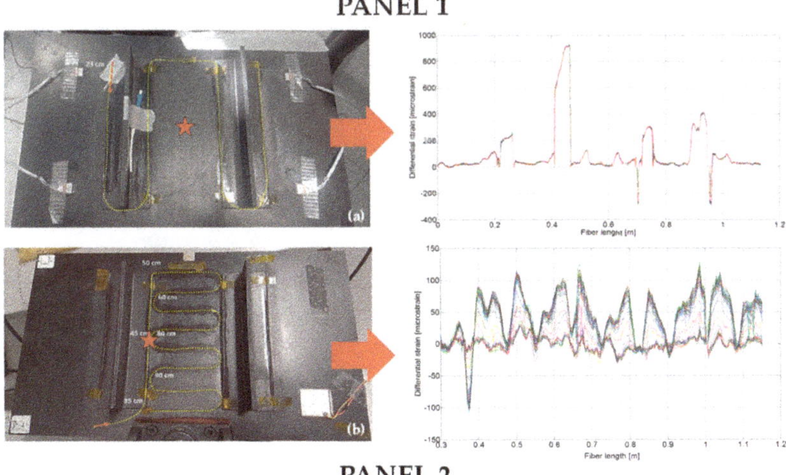

PANEL 2

Figure 5. Composite panels setup and differential strain distribution [46].

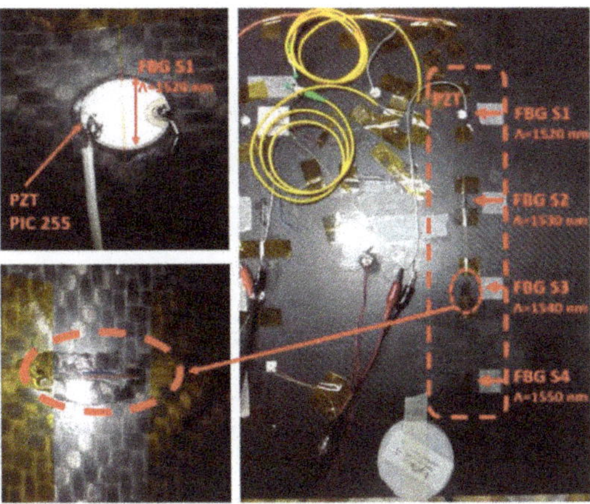

Figure 6. Composite panels setup and differential strain distribution [48].

Singh et al. presents an FBG sensor that may be surface mounted for simultaneous strain and temperature measurements. By inserting an FBG sensor into the laminate, local birefringence causes the bandwidth of the FBG spectrum to change with strain as well as temperature. Thermal–mechanical testing was performed to validate the sensor performance of FBG-composite assembly; the measurement errors, within one standard deviation, were found to be ±62 με and ±1.94 °C for the strain and temperature measurements, respectively [49]. A correlation between strain-gauge and FBG response has been investigated in [50,51] with reference to a composite sample representing a portion of a main landing gear bay for large aircraft in the contest of the ITEMB (Integrated Full Composite Main landing gear Bay Concept) Clean Sky 2 program, Figure 7. The experimental-numerical study [52] assesses the feasibility of the FBG sensor integrated into a composite tube structure. The results demonstrate that the FBG sensor can be successfully used for in-service SHM of the composite tube, showing that this category of sensors can be efficiently used, not only in purely plate-like structures. Looking at more innovative health monitoring tools for the next generation of composite structures, a nanotubes-based strain sensor has been realized in [53]. FBG measurements were performed to prove the feasibility of the sensor, Figure 8; a high correlation level was achieved between FBG response and nanotubes' electrical resistance change.

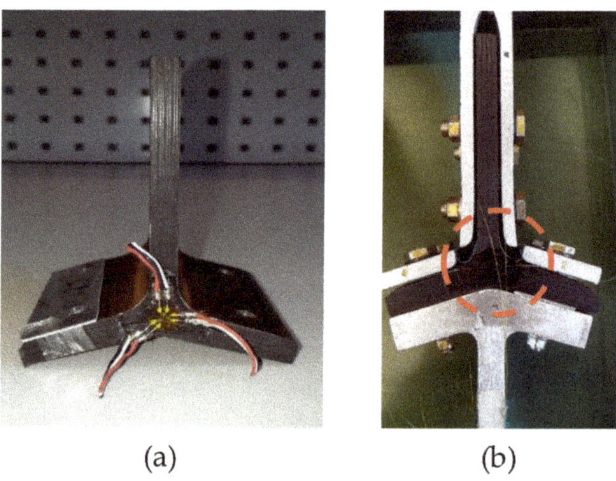

(a)   (b)

Figure 7. Tensile test on composite sample (UniNa laboratory) [50,51]: (**a**) Rosette; (**b**) FBG.

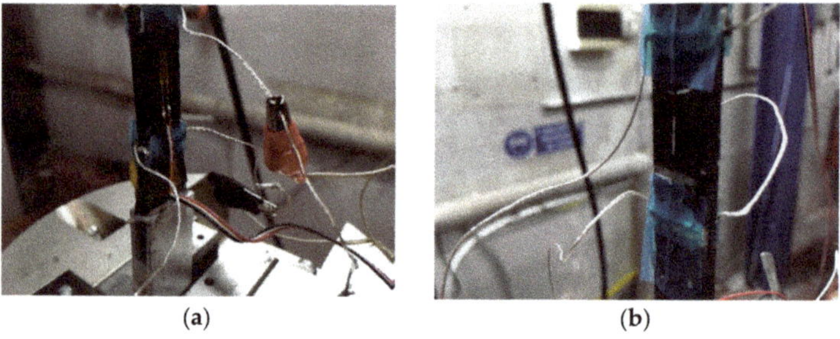

(a)   (b)

Figure 8. Tensile test on nanotubes-based coupon (UniNa laboratory) [53]. (**a**) Strain gauge; (**b**) FBG.

Another current FBG application concerns the debonding detection. Disbonding of the adhesive–adherent interface or cracks in the bond lead to a redistribution of the strain–stress profile in the bond layer. Therefore, research is investigating a method that allows continuously assessing the load distribution in the adhesive bond, thus getting data about the bond quality. The papers of [54,55] present the SHM technique of composite repair patches using small-diameter FBG sensors embedded into the adhesive layers. The debonding length was evaluated quantitatively by the monitoring of the form of the reflection spectrum. The authors of [56] have developed a novel technique to detect debonding in honeycomb sandwich structures using small-diameter chirped FBG sensors between the core and the facesheet during the curing process of the adhesive layer. The work outlines some key insights with reference to numerical modeling of debonding. Debonding can be detected with high sensitivity in real time from the recovery in the shape of the reflection spectrum. The development of a damage detection system using ultrasonic waves is described in [57,58] with reference to the skin/stringer debonding of an aeronautical CFRP panel. A novel damage index was investigated that could be acquired on the basis of the difference in the distribution of the wavelet transform coefficient. In particular, this damage index increased with an extension in the debonded area. Furthermore, it was experimentally verified in [59] that the pre-attachment and curing (PAC) technique has sufficient feasibility for the applications of debonding monitoring. Recent advances and applications of FBG sensors, to the structural health monitoring of composite aircraft structures, have been reviewed in [60]. The performance characteristics and intrinsic limitations of currently available fiber optic-based sensors were outlined deploying to many possible application cases. The authors highlighted that distributed sensing may offer several advantages over quasi-distributed sensors, where large rather than local areas have to be monitored, with high spatial resolution at relatively low frequency. The FBG sensor readings are mostly affected by both thermal and mechanical effects; therefore, these properties should be isolated from one another. The authors of [61] present a hybrid approach to thermally isolate FBGs addressed to cure monitoring of thermoset matrix composites. The results showed good correlation between the model and the experimental method of fibres encased in the glass capillary tube. A very recent application encompasses the embedded FBG as temperature sensors to assess the effects of laminate thickness on the manufacturing quality of composites through cure monitoring performed using microwave curing equipment [62]. FBG-based measurements allowed authors to analyze the impact of temperature overshoot on curing quality as well as the effect of thickness variations and size on the quality of microwave curing, as represented in Figure 9. Fiber optic temperature (FOT) sensors were inserted into the sample to track temperature gradient at each step of the curing process. A futher paper [63] presents the results of numerical simulations with reference to a polymer composite material with an embedded optical fiber, which can be surrounded by a resin pocket, Figure 10. The presented outcomes allowed for estimating the error of the strain values calculated on the basis of 1D stress state assumption of the optical fiber. The relationships between the data measured by the sensor and the strain in the Bragg grating have a unique solution under the condition of uniaxial stress state of the fiber, which are not fulfilled when the fiber is embedded in the material. The work of [64] focuses on implementing an SHM approach to detect structural debonding during the in-service operative life of a composite winglet, Figure 11. The short time fast Fourier transform is applied to acquiring a strain data set in real time during the cyclic load condition. Deviation of structural dynamic response from normal condition is observed, owing to bonding failure in the skin-spar. The layout definition for FBGs' path was selected on the basis of a validated FE model strain distribution. The paper [65] outlines the application study of smart composites embedding FBG sensors, with focus on two main aspects: sensor network has been used for checking the composite resin flow line and residual stress state during the curing process to improve the cure technics. Simultaneously, the embedded FBGs could monitor strain changes and damages in laminates by processing the amplitude and the shape of FBG wavelength. A measurement method using two FBGs is proposed in [66], which allows microdeformations and minimal temperature deviations to be detected with high sensitivity, while the measuring circuit remains quite simple in design and cheap. The transfer characteristic of the sensor is considered, and the optimal range of the sensor is determined to ensure a linear characteristic. A method of damage detection of three-dimensional and six-directional (3Dim 6Dir) braided composites by embedding FBG sensors is proposed in [67]. By analyzing FBG sensors' signals, it is found that different damages and different damage locations at the 3Dim 6Dir braided composites will cause the wavelength of FBG sensor signals at different positions to have unusual changes. This shows the feasibility to detect the damages of 3 Dim braided laminates using FBG sensors.

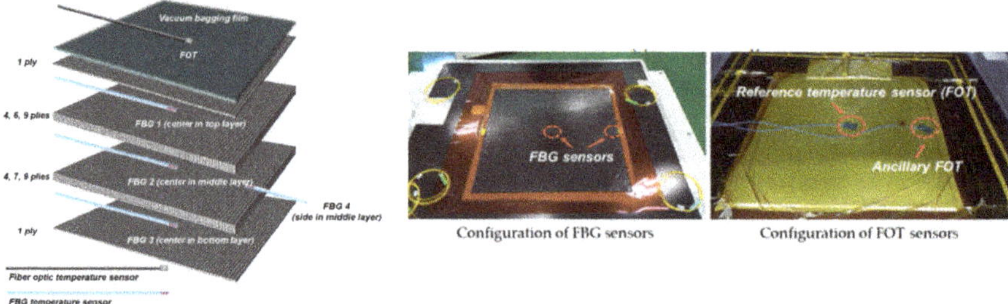

Figure 9. FBG and fiber optic temperature sensor (FOT) sensors configuration [62].

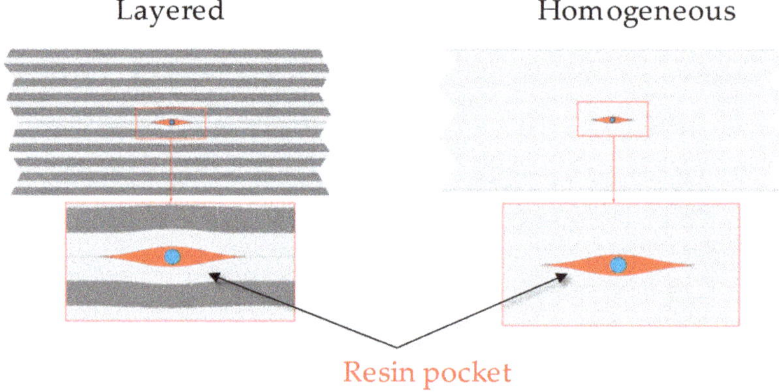

Figure 10. Representation of a polymer composite material with embedded optical fiber [63].

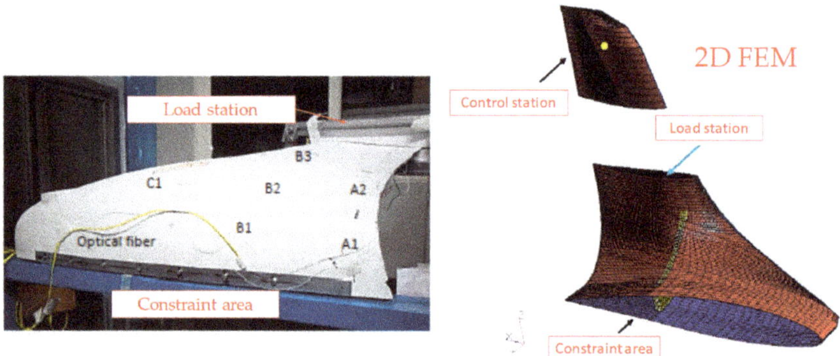

Figure 11. FBGs stations on composite winglet and reference numerical model [64].

For the real-time condition assessment, control, and safety monitoring of morphing aerospace structures, the FBG strain sensing technology represents a novel shape-sensing methodology [68,69]. The deformation of a shape memory polymer (SMP) skin at different temperature conditions is analyzed by FBGs array in order to establish the relationship between the deformation of the skin and pre-strain applied in the SMP skin. The relation of the strain and the deflection of the trailing-edge is established for the shape reconstruction [70]. A morphing wing prototype with polyimide thin film skin is constructed, and forty-eight FBG sensors are glued on the skin surface. The 3D shapes of the polyimide skin at different airfoil profiles are reconstructed. The 3D precise visual measurements are conducted using a digital photogrammetry system, and then the correctness of the shape reconstruction results is verified. The results prove that the maximum error between the 3D visual and FBG measurements is less than 5%, and the FBG sensing method is effective for the shape sensing of flexible morphing wing with polyimide skin [71,72]. In their study, the authors proposed a specially coated FBG, demonstrating its compatibility with in-flight conditions (fatigue, humidity, pressure cycling) of aerospace-grade composite materials [73]. The purpose of [74,75] is to present a conceptual design and modeling of an FBG-based distributed sensor system tailored to measure the span-wise and chord-wise variations of an adaptive trailing edge, thus allowing the complete deformed shape reconstruction, Figure 12.

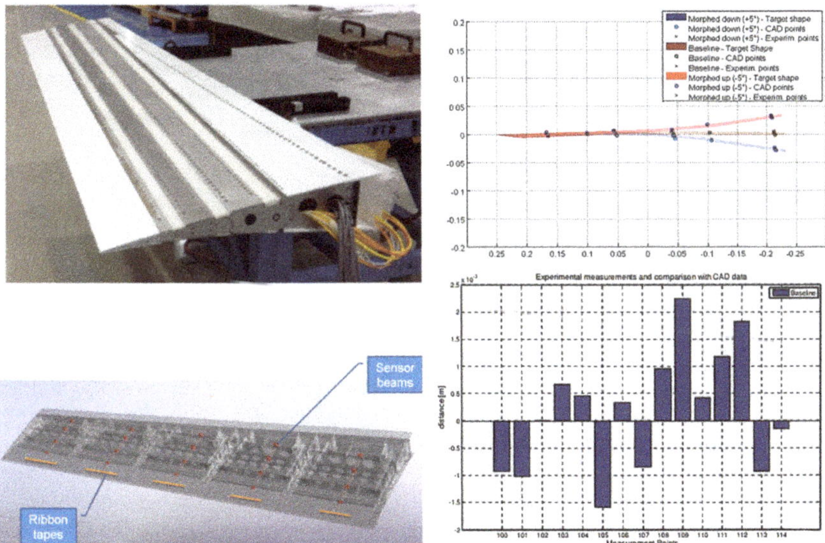

Figure 12. Shape reconstruction of a composite morphing wing segment [75].

2.2. Semiconductor Strain Gauges

The SG works as a bonded sensor; an ohmic resistor R is attached on an insulator I, fixed to a substrate S to be monitored. When S is strained, the length change is observed as ΔR gradient; the accuracy depends mainly on the bonding quality. The resistance of a conductor having length l, cross-sectional area A, and resistivity ρ is expressed by Equation (7):

$$R = \frac{\rho l}{A} \tag{7}$$

The deformation $\Delta l/l$ implies a resistance change according to Equation (8); or, in other terms, owing to a Poisson contraction in A, a resistivity change follows.

$$\frac{\Delta R}{R} = \left(1 + 2v + \frac{1}{\rho}\frac{\Delta\rho}{\Delta l}\right)\Delta l \tag{8}$$

The ratio $(\Delta R/R)/(\Delta l/l)$ represents the gauge factor. The term $(1 + 2v)$, where v is Poisson ratio, is about $1.6 \div 2$ for most metals when there is no resistivity change. Nickel-based alloys are usually used as the strain-sensitive conductor, notably nichrome (nickel-chromium) and constantan (copper-nickel). For backing insulator, instead, epoxy resins and polyamide films are used [76–79]. The application of standard SGs in composite structures is very wide [80–89]. Semiconductor-based SGs represent another category of contact devices for embedded structural monitoring applications. The sensors typically have a thin rectangular form, with a thickness of about 10 mm. The fiber-like form and material similarity to glass fibres make these transducers appealing to composite control. Their proportions are similar to the standard fiber; the length dimension is greater than the width or thickness. The basic principles of semiconductor SGs were described in 1964 by Higson [90]. Semiconductor SGs are much more sensitive than the metallic ones, and commonly used, but specific circuit arrangements are needed when high accuracy is required. The introduction of two new SG alloys provided the perspective of increasing the operative range from 400 °C up to about 1000 °C. The piezoresistive effect in silicon (Si) and germanium (Ge) was first studied by Smith in 1954 [91]. When exposed to an external force, the crystalline configuration of Ge and Si goes towards a reorganization, allowing more mobility to electronic charges, and thus resistivity changing. The gauge factor could be expressed as per Equation (9):

$$G = \frac{\Delta R}{R} / \frac{\Delta l}{l} = 1 + 2v + \frac{l}{\rho}\frac{\Delta \rho}{\Delta l} = 1 + 2v + \pi E \tag{9}$$

where π is the piezoresistive coefficient (m^2/N) and E is the elastic modulus (Pa, psi). In Equation (9), the metal SGs' behavior is explained by $(1 + 2v)$ contribution, but for semiconductor SGs, the piezoresistive parcel, πE, prevails [92]. The semiconductor material commonly silicon exhibits generally high piezoresistivity; the gauge factor can be up to 60 times higher compared with metallic foil gauges [93]. Moreover, they provide the possibility to be used as distributed wireless networks [94]. The linearity error of semiconductor SGs is in the range 10%–20%, while standard gauges have linearity error equal or less of 1%. The following references resume some characteristics of metal and semiconductor SGs [92,95–97]. The high fatigue life is one of the advantages of semiconductor SGs. Thanks to their single-crystal configuration, they do not present hysteresis or creep; in such a way, they are very suitable for long-term installations. High cycle loading of specimens with embedded and surface mounted semiconductor SGs is analysed in [98]. The base material comprised glass fiber laminate. Two different matrixes, that is, polyester and epoxy resins, were used. Hygrothermal loading was applied to weaken the sensor–matrix interface as well as to reproduce operative conditions. The embedded semiconductor SGs' performance was estimated using fatigue testing according to ASTM (American Society for Testing and Materials) E1949 standard [99]. One limit associated with semiconductor SGs is the sensibility to temperature change with a reduction of the signal-to-noise ratio. A possible mitigation could be given by integrating temperature compensation circuitry, commonly used in most commercially available sensors [100].

3. Noncontact-Based Techniques

3.1. Infrared Thermography

Infrared thermography (IRT) is well recognized as a technique for the inspection of composite materials, performed avoiding any surface contact. IR thermography detects surface temperature gradient as a direct function of thermal energy emitted by controlled bodies (in the infrared band of the electromagnetic spectrum) [101–105]. IR thermography is based on the physical evidence that, above the temperature of 0 K, any kind of body represents a thermal radiations source, owing to micromechanical entropy associated with the internal energy of the material. Constitutive components of thermal energy are the photons, generally classified as discrete particles with zero mass, no electric charge, indefinitely long lifetime, and moving in vacuum at the speed of light of $c \cong 3 \times 10^8$ m/s. The photon energy E_{ph} is equal to its frequency f_{ph} multiplied for the Planck's constant (h = 6.6×10^{-34} J s). A general law for such an energy level, according Einstein notation, is given by Equation (10):

$$E_{ph} = h f_{ph} = h\frac{c}{\lambda} \tag{10}$$

which states an inverse proportionality with respect wavelength λ of the considered radiation. To a higher energy level, will correspond a shorter wavelength. In order to express another general law for thermal radiation, it can be helpful to recall the black body concept. Planck's law in Equation (11) describes the spectral distribution of the radiation emitted by a black body:

$$E_{\lambda b} = \frac{2\pi h c^2}{\lambda^5 \left(e^{\frac{hc}{\lambda k_b T}} - 1 \right)} \tag{11}$$

where $E_{\lambda b}$ represents the black body (monochromatic) spectral radiation intensity, T is its absolute temperature (K), and k_b is Boltzmann's constant equal to 1.38×10^{-23} J/K. In particular, by differentiating Equation (11) with respect to λ, it is possible to evaluate, for a specific temperature value, the λ_{max} threshold equivalent to the maximum radiation intensity and given by Wien's displacement law, Equation (12):

$$\lambda_{max} = \frac{d_w}{T} \tag{12}$$

in which d_w is Wien's displacement constant, approximately equal to 2898 μm K. By integrating Planck's law over the whole bandwidth, the total hemispherical radiation intensity emitted by a black body can be obtained as per Equation (13):

$$E_b = \sigma T^4 \tag{13}$$

which is well-known as Stefan Boltzmann's law, where σ is in fact the Stefan Boltzmann constant ($\sigma - 5.67 \times 10^{-8}$ W/m²K⁴). Relying upon the black body concept, an important issue regarding IR measurement is the emissivity ε, which is the ability of surface to emit energy. Values of ε are between 0 (for a perfect reflector) up to 1 (for the black body). IR techniques are deployed to solve low or uneven emissivity concerns, which could perturb acquisitions by undesirable reflections. IR thermography is mostly classified as passive (steady) and active (unsteady) technique [106,107]. The passive approach is commonly applied when materials present a temperature gradient with respect to the environment in which they operate [108]. It could be appropriate for the cyclic loading test to monitor heat variations owing to the resulting hysteretic. On the other side, active IR thermography generates heat transfer in the inspected component by means of external sources such as optical radiation (e.g., halogen heat lamps and laser), electromagnetic actions (induced eddy currents and microwaves), and mechanical ultrasonic waves [109,110]. In active test conditions, an IR camera tracks and analyzes temperature gradient responses at the medium surface to provide details about the structure's integrity. Additionally, the heat response allows for detecting damages or material defects by the thermal waves discontinuities [111]. Thanks to its versatility, active IR tools represent a well-recognized experimental method in terms of reliability and costs, especially if compared with more sophisticated ones such as ultrasonic phased array and X-ray system. Scientific literature now offers a huge amount of information about for several application fields and types of materials/damage [112,113]. The work of [114] is one of the more recent reviews including a deep assessment of active IRT thermography specifically for the aerospace industry. Table 1 summarizes the main active IR methods according to their physical principle and thermal sources [114].

The generic reader will be able to deepen every single technique thanks to the bibliographic references provided. The manuscript will instead flow into an analysis of the applications in the field of composite materials in order to analyse the progress of the scientific community, thus opening the discussion towards new objectives. IRT-based non-destructive evaluation could be performed with different approaches that allow for identifying material defects and reconstructing their position in a 3D reference too. IRT methods have found in fact a large interest for detecting defects within adhesive bonds. Furthermore, IR technology could represent an advanced method for thermo-elastic stress analysis (TSA) purposes, monitoring the surface temperature gradient of a mechanical component. The paragraph is introducing the reader to these two main branches of application.

Table 1. Summary of active infrared thermography (IRT) methods [114].

Physical Principle	Thermal Source	Active IRT Terminology	
Optical radiation	Optical flash, lamp, and electrical heaters	Optically Stimulated Thermography (OST)	Pulsed Thermography [115] (or Flash Thermography)
			Lock-in Thermography [116] (or Amplitude Modulated Thermography)
			Step-Heating Thermography [117] and Long Pulse Thermography [118]
	Optical laser		Frequency Modulated Thermography [119]
			Laser-Spot Thermography [120] and Laser-Line Thermography [121]
Acoustic/ultrasonic wave propagation	Acoustic/ultrasonic horn, piezo-ceramic sensors, air-coupled transducers		Ultrasonic Stimulated Thermography [122] (or Thermosonics, Vibro-Thermography, and Sonic IR Thermography)
			Nonlinear Ultrasonic Stimulated Thermography [123]
Electromagnetic Radiation for dielectric materials	Induced eddy currents		Eddy Current Stimulated Thermography [124]
	Microwaves		Microwave Thermography [125]
Material enabled thermo-resistive radiation for composite materials	Electrical current applied to carbon fibres	Direct Material-Based Thermography (DMT)	Electrical Resistance Change Method (ERCM) coupled to thermography [126]
	Electrical current applied to embedded steel wires		
	Electrical current applied to embedded carbon nanotubes	Indirect Material-Based Thermography (IMT)	Metal-Based Thermography [127]
	Electrical current applied to embedded shape memory alloys wires		Carbon Nanotubes-Based Thermography [128]
			Shape Memory Alloys-Based Thermography [129]

3.1.1. Quality Control: Defects Detection

Thermography methods are not so recent. Infrared imaging had always offered a particularly attractive solution for the inspection of composite material structures [130]. The work of [131] explained practical examples of the remote thermal inspection as a non-destructive method to characterize subsurface features and defects. Avdelidis et al. developed a transient processing technique, useful for the inspection of failure inside different composite configurations representative of various aircraft panels [132]. The study of [133] discusses the development of a novel holistic inspection approach for aviation composite materials comprising three thermographic cameras, which operate at different wavelengths. A large number of different types of defects have been investigated, covering different manufacturing stages. An in-house methodology involving best fitting and normalisation techniques has been implemented to process the thermal images, increasing the results' quality. The works describes the raw thermal image obtained from a CFRP coupon: the non-uniform heating effect has been significantly improved thanks to the proposed processing algorithm. The idealized process demonstrates an increased signal-to-noise ratio, allowing for more indicative mapping results. The study of [134] focuses on the comparison of the results obtained by means of different camera devices: one is FLIR SC5000, while the other one is from the high-end ImageIR series, ImageIR 8300. Both of them belong to the medium wavelength infrared (MWIR) class. Additionally, a long wavelength infrared (LWIR) camera, an ImageIR 8800, was used too. The comparative study was carried out by inspecting three different calibrated and induced defect samples using similar excitation sources, so that the tests configuration and lay out are comparable with each other. A proposed processing to increase signal-to-noise ratio has been assessed. Optical pulsed thermography (OPT) was selected as one of inspection techniques; flash lamps and their corresponding generators were used as heat sources, thus exciting the specimens by heat pulses (close to theoretical Dirac delta signal during three milliseconds). Optical lock-in thermography (OLT) uses halogen lamps, which emitted radiation and can be modulated in both amplitude and frequency. D'Orazio et al. address the issue of developing an automatic signal processing system that performs a time–space denoising in a sequence of thermographic images, thus allowing the identification of undetected defects in composite laminates [135]. The original thermographic images were segmented both using a neural network classifier trained with a backpropagation algorithm and using the self organizing map (SOM) too. Such a method led to a signal-to-noise ratio optimization, recognizing different regions containing several types defects. In the study of [136], three different CFRP specimens having the same thickness and defects characteristics, but with a different shape (planar, trapezoid, and curved) were investigated under laboratory conditions. The qualitative analysis of the three samples indicated that techniques as thermographic signal reconstruction (TSR), pulsed phase thermography (PPT), and principal component thermography (PCT) can improve the perceptibility of the recorded pulsed thermographic signals and additionally enhance the thermal "footprints", which is critical in the case of deep and/or small flaws. De Angelis et al. [137] compared nondestructive systems detecting defects across vulnerable surfaces of a Typhoon air-cooling inlet composite panel. The research of [138] likewise provided a quantitative analysis of artificial defects in a glass fibre reinforced plastic (GFRP) composite plate comparing lock-in thermography (LIT) amplitude and phase images with shearograhy and linear ultrasounds. An inverse analysis based methodology was established to simultaneously identify the thermal diffusivities and subsurface defect of CFRP laminate through lock-in thermographic phase profile reconstruction [139]. Experimental and FE simulation of defective areas were used to assess the thermal diffusivities, recognizing the size and depth of subsurface defect. The outcomes showed a good agreement between estimation and reference values, with a relative deviation of less than 5%. IR thermography has represented an efficient tool during the manufacturing of graphene nanoplatelet (GNP)/polymer composites to measure void content and map void distribution [140]. Considering the size of each pixel (~100 μm), this method enables the NDI of flaws having a size of 200 μm approximately (Figure 13). Their impact on thermal conductivity has been assessed at nanoscale level by 3D FEA. The studies of [141–143] prove that long pulse thermography could forecast and characterize defects in basalt composites in reflection and transmission mode. During the test, both peak contrast derivative method and Parker method were used to determine the defects depth. The first one predicts the defects' depth with a better resolution, independent of their sizes and error percentage increases with increasing depth. The Parker method [144] could be used for estimating the thermal diffusivity if the sample depth is known, for few milliseconds heating time. In the Parker method, in fact, the error percentage is higher for a lesser depth.

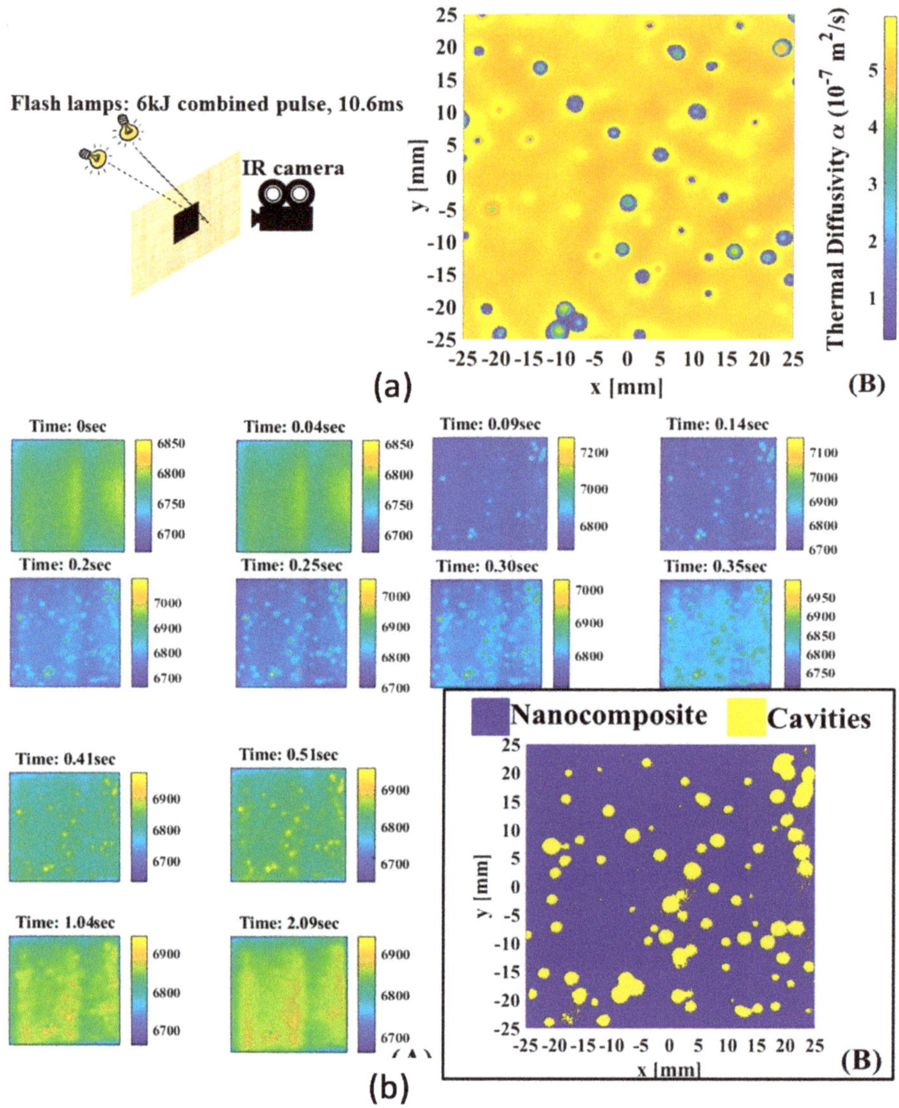

Figure 13. Thermal analysis on graphene nanoplatelet (GNP)/epoxy specimen. (**a**) Thermal diffusivity; (**b**) Temperature fields and void detection; [140].

3.1.2. Structural Qualification: Stress–Strain State Detection

The change in temperature is mostly used to perform thermo-elastic stress analysis tests [145–149]. In the assumption of reversible and adiabatic conditions, thus neglecting heat transfer within the body and to the environment and elastic regime as well, for isotropic materials, the temperature variation can be written as Equation (14):

$$\Delta T = -KT_a\Delta\sigma \tag{14}$$

where T_a is the absolute body temperature, $\Delta\sigma$ is the mean stress amplitude variation, and K is the material thermo-elastic constant. Practically, Equation (2) relates the temperature local variations to the volume variations. In particular, under adiabatic conditions, positive dilatation (tension) entails cooling of the material, and vice versa. In metals, the thermo-elastic limit is generally assumed [149] as an indication for the yielding point. In orthotropic materials, as CFRP, Equation (14) modifies as follows [146]:

$$\Delta T = -\frac{T_a}{\rho c_p}(\alpha_1\Delta\sigma_1 + \alpha_2\Delta\sigma_2) \tag{15}$$

with α_1 and α_2 representing the mean thermal expansion coefficients along the principal material directions. Generally, within the thermo-elastic stress analysis, the infrared imaging device is mainly used for mapping temperature variations under cyclic loading. ΔT depends on several quantities: the material elastic modulus E, the thermoelastic constant K, and some geometrical parameters. Luong [150] discussed on the use of IR thermography for spotting the intrinsic dissipation onset or as a damage indicator. The assumed procedure consisted of acquiring a thermal map after a fixed number of load cycles. The analysis was carried out towards the thermoplastic phase, where the temperature gradient is remarkable and then easy to be measured. A much more difficult concern deals with measurement of thermal variations experienced by the specimen during the elastic phase, that is, thermo-elastic effects. In this context, a common approach is to perform tensile tests with simple standard notched specimens. IRT bases its principle on the thermal energy emitted from bodies in the infrared band of the electromagnetic spectrum [151]. As recently demonstrated by the research group at the University of Naples "Federico II" [152–163], an IR-based imaging device could be really useful to monitor thermal effects evolving on the composite materials' surface when subjected to mechanical loads. Regardless of whether the type of load may be cyclic bending, quasi-static bending, or impact, the infrared camera is able to visualize thermal signatures, which may be related to what happened to the specimen under external load. This is possible because the hot spots that occur on the displayed surface are indicative of damage sites grown in the sample. Examples of images taken during online monitoring are shown in Figure 14a (impact events) and in Figure 14b (quasi-static bending). In particular, Figure 14a shows some thermal images recorded during impact with a modified Charpy pendulum, [159], on GFRP specimens with unidirectional E-glass fibers (300 g/m^2) and low viscosity epoxy resin (MATES® SX10, Italy). More specifically, the images of Figure 14a are ΔT images of specimen impacted at E = 8.3 J, accounting for the temperature rise over the initial ambient temperature. It is worth noting that an abrupt temperature rise is a symptom of fibres breakage, while light temperature variations are mostly linked with delamination. The ΔT images acquired at the lower frame rate by means of SC6000 camera have been additionally post-processed for the evaluation of the warm area extension: the noise correction reference (NCR) has been applied to improve the image quality (Figure 14a on the right). Figure 14b shows some thermal images of a specimen undergoing quasi-static bending at 5 mm/minute; the infrared camera views the specimen from its bottom surface and thickness contemporaneously and records sequences of images at 30 Hz, [163]. It is possible to see the initiation of the damage through the formation of the hot spot on the top right image and its successive enlargement and final collapse as the specimen is bent. Infrared thermography is also effective as a non-destructive testing technique to either ascertain the integrity of a structure before structural testing or detect the occurred damage. The research of [50,51], as already seen previously, provides some relevant outcomes concerning the design of a composite sample for the main landing gear bay of a large commercial airplane within Clean Sky 2 ITEMB project. The innovative method used to forecast the shape and the location of the breakdown showed in the filler of the ITEMB coupons was the acquisition of images from a thermal camera. The thermal camera measurements were useful to understand the crack triggering inside the sample. The frame sequence highlights a compression of the filler, and then its sudden and symmetrical failure, which expands on each side of the coupon, Figure 15. The durability and efficiency of the patch repair during both maintenance (off-line) and service (on-line) conditions were duly investigated by Grammatikos et al. with respect to an aluminium wing structure with CFRP composite patches [164]. The skin-to-core disbonds in a titanium alloy honeycomb structure were analysed in [165]; the numerical predictions by a detailed finite element matched in a very good correlation with experimental evidence. Recent studies have demonstrated the applicability for fatigue test monitoring. In [166], the authors proposed a procedure for tracking the damage of composite materials obtained by automatic fiber placement. Studies were carried out on three specimens of the reference fatigue curve used to evaluate the thermoelastic maximum and minimum limits compared with the extensometer values. The stress/strain redistribution and the consequent stiffness degradation were assessed by following the thermoelastic temperature signal.

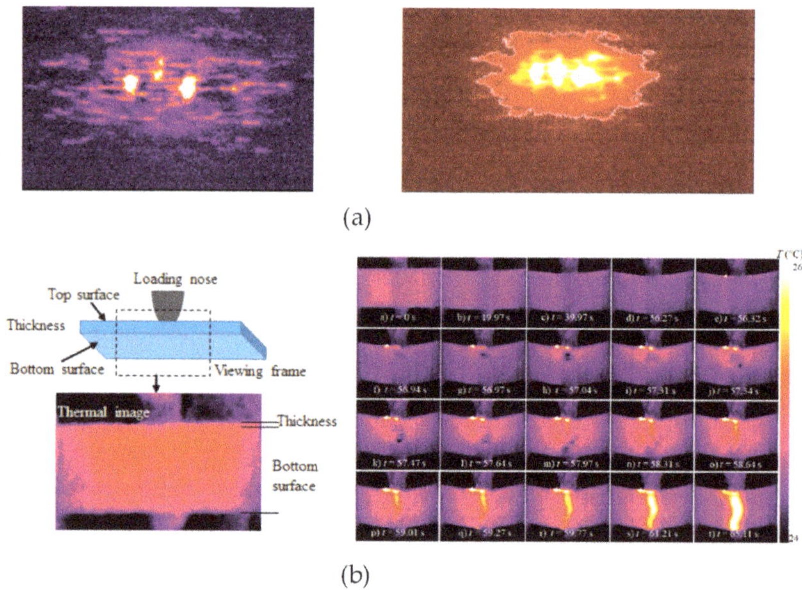

Figure 14. IR online monitoring of rising fiber fractures (UNINA laboratory) [163]. (**a**) Rising fiber fractures when impact occurred; (**b**) quasi-static bending tests.

The authors of [167] evaluated the static and fatigue performance of FRP specimens (a pre-preg IM7/8552 carbon fibre-epoxy and a glass-fibre reinforced epoxy laminates) under Mode II delamination, proposing an innovative approach given by the combined use of a modified transverse crack tension (MTCT) test coupon and IR thermography. The instable crack growth onset under monotonic loading as well as the delamination growth under fatigue cycling have been well achieved by thermographic, thermoelastic, and second harmonic signals. The study of [168] explored the feasibility of pulsed thermography technique for detecting damage owing to heat exposure, that is the case of aero-engine structures. Data indicated that the thermal degradation is mainly dependent on the material properties, environmental condition, and heat source. IR spectra were collected and analyzed in [169] using multivariate analysis techniques with reference to a short beam strength testing (ASTM D2344-16 [170]). The aim of the authors is to develop a physical model for predicting the onset and extent of damage as a function of interlaminar shear strength change. The study of [171] discusses on assessment of ageing for thin thermal barrier coatings (TBCs) by applying active IR thermography. As the material ages and the physical properties, that is, porosity, change over time, it becomes possible to evaluate the remaining useful life of coating ageing.

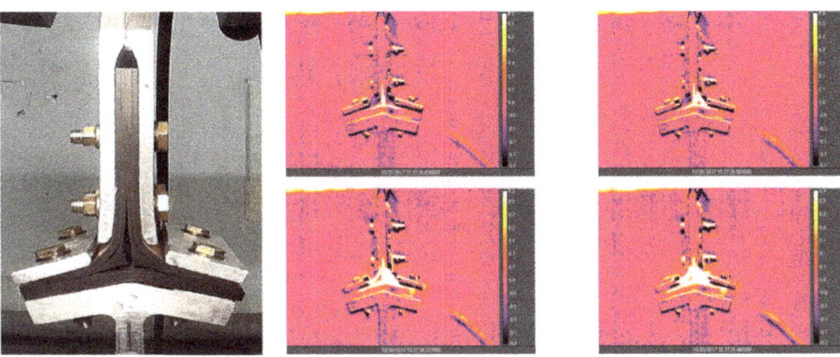

Figure 15. IR online monitoring of sample failure [50,51].

3.2. Digital Image Correlation (DIC)

Developed since the 1980s by a research group from the University of South Carolina [172], DIC is an original non-invasive optical technique for measuring displacements, strain, crack tip, and propagation by comparing digital photographs of a component or test piece at different loading steps [173]. Starting from a reference blocks of pixels, namely the region of interest (ROI), representative of the controlled map, the DIC-based system can quantify surface translation building up 2D and 3D deformation vector fields and strain contours. The surface strains could be determined with a very high spatial resolution, as well as with quite poor pixel resolution of the image. It requires no particular lighting and, in many cases, the clean surface to be controlled already has an acceptable texture for recording, thus there is the need for any special surface preparation. Moreover, images can be acquired from many sources including conventional Charge-Coupled Device (CCD) or commercial digital cameras, high-speed video, up to scanning electron and atomic force microscopes for particularly advanced applications. The DIC has been considered of extreme suitability in the characterization of strain distributions in different load conditions (tensile, torsion, bending, or combined tests). The study of [174] provided a review of surface deformation and strain measurement using 2D DIC. Figure 16 represents a schematic sketch of the 2D DIC windowing set-up with a focus on the reference measurement stage.

Using a stereoscopic transducer setup, the position of each point in the area of interest (Figure 16) is focused on a specific pixel in the camera plane. The average in-plane displacement of the examined surface is determined by mapping shape functions [175] of the image under loading conditions. These shape functions have zero-order for representing rigid body motions with respect xy frame. Therefore, the first-order shape functions are more adequate to represent the subsets subjected to a combination roto-translation, normal and shear strain and expressed in the form of Equation (16):

$$\varsigma_1 = u + \frac{\partial u}{\partial x}\Delta x + \frac{\partial u}{\partial y}\Delta y$$
$$\eta_1 = v + \frac{\partial v}{\partial x}\Delta x + \frac{\partial v}{\partial y}\Delta y \tag{16}$$

In the equations, ς_1, η_1 are the subset total displacements, while u and v are translations. The derivative terms ($\partial u/\partial x$; $\partial v/\partial y$) and ($\partial u/\partial y$; $\partial v/\partial x$) are the normal and shear strains, respectively. Δx and Δy are the incremental distances from the subset centre to an arbitrary point, within the same subset in x- and y-directions, respectively. Many higher-order interpolation schemes have been outlined in the years for the image processing, conceived to provide results with a reliable degree of accuracy [176–180]. Both industrial and scientific interests have been rapidly growing in several engineering fields including composite material characterization and analysis. DIC combined with a stereo camera system allows for reconstructing full-field deformation (in-plane and out-of-plane) and surface strains in many structural testing applications. The DIC can be used to measure 3D surface displacement fields of the specimen during dynamic or static loading [181]. 3D DIC utilizes images taken simultaneously from a pair of digital cameras to track a stochastic pattern on a structure in order to provide full-field displacement and strain information for a structure. This is performed by using the principle of stereo-photogrammetry and bundle adjustment along with the standard two dimensional technique of DIC. In summary, the method of stereo-photogrammetry assumes the relative position of a pair of digital cameras with respect to each other is known through a calibration process and remains constant throughout testing. By imaging a common point with both cameras, the location of that point, with respect to the camera pair, can be triangulated in three-dimensional space. In addition, deformation and strain analysis can be applied to fatigue tests, fracture mechanics, FEA validation, and much more. The potential of the 3D DIC method in measuring strain and displacement of large bodies subjected to quasi-static and cyclic loading is demonstrated by several studies [182–187]. The images taken from a pair of stereo cameras are used to determine surface geometry, deformation, and strain in three dimensions on a part of a blade [188]. The real-time monitoring of the loaded structure then allows to capture both damage onset and growth. DIC provides additional information on damage pattern as crack location and width and composite-to-substrate load transfer mechanism (effective bond length and local stress concentrations). In the work of [189], DIC is used in tensile and bond tests on composite reinforcements comprising different textiles and matrices. The results obtained were then validated by comparison with displacement and strain transducers. The 2D DIC method has been used to investigate superficial strain state of carbon and glass fabric-reinforced composites subjected to low load level tensile and compression test. Artificial through-delamination, inserted into specimens contained, was detected from a strain map by DIC in most cases [190]. The authors of [191,192] explain the use of DIC to monitor the surface strain of composite samples under both tensile and buckling loading conditions and analyse results from both FE models and practical experiments. The paper [193] studied the delaminations' consequences on strain maps, combining FE numerical analyses as a guideline. The well correlated results could be used to extract the orthotropic bending stiffnesses of the damaged and undamaged zones and to assess the interlaminar properties (location, extent, and depth of delamination damage). Always within the ITEMB project, [50,51], the authors implemented the DIC during tensile test for correlating numerical strain distribution, Figure 17.

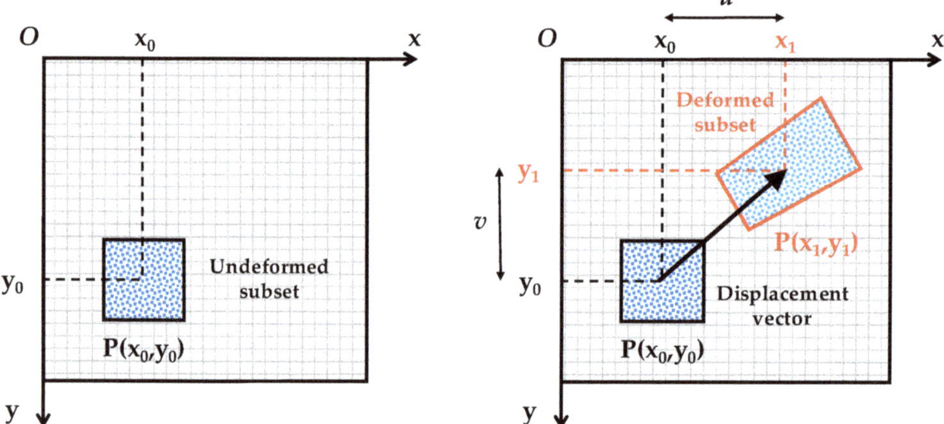

Figure 16. Schematic diagram of the image set-up.

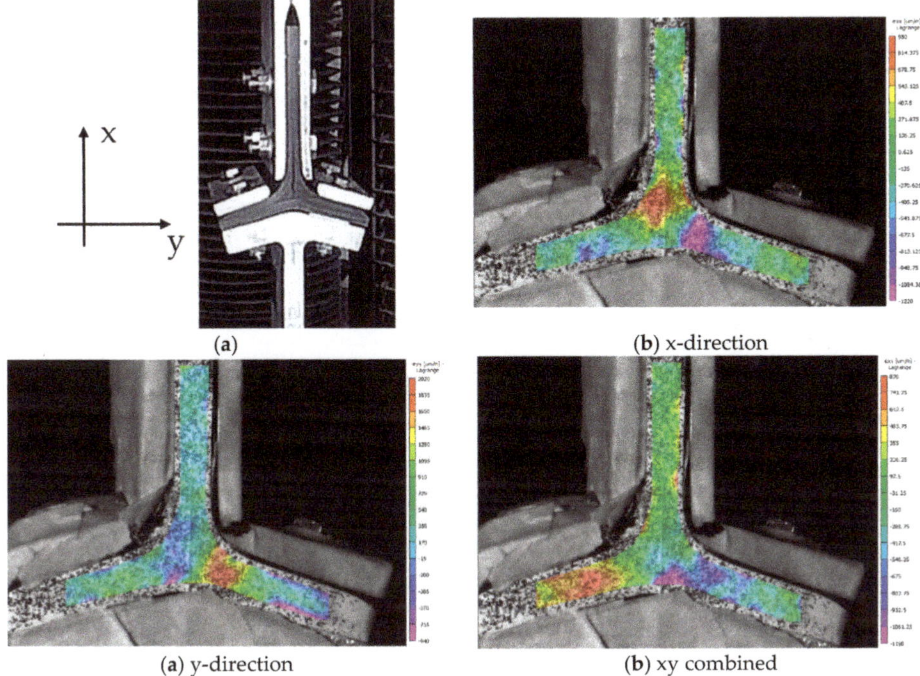

Figure 17. DIC online monitoring of plane strain components [50,51].

The full-field measurements through DIC are used as an efficient support for the experimental investigation on composite structures, as demonstrated in [194] to also fill the limits of standard techniques (LVDT sensor, strain gauges, acoustic emission). The additional evidence provided by DIC permitted to check the boundary conditions of the tests, such as detection of initial misalignment of the specimen in the experimental device, for instance, as well as to improve the damage and failure mechanisms characterization. It allowed for validating static and buckling behaviours predicted by FE simulations. Caminero et al. employed both 2D and 3D DIC techniques to obtain full-field surface strain measurements in notched carbon–fibre/epoxy M21/T700 composite plates. The same authors investigated the stiffness redistribution when adhesively bonded patch repairs are applied: DIC strain results were well correlated with X-ray measurements, providing high-quality information on the location and extent of damage identified [195]. The issue of damage evolution in adhesively bonded patch repair of carbon/epoxy unidirectional composite laminates was faced in [196] too; damage initiation and propagation in notched and repaired panel as well as patch debonding were recorded using 3D DIC. The real strain trend was successfully compared with a 3D finite element analysis. Image stereo-correlation was combined with IR thermography describing the deformation and thermal state of CFRP specimens in conditions of axial and off-axis tensile tests, respectively. The experimental exploration allowed to characterize the influence of the material initial anisotropy on damage growth, localization, and failure mode too [197]. A practical approach based on the evolution of the DIC strain field is proposed by [198] in order to detect local non-linearities leading to a local damage indicator for 3D carbon/epoxy composites. The authors addressed the development of filtering algorithm for post-processing stage too. Considerable key insight for the DIC data analysis has been recently discussed in [199]; a stochastic approach is proposed in order to extract the probabilistic density functions of the composite mechanical properties. Characterizing the mechanical properties of fibres is a crucial issue to acquire more insight about the composite mechanics. Although standardized, the current experimental methods could still lead to a significant error. A method is outlined in [200] for tensile test where direct strain measurement based on DIC eliminates the machine slippage effect at different gauge lengths thanks to the optical monitoring of the fiber deformation. Périé and Passieux led an interesting special issue covering recent developments of DIC based algorithms or new methodologies which could help to reduce the amount of measurement and identification uncertainties, [201]. The article [202] discusses about DIC methodologies to address for the inspection of the in-situ manufacturing of thermoplastic composite materials. Transient kinematic measurements have been performed in order to assess the DIC potentiality to detect gaps and overlaps, as shown in Figure 18. Just for evaluation purposes, two tows were inserted in such a way to intentionally create continuously decreasing gap. The authors found that thanks also to a fine random pattern quality, gaps and overlaps as small as about 0.4 mm size could be really tracked. A DIC-based inverse characterization was proposed in [203] for measurement of following properties with reference to a carbon/epoxy open-hole compression sample: interlaminar stress–strain, interlaminar tensile (ILT) strength, and ILT fatigue life, Figure 19.

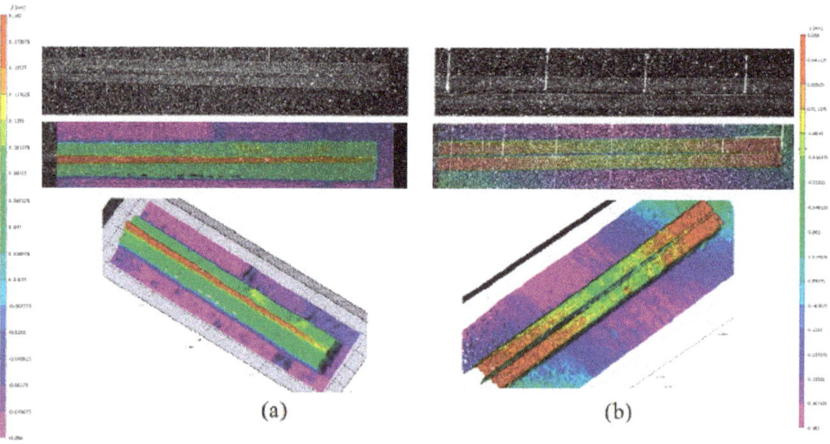

Figure 18. DIC ply inspection. (**a**) Overlap between two tow detections; (**b**) gap detection [202].

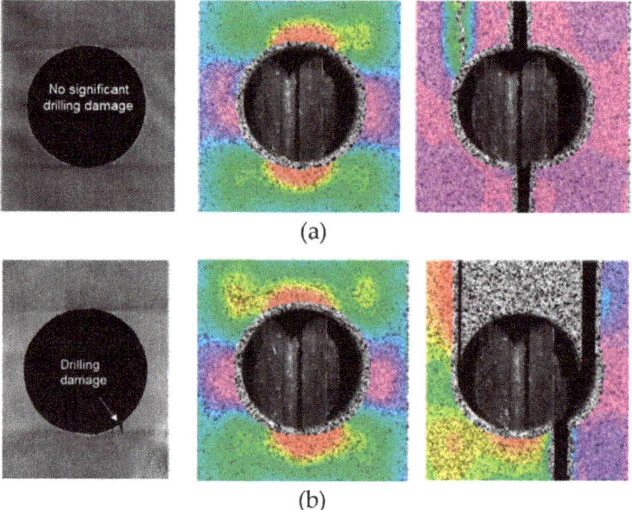

Figure 19. DIC inspection. (**a**) Strain distribution in an acceptable failure mode; (**b**) failure due to drilling imperfection [203].

Direct strain measurement could be advantageous for reducing the amount of measurements and, therefore, of prototypes if compared with indirect techniques. The feasibility has been demonstrated by the authors extending the study to natural fibres, technical flax, and bamboo fibres. 3D high-speed digital photogrammetry was used in [204] to catch displacements and strains distributions following low-velocity impact phenomenon on a composite back surface. Current advances in digital images processing allowed for extending the applicability of DIC to reconstruction of dynamic structural response. The study in [205] presents a comparison among accelerometer, laser vibrometry, and 3D DIC for the vibrations analysis; the results achieved by the three approaches, correlated well with the FEM predictions, demonstrated that the DIC approach could be suitable for full-field modal measurement. The authors of [206] proposed a modal expansion approach for dynamic strain reconstruction on structures containing several components, like rotor based systems, and subjected to complex loading conditions. In this case, time-dependent strains on a wind turbine prototype subjected to arbitrary excitations, that is, hammer and shaker, were successfully extracted, and demonstrated to be in extraordinary agreement with standard strain gages output. Strain measurement by the 2D DIC technique has been used for quantitative in situ investigations of damage nucleation and propagation in a large variety of materials [207,208]. The strain field can be obtained with high precision starting from displacement field. DIC, however, is confined to the surface of specimens. 3D DIC allowed for overcoming the limits of surface detection; tomography by digital volume correlation (DVC) can provide quantitative information of the damage level inside material [209–212]. The elastic strains are generally obtained by scattering techniques such as diffraction; even if widely applied to monolithic materials [213–215], some difficulties can be found into heterogeneous materials (such as graphite) [216–218]. DIC of X-ray radiographs and digital volume correlation of tomographs allows bulk elastic moduli measurement and inspection of the deformation heterogeneity at microstructure level. The authors of [219] studied in situ the mechanical response to tensile and bending loading of polygranular Gilsocarbon nuclear grade near-isotropic graphite (grade IM1-24) by neutron diffraction and synchrotron X-ray diffraction, Figure 20.

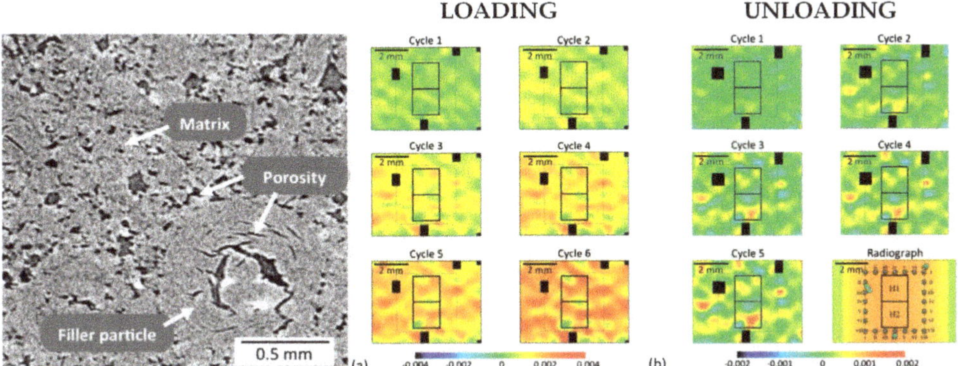

Figure 20. A section of the three-dimensional microstructure, observed by computed tomography and X-ray maps during tensile loading [219].

4. Discussion and Conclusions

The standard methodology for nondestructive characterization of mechanical properties, even if extensively safe in terms of structural load distribution, leads to many drawbacks owing to both the poor reliability for the inspection of damage quality and integrity, but especially to detecting the strain state, especially for composites. The development of different NDT procedures, with a robust methodology for virtual prediction, could lead to an interesting simplification of the whole sub-assembly with a potential advantage in terms of material characterization and cost saving. This paper has presented an overview of most and less recent techniques for the non-destructive evaluation of composite components. For the reliable application of composite materials, strain monitoring provides effective means to understand the material behaviour when loaded. Strain-gauges or strain-gauge-based extensometers have been commonly used sensors for strain monitoring for many ages. However, they have various drawbacks as they are sensitive to electromagnetic fields, have a surface limitation, and do not fit with embedding target inside composite layups. In addition, they have low fatigue resistance, which causes unreliable strain monitoring in fatigue. Taking into account their miniaturized dimensions and relative robustness, FBGs play an interesting role for integration into laminates. FBG transducers can be embedded discretely into laminates' layup, allowing for the analysis of local strain distribution and growing without compromising the structural integrity of the host material. They also have further key attributes such as light weight, multiplexing and absolute measurement capability, and high corrosion resistance, which make them ideal for strain and structural monitoring tasks. They are extremely useful when the structural state over large areas is required, thanks to the possibility of being able to obtain numerous gratings from a single optical wire. Thus, the greatest advantage for the application in composite materials is precisely the ability for meshing a dense sensing network with low structural impact. On the other side, their extreme fragility requires careful and expert installation capability. The authors have presented an overview of semiconductor SGs other possible contact transducers for composite monitoring. The main benefit with respect to traditional SGs is a higher fatigue life, thus fitting with long-term installations. They show high gauge factor, which provides high sensitivity; dynamic strains as small as 0.01 microstrains can be measured. Moreover, a small SG size (<1 mm) makes possible to measure highly localized strains, where a foil metal SG would be too large. Composite structures including various resistors in a Wheatstone bridge can be manufactured, in this case also leading to an embedded transducer configuration. Anyway, semiconductor SGs suffer noteworthy limitations such as non-linearity and strong thermal dependence. Where direct access to the structure is not possible, non-invasive measurement approaches could anyway represent a reliable alternative. As already said, IR thermography can be used in a wide range of applications; there is availability of a wide selection of infrared devices, which differentiate for weight, dimensions, shape, performance, and of course costs to fulfil the needs of users in a huge variety of applications. However, infrared thermography has been used mainly for nondestructive evaluation purposes to discern either manufacturing flaws, or artificially created defects, or impact damage within composite materials. On behalf of its working principle, different techniques have been developed: pulse, long, or step pulse; lock-in; pulse phase; ultrasound lock-in and so on. With regard to the impact damage, IR thermography can also be used to monitor impact tests and acquire information useful for identifying the impact damaging physics. In the recent years, the DIC technique has represented an effective tool for the displacement measurement in experimental solid mechanics. In comparison with the IR methods, the DIC system has been given significant attention as it does not require a rigorous experimental setup as well as a complex optical system. Furthermore, as it requires neither the fringe processing nor the phase analysis, the surface deformation measurement using the DIC method can be carried out effortlessly. The processing algorithms have been thus gradually improved in order to increase their computation efficacy and measurement accuracy. In this perspective, interesting applications of DIC in composite materials deal with the reconstruction of dynamic structural response; many research studies demonstrated that DIC approach could be suitable for full-field modal measurement. The testing methodologies discussed in this review are essential to provide to the transport industry novel highly integrated systems to ensure safety of complex structural configurations. Within this scenario, the main expected impacts of the current research will be as follows:

- to develop integrated sensing technology, which transforms passive composite elements into self-sensing smart structures by embedding FBGs;
- to achieve longer life-cycles and more accurate remnant life predictions for composite components through the use of structural monitoring data;

- to develop new design criteria for developing lighter composite structures;
- to improve the answer to FAA/EASA damage tolerance requirements with higher knowledge and accuracy in inspections;
- In addition to health monitoring purpose, these techniques providing a strong spatial resolution could be extended to dynamic applications even in service. The characterization of the modal parameters is an index of the right stiffness and inertial distribution, thus allowing for further evaluation of the structural integrity.

Author Contributions: Both authors have equally contributed to this review paper. All authors have read and agreed to the published version of the manuscript.

Funding: This research received no external funding.

Acknowledgments: The authors are extremely grateful to the colleagues of the Industrial Engineering Department of University of Naples "Federico II" for their collaboration in research activities concerning the development and application of the described methodologies.

Conflicts of Interest: The authors declare no conflict of interest.

References

1. Srivastava, A.N.; Mah, R.W.; Meyer, C. Integrated Vehicle Health Management. In *Automated Detection, Diagnosis, Prognosis to Enable Mitigation of Adverse Events, NASA Technical Plan 2009, Version 2.03*; NASA Aeronautics Research Mission Directorate Aviation Safety Program: Washington, DC, USA, 2009.

2. Chia, C.C.; Lee, J.-R.; Park, C.-Y.; Jeong, H.-M. Laser ultrasonic anomalous wave propagation imaging method with adjacent wave subtraction: Application to actual damages in composite wing. *Opt. Laser Technol.* **2012**, *44*, 428–440. [CrossRef]

3. Zagrai, A.; Doyle, D.; Arritt, B. Embedded nonlinear ultrasonics for structural health monitoring of satellite joints. In Proceedings of the SPIE 6935, Health Monitoring of Structural and Biological Systems 2008, San Diego, CA, USA, 27 March 2008; p. 6935.

4. Mao, Z.; Todd, M. A structural transmissibility measurements-based approach for system damage detection. In Proceedings of the SPIE 7650, Health Monitoring of Structural and Biological Systems 2010, San Diego, CA, USA, 8 April 2010; p. 7650.

5. Do, R.; Haynes, C.; Todd, M.; Gregory, W.; Key, C. Efficient Detection Methods on a Composite Plate with Interior Embedded Fiber Optic Sensors via Impact Test. In Proceedings of the International Workshop on Structural Health Monitoring, Stanford, CA, USA, 10–12 September 2013.

6. Takeda, N.; Minakuchi, S.; Nadabe, T. Distributed Strain Monitoring for Damage Evolution in CFRP Bolted Structures with Embedded Optical Fibers. *Key Eng. Mater.* **2013**, *558*, 252–259. [CrossRef]

7. Karbhari, V.M.; Kaiser, H.; Navada, R.; Ghosh, K.; Lee, L. *Methods for Detecting Defects in Composite Rehabilitated Concrete Structures*; Oregon Department of Transportation Research Unit: Salem, OR, USA, 2005.

8. Walter, P.L. Strain gage instrumentation. In *Harris' Shock and Vibration Handbook*; Piersol, A.G., Paez, T.L., Eds.; McGraw-Hill Companies Inc.: New York, NY, USA, 2002.

9. Starr, G.E. Basic Strain Gage Characteristic. In *Strain Gage Users' Handbook*; Hannah, R.L., Reed, S.E., Eds.; Chapman & Hall: London, UK, 1994.

10. Radojicic, A.; Bailey, S.; Brühwiler, E. *Consideration of the Serviceability Limit State in a Time Dependant Probabilistic Cost Model, Application of Statistics and Probability*; Balkema: Rotterdam, The Netherlands, 1999.

11. Kashyap, R. *Theory of Fiber Bragg Gratings-Chapter 4, Fiber Bragg Gratings*, 2nd ed.; Academic Press: Cambridge, MA, USA, 2010; pp. 119–187.

12. Othonos, A.; Kalli, K. *Fiber Bragg Gratings: Fundamentals and Applications in Telecommunications and Sensing*; Artech House: New York, NY, USA, 1999.

13. Kersey, A.; Davis, M.A.; Patrick, H.J.; Leblanc, M. Fiber grating sensors. *J. Light. Technol.* **1997**, *15*, 1442–1463. [CrossRef]

14. Luyckx, G.; Voet, E.; Lammens, N.; Degrieck, J. Strain measurements of composite laminates with embedded fibre Bragg gratings: Criticism and opportunities for research. *Sensors* **2011**, *11*, 384–408. [CrossRef] [PubMed]

15. Jensen, D.W.; Sirkis, J.S. Integrity of composite structures with embedded optical fibers. In *Fiber Optic Smart Structures*; Udd, E., Ed.; Wiley: New York, NY, USA, 1995; pp. 109–129.

16. Shivakumar, K.; Emmanwori, L. Mechanics of failure of composite laminates with an embedded fiber optic sensor. *J. Compos. Mater.* **2004**, *38*, 669–680. [CrossRef]

17. Kashyap, R. *Fiber Bragg Gratings*; Academic Press: San Diego, CA, USA, 1999.

18. Ye, X.W.; Su, Y.H.; Han, J.P. Structural Health Monitoring of Civil Infrastructure Using Optical Fiber Sensing Technology: A Comprehensive Review. *Sci. World J.* **2014**, 652329. [CrossRef] [PubMed]

19. Measures, R.M. Structural Monitoring with Fiber Optic Technology. *Meas. Sci. Technol.* **2001**, 12. [CrossRef]

20. Gagliardi, G.; Salza, M.; Avino, S.; Ferraro, P.; Natale, P.D. Probing the Ultimate Limit of Fiber-Optic Strain Sensing. *Science* **2010**, *330*, 1081–1084. [CrossRef]

21. Narasimhamurty, T.S. *Photoelastic and Electro-Optic Properties of Crystals*; Plenum Press: New York, NY, USA, 1981.

22. Othonos, A. Fiber Bragg gratings. *Rev. Sci. Instrum.* **1997**, *68*, 4309–4341. [CrossRef]

23. Campanella, C.E.; Cuccovillo, A.; Campanella, C.; Yurt, A.; Passaro, V.M.N. Fibre Bragg Grating Based Strain Sensors: Review of Technology and Applications. *Sensors* **2018**, *18*, 3115. [CrossRef]

24. Ferdinand, P.; Magne, S.; Dewynter-Marty, V.; Rougeault, S.; Maurin, L. Applications of Fiber Bragg Grating sensors in the composite industry. *MRS Bull.* **2002**, *27*, 400–407. [CrossRef]

25. Dewynter-Marty, V.; Rougeault, S.; Ferdinand, P.; Bugaud, M.; Brion, P.; Marc, G.; Plouvier, P. Contrôle de la fabrication de pieces en matériaux composites réalisées par proceed RTM à l'aide de capteurs à réseaux de Bragg. In Proceedings of the 20th Journées Nationales d'Optique Guidée (JNOG 2000), Toulouse, France, 20–22 November 2000.

26. Green, A.K.; Shafir, E. Termination and connection methods for optical fibres embedded in aerospace composite components. *Smart Mater. Struct.* **1999**, *8*. [CrossRef]

27. Luyckx, G.; Kinet, D.; Lammens, N.; Chah, K.; Caucheteur, C.; Mégret, P.; Degrieck, J. Temperature-insensitive cure cycle monitoring of cross-ply composite laminates using the polarization dependent loss property of FBG. In Proceedings of the 15th European Conference on Composite Materials, Venice, Italy, 24–28 June 2012.

28. Lammens, N.; Kinet, D.; Chah, K.; Luyckx, G.; Caucheteur, C.; Degrieck, J.; Mégret, P. Residual strain monitoring of out-of-autoclave cured parts by use of polarization dependent loss measurements in embedded optical fiber Bragg gratings. *Compos. Part A* **2013**, *52*, 38–44. [CrossRef]

29. Luyckx, G.; Voet, E.; de Waele, W.; Degrieck, J. Multi-axial strain transfer from laminated CFRP composites to embedded Bragg sensor: I. Parametric study. *Smart Mater. Struct.* **2010**, *19*, 105017. [CrossRef]

30. Voet, E.; Luyckx, G.; De Waele, W.; Degrieck, J. Multi-axial strain transfer from laminated CFRP composites to embedded Bragg sensor: II. Experimental validation. *Smart Mater. Struct.* **2010**, *19*, 105018. [CrossRef]

31. Takeda, S.; Okabe, Y.; Takeda, N. Delamination detection in CFRP laminates with embedded small-diameter fiber Bragg grating sensors. *Compos. Part A* **2002**, *33*, 971–980. [CrossRef]

32. Studer, M.; Peters, K.; Botsis, J. Method for determination of crack bridging parameters using long optical fiber Bragg grating sensors. *Compos. Part B* **2003**, *34*, 347–359. [CrossRef]

33. Okabe, Y.; Yashiro, S.; Kosaka, T.; Takeda, N. Detection of transverse cracks in CFRP composites using embedded fiber Bragg grating sensors. *Smart Mater. Struct.* **2000**, *9*. [CrossRef]

34. Kang, D.H.; Park, S.O.; Hong, C.S.; Kim, C.G. The signal characteristics of reflected spectra of fiber Bragg grating sensors with strain gradients and grating lengths. *NDT E Int.* **2005**, *38*, 712–718. [CrossRef]

35. Huang, S.; Ohn, M.M.; LeBlanc, M.; Measures, R.M. Continuous arbitrary strain profile measurements with fiber Bragg gratings. *Smart Mater. Struct.* **1998**, *7*. [CrossRef]

36. Sorensen, L.; Botsis, J.; Gmür, T.; Cugnoni, J. Delamination detection and characterization of bridging tractions using long FBG optical sensors. *Compos. Part A* **2007**, *38*, 2087–2096. [CrossRef]

37. Daggumati, S.; Voet, E.; Van Paepegem, W.; Degrieck, J.; Xu, J.; Lomov, S.V.; Verpoest, I. Local strain in a 5-harness satin weave composite under static tension: Part I—Experimental analysis. *Compos. Sci. Technol.* **2011**, *71*, 1171–1179. [CrossRef]

38. Chiesura, G.; Luyckx, G.; Lammens, N.; van Paepegem, W.; Degriecj, J.; Dierick, M.; van Hoorebeke, L. A micro-computer tomography technique to study the interaction between the composite material and an embedded optical fiber sensor. In Proceedings of the International Conference on Composites Testing and Model Identification, Aalborg, Denmark, 22–24 April 2013.

39. Peairs, D.M.; Sterner, L.; Flanagan, K.; Kochergin, V. Fiber optic monitoring of structural composites using optical backscatter reflectometry. In Proceedings of the 41st International SAMPE Technical Conference, Wichita, KS, USA, 19–22 October 2009.

40. Beukema, R.P. Embedding technologies of FBG sensors in composites: Technologies, applications and practical use. In Proceedings of the 6th European Workshop on Structural Health Monitoring, Dresden, Germany, 3 July 2012.

41. Kinet, D.; Garray, D.; Wuilpart, M.; Dortu, F.; Dusermont, X.; Giannone, D.; Mégret, P. Behaviour of optical fibre Bragg grating sensors embedded into composite material under flexion. In Proceedings of the 14th European Conference on Composite Materials, Budapest, Hungary, 7–10 June 2010; pp. 289–292.

42. Kinet, D.; Mégret, P.; Goossen, K.W.; Qiu, L.; Heider, D.; Caucheteur, C. Fiber Bragg Grating Sensors toward Structural Health Monitoring in Composite Materials: Challenges and Solutions. *Sensors* **2014**, *14*, 7394–7419. [CrossRef]

43. Emmons, M.C.; Karnani, S.; Trono, S.; Mohanchandra, K.P.; Richards, W.L.; Carman, G.P. Strain Measurement Validation of Embedded Fiber Bragg Gratings. *Int. J. Optomechatron.* **2010**, *4*, 22–33. [CrossRef]

44. Wang, H.; Dai, J.-G. Strain transfer analysis of fiber Bragg grating sensor assembled composite structures subjected to thermal loading. *Compos. Part B Eng.* **2019**, *162*, 303–313. [CrossRef]

45. Kocaman, E.S.; Akay, E.; Yılmaz, C.; Türkmen, H.S.; Misirlioglu, I.B.; Suleman, A.; Yildiz, M. Local Strain Monitoring of Fiber Reinforced Composites under High Strain Fatigue using Embedded FBG Sensors. In Proceedings of the 8th European Workshop On Structural Health Monitoring (EWSHM 2016), Spain, Bilbao, 5–8 July 2016.

46. Ciminello, M.; Boffa, N.D.; Concilio, A.; Galasso, B.; Romano, F.; Monaco, E. Damage Detection of CFRP Stiffened Panels by Using Cross-Correlated Spatially Shifted Distributed Strain Sensors. *Appl. Sci.* **2020**, *10*, 2662. [CrossRef]

47. Boffa, N.D.; Monaco, E.; Ricci, F.; Lecce, L.; Barile, M.; Memmolo, V. Hybrid structural health monitoring on a composite plate manufactured with automatic fibers placement including embedded fiber Bragg gratings and bonded piezoelectric patches. In Proceedings of the SPIE, Health Monitoring of Structural and Biological Systems XIII, 109720U, Denver, CO, USA, 1 April 2019; p. 10972.

48. Boffa, N.D.; Monaco, E.; Ricci, F.; Memmolo, V. Hybrid Structural Health Monitoring on composite plates with embedded and secondary bonded Fiber Bragg Gratings arrays and piezoelectric patches. In Proceedings of the 11th International Symposium NDT in Aerospace (AeroNDT 2019), Paris-Saclay, France, 13–15 November 2019.

49. Singh, A.K.; Berggren, S.; Zhu, Y.; Han, M.; Huang, H. Simultaneous strain and temperature measurement using a single fiber Bragg grating embedded in a composite laminate. *Smart Mater. Struct.* **2017**, *26*, 115025. [CrossRef]

50. Viscardi, M.; Arena, M.; Ciminello, M.; Guida, M.; Meola, C.; Cerreta, P. Experimental Technologies Comparison for Strain Measurement of a Composite Main Landing Gear Bay Specimen. In Proceedings of the SPIE, Nondestructive Characterization and Monitoring of Advanced Materials, Aerospace, Civil Infrastructure, and Transportation XII, 105990N, Denver, CO, USA, 27 March 2018.

51. Viscardi, M.; Arena, M.; Cerreta, P.; Iaccarino, P.; Inserra, S.I. Manufacturing and Validation of a Novel Composite Component for Aircraft Main Landing Gear Bay. *J. Mater. Eng. Perform.* **2019**, *28*, 3292–3300. [CrossRef]

52. Chen, Y.C.; Hsieh, C.C.; Lin, C.C. Strain measurement for composite tubes using embedded, fiber Bragg grating sensor. *Sens. Actuators A Phys.* **2011**, *167*, 63–69. [CrossRef]

53. Viscardi, M.; Arena, M.; Barra, G.; Vertuccio, L.; Ciminello, M.; Guadagno, L. Piezoresistive strain sensing of carbon nanotubes-based composite skin for aeronautical morphing structures. In Proceedings of the SPIE, Nondestructive Characterization and Monitoring of Advanced Materials, Aerospace, Civil Infrastructure, and Transportation XII, 105991C, Denver, CO, USA, 27 March 2018.

54. Takeda, S.; Yamamoto, T.; Okabe, Y.; Takeda, N. Debonding monitoring of a composite repair patch using small-diameter FBG sensors. In Proceedings of the SPIE 5390, Smart Structures and Materials 2004: Smart Structures and Integrated Systems, San Diego, CA, USA, 26 July 2004.

55. Takeda, S.; Yamamoto, T.; Okabe, Y.; Takeda, N. Debonding monitoring of a composite repair patch using embedded small-diameter FBG sensors. *Smart Mater. Struct.* **2007**, *16*, 763–770. [CrossRef]

56. Minakuchi, S.; Okabe, Y.; Takeda, N. Real-time Detection of Debonding between Honeycomb Core and Facesheet using a Small-diameter FBG Sensor Embedded in Adhesive Layer. *J. Sandw. Struct. Mater.* **2007**, *9*, 9–33. [CrossRef]

57. Okabe, Y.; Kuwahara, J.; Takeda, N.; Ogisu, T.; Kojima, S.; Komatsuzaki, S. Evaluation of debonding progress in composite bonded structures by ultrasonic wave sensing with fiber Bragg grating sensors. In Proceedings of the SPIE, Advanced Sensor Technologies for Nondestructive Evaluation and Structural Health Monitoring II, 61790G, San Diego, CA, USA, 16 March 2006.

58. Okabe, Y.; Kuwahara, J.; Natori, K.; Takeda, N.; Ogisu, T.; Kojima, S.; Komatsuzaki, S. Evaluation of debonding progress in composite bonded structures using ultrasonic waves received in fiber Bragg grating sensors. *Smart Mater. Struct.* **2007**, *16*, 1370. [CrossRef]

59. Kim, S.; Yoo, S.; Kim, E.; Lee, I.; Kwon, I.; Yoon, D. Embedding techniques of FBG sensors in adhesive layers of composite structures and applications. In Proceedings of the ICCM International Conferences on Composite Materials, Jeju Island, Korea, 21–26 August 2011.

60. Sante, R.D. Fibre Optic Sensors for Structural Health Monitoring of Aircraft Composite Structures: Recent Advances and Applications. *Sensors* **2015**, *15*, 18666–18713. [CrossRef] [PubMed]

61. Boateng, E.K.G.; Schubel, P.; Umer, R. Thermal Isolation of FBG Optical Fibre Sensors for Composite Cure Monitoring. *Sens. Actuators A Phys.* **2019**, *287*, 158–167. [CrossRef]

62. Kim, H.; Kang, D.; Kim, M.; Jung, M.H. Microwave Curing Characteristics of CFRP Composite Depending on Thickness Variation Using FBG Temperature Sensors. *Materials* **2020**, *13*, 1720. [CrossRef] [PubMed]

63. Kosheleva, N.A.; Matveenko, V.; Serovaev, G.; Fedorov, A. Numerical analysis of the strain values obtained by FBG embedded in a composite material using assumptions about uniaxial stress state of the optical fiber and capillary on the Bragg grating. *Frat. Integrità Strutt.* **2019**, *13*, 177–189. [CrossRef]

64. Ciminello, M.; De Fenza, A.; Dimino, I.; Pecora, R. Skin-spar failure detection of a composite winglet using FBG sensors. *Arch. Mech. Eng.* **2017**, *64*, 287–300. [CrossRef]

65. Wang, W.; Song, H.; Xue, J.; Guo, W.; Zheng, M. Overview of intelligent composites with embedded FBG sensors. In Proceedings of the 2011 IEEE 5th International Conference on Cybernetics and Intelligent Systems (CIS), Qingdao, China, 17–19 September 2011; pp. 47–51.

66. Urakseev, M.A.; Vazhdaev, K.V.; Sagadeev, A.R. Differential Fiber Optic Sensor Based on Bragg Gratings. In Proceedings of the 2019 International Multi-Conference on Industrial Engineering and Modern Technologies (FarEastCon), Vladivostok, Russia, 1–4 October 2019; pp. 1–4.

67. Li, P.; Wan, Z. Study on the damages detection of 3 dimensional and 6 directional braided composites using FBG sensor. In Proceedings of the 2019 IEEE Symposium Series on Computational Intelligence (SSCI), Xiamen, China, 6–9 December 2019; pp. 3272–3274.

68. Derkevorkian, A.; Masri, S.F.; Alvarenga, J.; Boussalis, H.; Bakalyar, J.; Richards, W.L. Strain-Based Deformation Shape-Estimation Algorithm for Control and Monitoring Applications. *AIAA J.* **2013**, *51*, 2231–2240. [CrossRef]

69. Pak, C.G. Wing Shape Sensing from Measured Strain. *AIAA J.* **2016**, *54*, 1068–1077. [CrossRef]

70. Yin, W.; Fu, T.; Liu, J.; Leng, J. Structural shape sensing for variable camber wing using FBG sensors. In Proceedings of the SPIE 7292, Sensors and Smart Structures Technologies for Civil, Mechanical, and Aerospace Systems 2009, 72921H, San Diego, CA, USA, 30 March 2009.

71. Sun, G.; Wu, Y.; Li, H.; Zhu, L. 3D shape sensing of flexible morphing wing using fiber Bragg grating sensing method. *Optik* **2018**, *156*, 83–92. [CrossRef]

72. Sun, G.; Li, H.; Dong, M.; Lou, X.; Zhu, L. Optical fiber shape sensing of polyimide skin for a flexible morphing wing. *Appl. Opt.* **2017**, *56*, 9325–9332. [CrossRef] [PubMed]

73. Goossens, S.; De Pauw, B.; Geernaert, T.; Salmanpour, M.S.; Khodaei, Z.S.; Karachalios, E.; Saenz-Castillo, D.; Thienpont, H.; Berghmans, F. Aerospace-grade surface mounted optical fibre strain sensor for structural health monitoring on composite structures evaluated against in-flight conditions. *Smart Mater. Struct.* **2019**, *28*, 065008. [CrossRef]

74. Ciminello, M.; Ameduri, S.; Concilio, A.; Dimino, I.; Bettini, P. Fiber optic shape sensor system for a morphing wing trailing edge. *Smart Struct. Syst.* **2017**, *20*, 441–450.

75. Dimino, I.; Ciminello, M.; Concilio, A.; Gratias, A.; Shueller, M.; Pecora, R. Control system design for a morphing wing trailing edge. In Proceedings of the 7th ECCOMAS thematic conference on smart structures and materials (SMART 2015), Ponta Delgada, Azores, Portugal, 3–6 June 2015.

76. Holister, G.S. *Experimental Stress Analysis, Principles and Methods*; Cambridge University Press: Cambridge, UK, 1967.
77. Window, A.L.; Holister, G.S. (Eds.) *Strain Gauge Technology*; Allied Science Publishers: London, UK, 1982.
78. Keil, S. *Technology and Practical Use of Strain Gages: With Particular Consideration of Stress Analysis Using Strain Gages*, 1st ed.; Ernst & Sohn GmbH & Co. KG.: Berlin, Germany, 2017.
79. Noltingk, B.E. *Chapter 9-Measurement of Strain, Instrumentation Reference Book*, 4th ed.; Butterworth-Heinemann: Oxford, UK, 2010; pp. 93–101.
80. Scowen, G.D. Measurement of surface strain in glass reinforced products. *Strain* **1982**, *18*, 99–104. [CrossRef]
81. Tuttle, M.E.; Brinson, H.F. Resistance-Foil Strain Gage Technology as Applied to Composite Materials. *Exp. Mech.* **1984**, *24*, 54–65. [CrossRef]
82. Perry, C.C. Strain gauge measurement on plastics and composites. *Strain* **1987**, *23*, 155–156. [CrossRef]
83. Whitehead, R.J. *Strain Gage Applications on Composites, Application Note*; Measurement Group Raleigh: Raleigh, NC, USA, 1988.
84. Slaminko, R. Strain Gages on Composites—Gage-Selection Criteria. In *Manual on Experimental Methods for Mechanical Testing of Composites*; Pendleton, R.L., Tuttle, M.E., Eds.; Springer: Berlin/Heidelberg, Germany, 1989.
85. Salzano, T.B.; Calder, C.A.; DeHart, D.W. Embedded-strain-sensor development for composite smart structures. *Exp. Mech.* **1992**, *32*, 225–229. [CrossRef]
86. Tairova, L.P.; Tsvetkov, S.V. Specific features of strain measurement in composite materials. *Mech. Compos. Mater.* **1993**, *28*, 489–493. [CrossRef]
87. Ifju, P.G. Composite materials. In *Handbook of Experimental Solid Mechanics*; Sharpe, W.N., Ed.; Springer: Berlin/Heidelberg, Germany, 2008; pp. 97–124.
88. Kanerva, M.; Antunes, P.; Sarlin, E.; Orell, O.; Jokinen, J.; Wallin, M.; Brander, T.; Vuorinen, J. Direct measurement of residual strains in CFRP-tungsten hybrids using embedded strain gauges. *Mater. Des.* **2017**, *127*, 352–363. [CrossRef]
89. Belhouideg, S.; Lagache, M. Effect of Embedded Strain Gage on the Mechanical Behavior of Composite Structures. *J. Mod. Mater.* **2018**, *5*, 1–7. [CrossRef]
90. Higson, G.R. Recent advances in strain gauges. *J. Sci. Instrum.* **1964**, *41*, 405. [CrossRef]
91. Smith, C.S. Piezoresistance Effect in Germanium and Silicon. *Phys. Rev.* **1954**, *94*, 42. [CrossRef]
92. Cobbold, R.S.C. *Transducers for Biomedical Measurements: Principles and Applications*; Wiley Chap: Hoboken, NJ, USA, 1974.
93. Watson, R.B.; Sharape, W.N. *Springer Handbook of Experimental Solid Mechanics: Bonded Electrical Resistance Strain Gages*; Springer: Berlin/Heidelberg, Germany, 2008.
94. Choi, H.; Choi, S.; Cha, H. Structural Health Monitoring System Based on Strain Gauge Enabled Wireless Sensor Nodes. In Proceedings of the 5th International Conference on Networked Sensing Systems, Kanazawa, Japan, 17–19 June 2008; pp. 211–214.
95. Dally, J.W.; Riley, W.F.; McConnell, K.G. *Instrumentation for Engineering Measurements*, 2nd ed.; John Wiley & Sons: Hoboken, NJ, USA, 1993.
96. Geddes, L.A.; Baker, L.E. *Principles of Applied Biomedical Instrumentation*; John Wiley & Sons: Hoboken, NJ, USA, 1968.
97. Webster, J. (Ed.) Medical instrumentation. In *Application and Design*, 4th ed.; Wiley & Sons, Inc.: Hoboken, NJ, USA, 2010.
98. Herranen, H.; Saar, T.; Gordon, R.; Pohlak, M.; Lend, H. Fatigue Performance of Semiconductor Strain Gauges in GFRP Laminate. *Adv. Mater. Res.* **2014**, *905*, 244–248. [CrossRef]
99. *ASTM E1949-03e1, Standard Test Method for Ambient Temperature Fatigue Life of Metallic Bonded Resistance Strain Gages*; ASTM International: West Conshohocken, PA, USA, 2009.
100. Zhang, J.X.J.; Hoshino, K. *Molecular Sensors and Nanodevices*, 2nd ed.; Academic Press: Cambridge, MA, USA, 2019.
101. Planck, M. *The Theory of Heat Radiation*; P. Blakiston's Son & Co.: Philadelphia, PA, USA, 1914.
102. Maldague, X. *Nondestructive Evaluation of Materials by Infrared Thermography*; Springer: Berlin/Heidelberg, Germany, 1994.
103. Kaplan, H. *Practical Applications of Infrared Thermal Sensing and Imaging Equipment*; SPIE: Bellingham, WA, USA, 1993.

104. Maldague, X. (Ed.) *Infrared Methodology and Technology*; Gordon and Breach: New York, NY, USA, 1994; p. 525.

105. Cuevas, J.C.; García-Vidal, F.J. Radiative heat transfer. *ACS Photonics* **2018**, *5*, 3896–3915. [CrossRef]

106. Ibarra-Castanedo, C.; Galmiche, F.; Darabi, A.; Pilla, M.; Klein, M.; Ziadi, A.; Maldague, X.P. Thermographic nondestructive evaluation: Overview of recent progress. *AeroSense* **2003**, *5073*, 450–459.

107. Balageas, D.; Maldague, X.; Burleigh, D.; Vavilov, V.P.; Oswald-Tranta, B.; Roche, J.M.; Carlomagno, G.M. Thermal (IR) and Other NDT Techniques for Improved Material Inspection. *J. Nondestruct. Eval.* **2016**, *35*, 1–17. [CrossRef]

108. Montesano, J.; Fawaz, Z.; Bougherara, H. Use of infrared thermography to investigate the fatigue behaviour of a carbon fiber reinforced polymer composite. *Compos. Struct.* **2013**, *97*, 76–83. [CrossRef]

109. Bates, D.; Smith, G.; Lu, D.; Hewitt, J. Rapid thermal non-destructive testing of aircraft components. *Compos. Part B Eng.* **2000**, *31*, 175–185. [CrossRef]

110. Vavilov, V.P.; Pawar, S.S. A novel approach for one-sided thermal nondestructive testing of composites by using infrared thermography. *Polym. Test.* **2015**, *44*, 224–233. [CrossRef]

111. Pickering, S.; Almond, D.-P. Matched excitation energy comparison of the pulse and lock-in thermography. *NDT E Int.* **2008**, *41*, 501–509. [CrossRef]

112. Ibarra-Castanedo, C.; Maldague, X.P.V. Infrared Thermography. In *Handbook of Technical Diagnostics*; Czichos, H., Ed.; Springer: Berlin/Heidelberg, Germany, 2013; pp. 175–220.

113. Yang, R.; He, Y. Optically and non-optically excited thermography for composites: A review. *Infrared Phys. Technol.* **2016**, *75*, 26–50. [CrossRef]

114. Ciampa, F.; Mahmoodi, P.; Pinto, F.; Meo, M. Recent Advances in Active Infrared Thermography for Non-Destructive Testing of Aerospace Components. *Sensors* **2018**, *18*, 609. [CrossRef] [PubMed]

115. Maldague, X.P. *Theory and Practice of Infrared Technology for Nondestructive Testing*; John Wiley Interscience: New York, NY, USA, 2001.

116. Bai, W.; Wong, B.S. Nondestructive evaluation of aircraft structure using lockin thermography. In Proceedings of the SPIE's 5th Annual International Symposium on Nondestructive Evaluation and Health Monitoring of Aging Infrastructure, Newport Beach, CA, USA, 6–8 March 2000; International Society for Optics and Photonics: Bellingham, WA, USA, 2000; pp. 37–46.

117. Badghaish, A.A.; Fleming, D.C. Non-destructive inspection of composites using step heating thermography. *J. Compos. Mater.* **2008**, *42*, 1337–1357. [CrossRef]

118. Almond, D.P.; Angioni, S.L.; Pickering, S.G. Long pulse excitation thermographic non-destructive evaluation. *NDT E Int.* **2017**, *87*, 7–14. [CrossRef]

119. Mulaveesala, R.; Tuli, S. Theory of frequency modulated thermal wave imaging for nondestructive subsurface defect detection. *Appl. Phys. Lett.* **2006**, *89*, 191913. [CrossRef]

120. Li, T.; Almond, D.P.; Rees, D.A.S. Crack imaging by scanning pulsed laser spot thermography. *NDT E Int.* **2011**, *44*, 216–225. [CrossRef]

121. Woolard, D.F.; Cramer, K.E. Line scan versus flash thermography: Comparative study on reinforced carbon-carbon. In Proceedings of the SPIE 5782, Thermosense XXVII, Orlando, FL, USA, 28 March 2005; International Society for Optics and Photonics: Bellingham, WA, USA, 2005; Volume 5782, pp. 315–324.

122. Zweschper, T.; Riegert, G.; Dillenz, A.; Busse, G. Ultrasound excited thermography-advances due to frequency modulated elastic waves. *Quant. InfraRed Thermogr. J.* **2005**, *2*, 65–76. [CrossRef]

123. Fierro, G.P.M.; Ginzburg, D.; Ciampa, F.; Meo, M. Imaging of barely visible impact damage on a complex composite stiffened panel using a nonlinear ultrasound stimulated thermography approach. *J. Nondestruct. Eval.* **2017**, *36*, 69. [CrossRef]

124. Wilson, J.; Tian, G.Y.; Abidin, I.Z.; Yang, S.; Almond, D. Modelling and evaluation of eddy current stimulated thermography. *Nondestruct. Test. Eval.* **2010**, *25*, 205–218. [CrossRef]

125. Levesque, P.; Deom, A.; Balageas, D. Non destructive evaluation of absorbing materials using microwave stimulated infrared thermography. In *Review of Progress in Quantitative Nondestructive Evaluation*; Springer: Berlin/Heidelberg, Germany, 1993; pp. 649–654.

126. Sakagami, T.; Ogura, K. A New flaw inspection technique based on infrared thermal images under Joule effect heating. *JSME Int. J. Ser. A* **1994**, *37*, 380–388. [CrossRef]

127. Ahmed, T.; Nino, G.; Bersee, H.; Beukers, A. Heat emitting layers for enhancing NDE of composite structures. *Compos. Part A Appl. Sci. Manuf.* **2008**, *39*, 1025–1036. [CrossRef]

128. De Villoria, R.G.; Yamamoto, N.; Miravete, A.; Wardle, B.L. Multi-physics damage sensing in nano-engineered structural composites. *Nanotechnology* **2011**, *22*, 185502. [CrossRef] [PubMed]

129. Pinto, F.; Ciampa, F.; Meo, M.; Polimeno, U. Multifunctional SMArt composite material for in situ NDT/SHM and de-icing. *Smart Mater. Struct.* **2012**, *21*, 105010. [CrossRef]

130. Jones, T.S. Infrared Thermographic evaluation of marine composite structures. *Nondestruct. Eval. Aging Infrastruct.* **1995**, *2459*, 42–51.

131. Wu, D.; Busse, G. Lock-in thermography for nondestructive evaluation of materials. *Rev. Gen. Therm.* **1998**, *37*, 693–703. [CrossRef]

132. Avdelidis, N.P.; Hawtin, B.C.; Almond, D.P. Transient thermography in the assessment of defects of aircraft composites. *NDT E Int.* **2003**, *36*, 433–439. [CrossRef]

133. Jorge Aldavea, I.; Venegas Bosom, P.; Vega González, L.; López de Santiago, I.; Vollheim, B.; Krausz, L.; Georges, M. Review of thermal imaging systems in composite defect detection. *Infrared Phys. Technol.* **2013**, *61*, 167–175. [CrossRef]

134. Moustakidis, S.; Anagnostis, A.; Karlsson, P.; Hrissagis, K. Non-destructive inspection of aircraft composite materials using triple IR imaging. *IFAC-PapersOnLine* **2016**, *49*, 291–296. [CrossRef]

135. D'Orazio, T.; Leo, M.; Guaragnella, C.; Distante, A. Analysis of Image Sequences for Defect Detection in Composite Materials. *Adv. Concepts Intell. Vision Syst.* **2007**, *4678*, 855–864.

136. Theodorakeas, P.; Avdelidis, N.P.; Ibarra-Castanedo, C.; Koui, M.; Maldague, X. Pulsed thermographic inspection of CFRP structures: Experimental results and image analysis tools. In Proceedings of the SPIE 9062, Smart Sensor Phenomena, Technology, Networks, and Systems Integration, San Diego, CA, USA, 8 March 2014.

137. De Angelis, G.; Dati, E.; Bernabei, M.; Betti, L.; Menner, P. Non Destructive Investigation of disbonding damage in air cooling inlet composite panel using active thermography and shearography. *MAYFEB J. Mater. Sci.* **2016**, *1*, 10–20.

138. Ranjit, S.; Choi, M.; Kim, W. Quantification of defects depth in glass fiber reinforced plastic plate by infrared lock-in thermography. *J. Mech. Sci. Technol.* **2016**, *30*, 1111–1118. [CrossRef]

139. Junyan, L.; Fei, W.; Yang, L.; Yang, W. Inverse methodology for identification the thermal diffusivity and subsurface defect of CFRP composite by lock-in thermographic phase (LITP) profile reconstruction. *Compos. Struct.* **2016**, *138*, 214–226. [CrossRef]

140. Manta, A.; Gresil, M.; Soutis, C. Infrared thermography for void mapping of a graphene/epoxy composite and its full-field thermal simulation. *Fatigue Fract Eng Mater Struct.* **2019**, *42*, 1441–1453. [CrossRef]

141. Kalyanavalli, V.; Ramadhas, T.K.A.; Sastikumar, D. Long pulse thermography investigations of basalt fiber reinforced composite. *NDT E Int.* **2018**, *100*, 84–91. [CrossRef]

142. Kalyanavalli, V.; Ramadhas, T.K.A.; Sastikumar, D. Determination of thermal diffusivity of Basalt fiber reinforced epoxy composite using infrared thermography. *Measurement* **2019**, *134*, 673–678. [CrossRef]

143. Kalyanavalli, V.; Mithun, P.M.; Sastikumar, D. Analysis of long-pulse thermography methods for defect depth prediction in transmission mode. *Meas. Sci. Technol.* **2019**, *31*, 014002. [CrossRef]

144. Parker, W.J.; Jenkins, R.J.; Butler, C.P.; Abbott, G.L. Flash method of determining thermal diffusivity, heat capacity, and thermal conductivity. *J. Appl. Phys.* **1961**, *32*, 1679–1684. [CrossRef]

145. Dulieu-Barton, J.M.; Stanley, P. Development and applications of thermoelastic stress analysis. *J. Strain. Anal.* **1998**, *33*, 93–104. [CrossRef]

146. Stanley, P.; Chan, W.K. The application of thermoelastic stress analysis to composite materials. *J. Strain. Anal.* **1988**, *23*, 137–142. [CrossRef]

147. Galietti, U.; Modugno, D.; Spagnolo, L. A novel signal processing method for TSA applications. *Meas. Sci. Technol.* **2005**, *16*, 2251–2260. [CrossRef]

148. Emery, T.R.; Dulieu-Barton, J.M.; Earl, J.S.; Cunningham, P.R. A generalised approach to the calibration of orthotropic materials for thermoelastic stress analysis. *Compos. Sci. Technol.* **2008**, *68*, 743–752. [CrossRef]

149. Beghi, M.G.; Bottani, C.E.; Caglioti, G. Irreversible thermodynamics of metals understress. *Res. Mech.* **1986**, *19*, 365–379.

150. Luong, M.P. Fatigue limit evaluation of metals using an infrared thermographic technique. *Mech. Mater.* **1998**, *28*, 155–163. [CrossRef]

151. Meola, C.; Carlomagno, G.M. Recent advances in the use of infrared thermography. *Meas. Sci. Technol.* **2004**, *15*, 27–58. [CrossRef]

152. Meola, C.; Carlomagno, G.M. Impact damage in GFRP: New insights with Infrared Thermography. *Compos. Part A Appl. Sci. Manuf.* **2010**, *41*, 1839–1847. [CrossRef]

153. Meola, C.; Carlomagno, G.M. Infrared thermography to evaluate impact damage in glass/epoxy with manufacturing defects. *Int. J. Impact Eng.* **2014**, *67*, 1–11. [CrossRef]

154. Meola, C.; Carlomagno, G.M.; Boccardi, S.; Simeoli, G.; Acierno, D.; Russo, P. Infrared Thermography to Monitor Thermoplastic-matrix Composites Under Load. In Proceedings of the 11th ECNDT, Prague, Czech Republic, 6–10 October 2014.

155. Meola, C.; Boccardi, S.; Carlomagno, G.M. Measurements of very small temperature variations with LWIR QWIP infrared camera. *Infrared Phys. Technol.* **2015**, *72*, 195–203. [CrossRef]

156. Meola, C.; Boccardi, S.; Carlomagno, G.M.; Boffa, N.D.; Monaco, E.; Ricci, F. Nondestructive evaluation of carbon fibre reinforced composites with infrared thermography and ultrasonics. *Compos. Struct.* **2015**, *134*, 845–853. [CrossRef]

157. Boccardi, S.; Carlomagno, G.M.; Meola, C.; Simeoli, G.; Russo, P. Infrared thermography to evaluate thermoplastic composites under bending load. *Compos. Struct.* **2015**, *134*, 900–904. [CrossRef]

158. Boccardi, S.; Carlomagno, G.M.; Meola, C. Basic temperature correction of QWIP cameras in thermo-elastic-plastic tests of composite materials. *Appl. Opt.* **2016**, *55*, 87–94. [CrossRef] [PubMed]

159. Meola, C.; Boccardi, S.; Boffa, N.D.; Ricci, F.; Simeoli, G.; Russo, P.; Carlomagno, G.M. New perspectives on impact damaging of thermoset- and thermoplastic-matrix composites from thermographic images. *Compos. Struct.* **2016**, *152*, 746–754. [CrossRef]

160. Meola, C.; Boccardi, S.; Carlomagno, G.M. *Infrared Thermography in the Evaluation of Aerospace Composite Materials*; Woodhead Publishing Print Book: Amsterdam, The Netherlands, 2016.

161. Bonavolontà, C.; Aramo, C.; Valentino, M.; Pepe, G.P.; De Nicola, S.; Carotenuto, G.; Longo, A.; Palomba, C.M.; Boccardi, S.; Meola, C. Graphene polymer coating for the realization of strain sensors. *Beilstein J. Nanotechnol.* **2017**, *8*, 21–27. [CrossRef] [PubMed]

162. Meola, C.; Boccardi, S.; Carlomagno, G.M.; Boffa, N.D.; Ricci, F.; Simeoli, G.; Russo, P. Impact damaging of composites through online monitoring and non-destructive evaluation with infrared thermography. *NDT E Int.* **2017**, *85*, 34–42. [CrossRef]

163. Meola, C.; Boccardi, S.; Carlomagno, G.M. Infrared Thermography for Inline Monitoring of Glass/Epoxy under Impact and Quasi-Static Bending. *Appl. Sci.* **2018**, *8*, 301. [CrossRef]

164. Grammatikos, S.A.; Kordatos, E.Z.; Matikas, T.E.; Paipetis, A.S. Service and maintenance damage assessment of composite structures using various modes of infrared thermography. In Proceedings of the IOP Conference Series: Materials Science and Engineering, Macau, China, 3–6 August 2015; IoP Publishing: Bristol, UK, 2015; Volume 74, p. 012006.

165. Zhao, H.; Zhou, Z.; Fan, J.; Li, G.; Sun, G. Application of lock-in thermography for the inspection of disbonds in titanium alloy honeycomb sandwich structure. *Infrared Phys. Technol.* **2016**, *81*, 69–78. [CrossRef]

166. De Finis, R.; Palumbo, D.; Galietti, U. Fatigue damage analysis of composite materials using thermography-based techniques. *Procedia Struct. Integr.* **2019**, *18*, 781–791. [CrossRef]

167. Pitarresi, G.; Scalici, T.; Catalanotti, G. Infrared Thermography assisted evaluation of static and fatigue Mode II fracture toughness in FRP composites. *Compos. Struct.* **2019**, *226*, 111220. [CrossRef]

168. Addepalli, S.; Zhao, Y.; Roy, R.; Galhenege, W.; Colle, M.; Yu, J.; Ucur, A. Non-destructive evaluation of localised heat damage occurring in carbon composites using thermography and thermal diffusivity measurement. *Measurement* **2019**, *131*, 706–713. [CrossRef]

169. Toivola, R.; Afkhami, F.; Baker, S.; McClure, J.; Flinn, B.D. Detection of incipient thermal damage in carbon fiber-bismaleimide composites using hand-held FTIR. *Polym. Test.* **2018**, *69*, 490–498. [CrossRef]

170. ASTM D2344/D2344M-16. *Standard Test Method for Short-Beam Strength of Polymer Matrix Composite Materials and Their Laminates*; ASTM International: West Conshohocken, PA, USA, 2016.

171. Tinsley, L.; Chalk, C.; Nicholls, J.; Mehnen, J.; Roy, R. A study of pulsed thermography for life assessment of thin EB-PVD TBCs undergoing oxidation ageing. *NDT E Int.* **2017**, *92*, 67–74. [CrossRef]

172. Pan, B.; Qian, K.; Xie, H.; Asundi, A. Two-dimensional digital image correlation for in-plane displacement and strain measurement: A review. *Meas. Sci. Technol.* **2009**, *20*, 062001. [CrossRef]

173. Sutton, M.; Orteu, J.J.; Schreier, H. *Image Correlation for Shape, Motion and Deformation Measurements*; Springer: Berlin/Heidelberg, Germany, 2009.

174. Khoo, S.W.; Karuppanan, S.; Tan, C.S. A review of surface deformation and strain measurement using two-dimensional digital image correlation. *Metrol. Meas. Syst.* **2016**, *23*, 461–480. [CrossRef]

175. Sutton, M.A.; Orteu, J.J.; Schreier, H.W. *Image Correlation for Shape, Motion and Deformation Measurements: Basic Concepts, Theory and Applications*; Springer: Berlin/Heidelberg, Germany, 2009; p. 88.

176. Sutton, M.A.; Wolters, W.J.; Peters, W.H.; Ranson, W.F.; McNeill, S.R. Determination of displacements using an improved digital image correlation method. *Image Vis. Comput* **1983**, *1*, 133–139. [CrossRef]

177. Chu, T.C.; Ranson, W.F.; Sutton, M.A.; Peters, W.H. Applications of digital image correlation techniques to experimental mechanics. *Exp. Mech.* **1985**, *25*, 232–244. [CrossRef]

178. Bruck, H.A.; McNeill, S.R.; Sutton, M.A.; Peters, W.H. Digital image correlation using Newton- Raphson method of Partial Differential Correction. *Exp. Mech.* **1989**, *29*, 261–267. [CrossRef]

179. Lu, H.; Cary, P.D. Deformation measurements by digital image correlation: Implementation of a second-order displacement gradient. *Exp. Mech.* **2000**, *40*, 393–400. [CrossRef]

180. Cheng, P.; Sutton, M.A.; Schreier, H.W.; McNeill, S.R. Full-field speckle pattern image correlation with B-Spline deformation function. *Exp. Mech.* **2002**, *42*, 344–352. [CrossRef]

181. LeBlanc, B.; Niezrecki, C.; Avitabile, P.; Sherwood, J.; Chen, J. Surface Stitching of a Wind Turbine Blade Using Digital Image Correlation. In Proceedings of the IMAC-XXX, Jacksonville, FL, USA, 30 January–2 February 2012.

182. LeBlanc, B.; Niezrecki, C.; Hughes, S.; Avitabile, P.; Chen, J.; Sherwood, J. Full Field Inspection of a Wind Turbine Blade Using 3D Digital Image Correlation. In Proceedings of the SPIE Symposium on Smart Structures & Materials/NDE for Health Monitoring, San Diego, CA, USA, 7–11 March 2011.

183. Leblanc, B. Non-Destructive Testing of Wind Turbine Blades with Three Dimensional Digital Image Correlation. Master's Thesis, University of Massachusetts Lowell, Lowell, MA, USA, 2011.

184. Grédiac, M. The use of full-field measurement methods in composite material characterization: Interest and limitations. *Compos. Part A Appl. Sci. Manuf.* **2004**, *35*, 751–761. [CrossRef]

185. Puri, A.; Dear, J.P.; Morris, A.; Jensen, F.M. Analysis of wind turbine material using digital image correlation. In Proceedings of the XIth International Congress and Exposition, Orlando, FL, USA, 2–5 June 2008; pp. 2126–2133.

186. Carr, J.; Baqersad, J.; Niezrecki, C.; Avitabile, P.; Slattery, M. Dynamic stress-strain on turbine blades using digital image correlation techniques Part 2: Dynamic measurements. In *Topics in Experimental Dynamics Substructuring and Wind Turbine Dynamics*; Springer: Berlin/Heidelberg, Germany, 2012; pp. 221–226.

187. Johnson, J.T.; Hughes, S.; van Dam, J. A stereo-videogrammetry system for monitoring wind turbine blade surfaces during structural testing. *ASME Early Career Tech. J.* **2009**, *8*, 1–10.

188. Poozesh, P.; Baqersad, J.; Niezrecki, C. *Full Field Inspection of a Utility Scale Wind Turbine Blade Using Digital Image Correlation*; CAMX: Orlando, FL, USA, 2014.

189. Tekieli, M.; De Santis, S.; de Felice, G.; Kwiecień, A.; Roscini, F. Application of Digital Image Correlation to composite reinforcements testing. *Compos. Struct.* **2017**, *160*, 670–688. [CrossRef]

190. Szebényi, G.; Hliva, V. Detection of Delamination in Polymer Composites by Digital Image Correlation—Experimental Test. *Polymers* **2019**, *11*, 523. [CrossRef] [PubMed]

191. Bataxi, C.X.; Yu, Z.; Wang, H.; Bil, C. Strain monitoring on damaged composite laminates using digital image correlation. *Procedia Eng.* **2015**, *99*, 353–360. [CrossRef]

192. Bataxi; Yu, Z.; Wang, H. Surface strain analysis of composite laminate with impact damage. In Proceedings of the 17th National Conference on Composite Materials, Beijing, China, 12–15 October 2012.

193. Devivier, C.; Thompson, D.; Pierron, F.; Wisnom, M.R. Correlation between full-field measurements and numerical simulation results for multiple delamination composite specimens in bending. *Appl. Mech. Mater.* **2010**, *24–25*, 109–114. [CrossRef]

194. Laurin, F.; Charrier, J.-S.; Lévêque, D.; Maire, J.-F.; Mavel, A.; Nuñez, P. Determination of the properties of composite materials thanks to digital image correlation measurements. *Procedia IUTAM* **2012**, *4*, 106–115. [CrossRef]

195. Caminero, M.; Lopez-Pedrosa, M.; Pinna, C.; Soutis, C. Damage monitoring and analysis of composite laminates with an open hole and adhesively bonded repairs using digital image correlation. *Compos. Part B Eng.* **2013**, *53*, 76–91. [CrossRef]

196. Kashfuddoja, M.; Ramji, M. Whole-field strain analysis and damage assessment of adhesively bonded patch repair of CFRP laminates using 3D-DIC and FEA. *Compos. Part B Eng.* **2013**, *53*, 46–61. [CrossRef]

197. Goidescu, C.; Welemane, H.; Garnier, C.; Fazzini, M.; Brault, R.; Péronnet, E.; Mistou, S. Damage investigation in CFRP composites using full-field measurement techniques: Combination of digital image stereo-correlation, infrared thermography and X-ray tomography. *Compos. Part B Eng.* **2013**, *48*, 95–105. [CrossRef]

198. Feissel, P.; Schneider, J.; Aboura, Z.; Villon, P. Use of diffuse approximation on DIC for early damage detection in 3D carbon/epoxy composites. *Compos. Sci. Technol.* **2013**, *88*, 16–25. [CrossRef]

199. Garcia-Martin, R.; Bautista-De Castro, Á.; Sánchez-Aparicio, L.J.; Fueyo, J.G.; Gonzalez-Aguilera, D. Combining digital image correlation and probabilistic approaches for the reliability analysis of composite pressure vessels. *Arch. Civ. Mech. Eng.* **2019**, *19*, 224–239. [CrossRef]

200. Depuydt, D.; Hendrickx, K.; Biesmans, W.; Ivens, J.; Willem, A.; Van Vuure, W. A Digital image correlation as a strain measurement technique for fibre tensile tests. *Compos. Part A Appl. Sci. Manuf.* **2017**, *99*, 76–83. [CrossRef]

201. Périé, J.-N.; Passieux, J.-C. Special Issue on Advances in Digital Image Correlation (DIC). *Appl. Sci.* **2020**, *10*, 1530. [CrossRef]

202. Shadmehri, F.; Hoa, S.V. Digital Image Correlation Applications in Composite Automated Manufacturing, Inspection, and Testing. *Appl. Sci.* **2019**, *9*, 2719. [CrossRef]

203. Seon, G.; Makeev, A.; Schaefer, J.D.; Justusson, B. Measurement of Interlaminar Tensile Strength and Elastic Properties of Composites Using Open-Hole Compression Testing and Digital Image Correlation. *Appl. Sci.* **2019**, *9*, 2647. [CrossRef]

204. Flores, M.; Mollenhauer, D.; Runatunga, V.; Beberniss, T.; Rapking, D.; Pankow, M. High-speed 3D digital image correlation of low-velocity impacts on composite plates. *Compos. Part B Eng.* **2017**, *131*, 153–164. [CrossRef]

205. Helfrick, M.N.; Niezrecki, C.; Avitabile, P.; Schmidt, T. 3D digital image correlation methods for full-field vibration measurement. *Mech. Syst. Signal Process.* **2011**, *25*, 917–927. [CrossRef]

206. Baqersad, J.; Niezrecki, C.; Avitabile, P. Extracting full-field dynamic strain on a wind turbine rotor subjected to arbitrary excitations using 3D point tracking and a modal expansion technique. *J. Sound Vib.* **2015**, *352*, 16–29. [CrossRef]

207. Mostafavi, M.; Marrow, T.J. In situ observation of crack nuclei in poly-granular graphite under ring-on-ring equi-biaxial and flexural loading. *Eng. Fract. Mech.* **2011**, *78*, 1756–1770. [CrossRef]

208. Cook, A.B.; Duff, J.; Stevens, N.; Lyon, S.; Sherry, A.; Marrow, J. Preliminary evaluation of digital image correlation for in-situ observation of low temperature atmospheric-induced chloride stress corrosion cracking in austenitic stainless steels. *ECS Trans.* **2010**, *25*, 119–132.

209. Bay, B.K.; Smith, T.S.; Fyhrie, D.P.; Saad, M. Digital volume correlation: Three-dimensional strain mapping using X-ray tomography. *Exp. Mech.* **1999**, *39*, 217–226. [CrossRef]

210. Forsberg, F.; Mooser, R.; Arnold, M.; Hack, E.; Wyss, P. 3D micro-scale de-formations of wood in bending: Synchrotron radiation muCT data analysed with digital volume correlation. *J. Struct. Biol.* **2008**, *164*, 255–262. [CrossRef] [PubMed]

211. Limodin, N.; Réthoré, J.; Buffière, J.-Y.; Hild, F.; Roux, S.; Ludwig, W.; Rannou, J.; Gravouil, A. Influence of closure on the 3D propagation of fatigue cracks in a nodular cast iron investigated by X-ray tomography and 3D volume correlation. *Acta Mater.* **2010**, *58*, 2957–2967. [CrossRef]

212. Mostafavi, M.; Collins, D.M.; Cai, B.; Bradley, R.; Atwood, R.C.; Reinhard, C.; Jiang, X.; Galano, M.; Lee, P.D.; Marrow, T.J. Yield behavior beneath hardness indentations in ductile metals, measured by three-dimensional computed X-ray tomography and digital volume correlation. *Acta Mater.* **2015**, *82*, 468–482. [CrossRef]

213. Marrow, T.J.; Steuwer, A.; Mohammed, F.; Engelberg, D.; Sarwar, M. Measurement of crack bridging stresses in environment-assisted cracking of duplex stainless by synchrotron diffraction. *Fatigue Fract. Eng. Mater. Struct.* **2006**, *29*, 464–471. [CrossRef]

214. Withers, P.J.; Bhadeshia, H.K.D.H. Residual stress part 1-measurement techniques. *Mater. Sci. Technol.* **2001**, *17*, 355–365. [CrossRef]

215. Owen, R. Neutron and synchrotron measurements of residual strain in TIG welded aluminium alloy 2024. *Mater. Sci. Eng. A* **2003**, *346*, 159–167. [CrossRef]

216. Withers, P.J.; Clarke, A.P. A neutron diffraction study of load partitioning incontinuous Ti/SiC composites. *Acta Mater.* **1998**, *46*, 6585–6598. [CrossRef]

217. Biernacki, J.J.; Parnham, C.J.; Watkins, T.R.; Hubbard, C.R.; Bai, J. Phase-resolved strain measurements in hydrated ordinary portland cement using synchrotron X-rays. *J. Am. Ceram. Soc.* **2006**, *89*, 2853–2859. [CrossRef]
218. Schulson, E.M.; Swainson, I.P.; Holden, T.M. Internal stress within hardened cement paste induced through thermal mismatch. *Cem. Concr. Res.* **2001**, *31*, 1785–1791. [CrossRef]
219. Marrow, T.J.; Liu, D.; Barhli, S.M.; Saucedo Mora, L.; Vertyagina, Y.; Collins, D.M.; Reinhard, C.; Kabra, S.; Flewitt, P.E.J.; Smith, D.J. In situ measurement of the strains within a mechanically loaded polygranular graphite. *Carbon* **2016**, *96*, 285–302. [CrossRef]

Journal of
Composites Science

Article

Damage Detection in Glass/Epoxy Laminated Composite Plates Using Modal Curvature for Structural Health Monitoring Applications

Mahendran Govindasamy [1,*], **Gopalakrishnan Kamalakannan** [1], **Chandrasekaran Kesavan** [1] and **Ganesh Kumar Meenashisundaram** [2,*]

[1]   Department of Mechanical Engineering, RMK Engineering College, RSM Nagar, Kavaraipettai, Tamil Nadu 601206, India; krishgopz@gmail.com (G.K.); kesavan.chandrasekaran@gmail.com (C.K.)
[2]   Singapore Institute of Manufacturing Technology, 73 Nanyang Drive, Singapore 637662, Singapore
*   Correspondence: gmn.mech@rmkec.ac.in (M.G.); mganesh@simtech.a-star.edu.sg (G.K.M.)

Received: 9 September 2020; Accepted: 9 December 2020; Published: 14 December 2020

Abstract: This paper deals with detection of macro-level crack type damage in rectangular E-Glass fiber/Epoxy resin (LY556) laminated composite plates using modal analysis. Composite plate-like structures are widely found in aerospace and automotive structural applications which are susceptible to damages. The formation of cracks in a structure that undergoes vibration may lead to catastrophic events such as structural failure, thus detection of such occurrences is considered necessary. In this research, a novel technique called as node-releasing technique in Finite Element Analysis (FEA), which was not attempted by the earlier researchers, is used to model the perpendicular cracks (the type of damage mostly considered in all the pioneering research works) and also slant cracks (a new type of damage considered in the present work) of various depths and lengths for Unidirectional Laminate (UDL) ($[0]_S$ and $[45]_S$) composite layered configurations using commercial FE code Ansys, thus simulating the actual damage scenario. Another novelty of the present work is that the crack is modeled with partial depth along the thickness of the plate, instead of the through the thickness crack which has been of major focus in the literature so far, in order to include the possibility of existence of the crack up to certain layers in the laminated composite structures. The experimental modal analysis is carried out to validate the numerical model. Using central difference approximation method, the modal curvature is determined from the displacement mode shapes which are obtained via finite element analysis. The damage indicators investigated in this paper are Normalized Curvature Damage Factor (NCDF) and modal strain energy-based methods such as Strain Energy Difference (SED) and Damage Index (DI). It is concluded that, all the three damage detection algorithms detect the transverse crack clearly. In addition, the damage indicator NCDF seems to be more effective than the other two, particularly when the detection is for damage inclined to the longitudinal axis of the plate. The proposed method will provide the base data for implementing online structural health monitoring of structures using technologies such as Machine Learning, Artificial Intelligence, etc.

Keywords: damage detection; laminated composite plates; modal analysis; curvature mode shape; strain energy

1. Introduction

In high-speed mechanical applications such as turbomachinery and aerospace structures, there is a pressing need for high strength lightweight materials. The usage of non-homogenous fiber reinforced composites has increased widely due to its tailorable direction-dependent properties, high stiffness and increased corrosion resistance. However, composites are susceptible to unnoticeable damages as they experience various loading conditions in-service such as fatigue, bird impacts, lightning strikes etc. which can alter its dynamic characteristics ultimately leading to failure [1]. Some of the types of damages that occur in laminated composites are delamination, fiber-matrix debonding, matrix cracking, inter-laminar failure and fiber breakage [2].

Most of the non-destructive testing (NDT) methods for detecting damage in structures may be characterised as either local or global methods. In the past, researchers studied various local non-destructive techniques for damage detection in composites like X-Ray [3], microwave [4], acoustic emission [5], infrared thermography [6], ultrasonic [7] and magnetic-field methods [8]. Although these techniques hold some merits, it was reported that they were found to be somewhat difficult for applying in in the stuructures where the accessibility is very difficult like offshore oil platforms and the flying machines like aerospace vehicles as it require disassembly of parts and the locality of the damage is known prior to the testing which may be difficult in complex structures. In particular, the matrix cracks in composites lying perpendicular to the surface are hard to detect as most of the afore-mentioned methods require a wide enough reflecting surface which is the case in delamination type damages [9]. Thus, damage detection technique which is global in nature such as vibration-based methods may be found more useful, individually or in combination with other local methods, for successful non-destructive evaluation of laminated composite structures.

The fundamental basis of such a technique is based on the fact that, the changes in the structure's inherent parameters such as mass, stiffness and damping reflect in its modal properties such as natural frequency, mode shape and modal damping respectively [10]. In the past few decades, there has been substantial research carried out in the field of vibration-based damage detection across the world. Dimarogonas [11] has contributed a review of the studies carried out on cracked structures such as beams, plates and shells until 1996. This paper provided an excellent overview of the scope of crack study in plate-like structures and highlighted the topics of interest for future research. In their respective summary reviews of various vibration-based methods for damage detection, Doebling et al. [12] focussed on structural and mechanical systems, and Wei Fan [13] emphasized on algorithms and signal processing methods. The reduction in natural frequency is considered as an easily observable change which has been explored by researchers in the past. The determination of crack location in varying-cross section slender beams using natural frequencies was investigated by Chinchalkar [14]. A description of a numerical-based finite element approach was made and the results obtained in the past through semi-analytical approach using Frobenius method are compared with the current. The proposed method had an advantage which allowed to model different depths and boundary conditions of the beam. However, its limitation lays in the fact that in case of large-sized damages, slight change in the natural frequencies may be observed which may go unnoticed.

Some of the authors investigated the displacement mode shape [15] and its rotation [16] as damage indicators from dynamic behaviour of beam and plate structures respectively. It was stated that, the former was less sensitive and localized to damage than the latter. In addition, when the damage was in proximity to the node point, the displacement shape showed very little change in comparison with that of intact structure. On the contrary, the first order derivative of the mode shapes highlighted damages of various sizes at multiple locations within the structure. Nevertheless, there are some drawbacks for considering mode shapes for damage detection. Firstly, the vibration testing carried out for large structures usually considers lower fundamental modes and its corresponding mode shapes may not be affected significantly due to damage. Secondly, the number of sensors placed on the structure and noise effects play a very important role in the damage diagnosis procedure.

Pandey et al. [17] investigated a novel parameter called Curvature Mode Shape (CMS) for identification and localization of damages in structures. A cantilever and simply supported beam models were used and the results exhibited that the modal curvature was localized at damage location. Lestari et al. [18] have applied curvature mode shape as a damage indicator for carbon/epoxy laminated composite beam with an assumed damage effect in the form of stiffness loss in the analytical model for the damaged structure. The discontinuities in the modal curvature due to the presence of damage have been observed in the numerical simulation and with the use of surface-bonded Polyvinylidene fluoride (PVDF) film as sensors, the curvature modes have been measured experimentally. However, for higher modes, the modal curvature showed peaks at locations other than the damage which may lead to misinterpretation of results. Abdel Wahab et al. [19] applied the method of changes in curvature mode shapes for damage localization in simply supported and continuous beam models using finite element simulation and to a real-time measured data obtained from pre-stressed Z24 concrete bridge. An attempt has been made to minimize the misleading peaks by an indicator called as Curvature Damage Factor (CDF) and concluded that it functioned to a certain extent, though not completely. Hence, to improve the accuracy of damage detection, an indicator called NCDF (Normalized Curvature Damage Factor) has been utilized by Lu et al. [20] for identification of crack in beam structures. Qiao et al. [18] carried out numerical and experimental modal analysis to determine curvature mode shape for detection, localization and quantification of delamination in laminated composite plates. It was stated that, the dynamic response data measured experimentally cannot be readily used for detection of damage and are often used in conjunction with damage detection algorithms. The Gapped Smoothing Method (GSM), Generalized Fractal Dimension (GFD) and Strain Energy Method (SEM) were the algorithms used to analyse the modal curvature data and concluded that, the GSM located delamination better than GFD and SEM which showed extra peaks at locations other than the damage.

Cornwell et al. [21] extended the application of strain energy method to plate-like structures which was initially developed for one-dimensional beam structures. They established a damage detection algorithm called Damage Index Method (DIM) based on modal strain energy and concluded that, it was successful in detecting damages in areas up to 10% reduction in stiffness value. Huiwen Hu et al. [22] presented an approach based on strain energy to detect surface crack type damage in carbon/epoxy composite laminated plate for four different layups. A combination of experimental and numerical methods was used to determine the mode shapes for calculating strain energies based on Differential Quadrature Method (DQM). They concluded that, the developed damage index successfully identified the presence of damage and required only a few mode shapes before and after damage. Wang et al. [23] investigated the modal characteristics of damaged laminated composite plates using analytical and numerical approaches. The degradation of stiffness of the structure was considered as damage and concluded that, the modal strain energy and curvature mode shape had higher sensitivity to damage than natural frequencies and displacement mode shapes. The effect of damage in a structure had been modelled through various methods previously. Many researchers have considered the damage to be a reduction in the elemental Young's modulus in numerical finite element model or reduction in the second moment of area of cross-section or modeled the crack as a V-shaped groove by considering a solid element in finite element modelling as summarised in Table 1.

Table 1. Summary of methods used for modeling of cracks in finite element/analytical approach.

S. No.	Reference Cited/Year	Structure Considered	Method of Damage Detection	Method of Crack Modeling in Finite Element/Analytical Approach
01	Pandey et al. [17]/1991	Isotropic Beam	Change in Curvature Mode Shape	Approximate reduction in the elemental Young's modulus at the damaged location
02	Narkis [24]/1992	Isotropic Beam	Change in Natural Frequencies	For analytical methods of beam like structures, the beam was separated to two halves and damage was considered to be a massless rotational spring where the stiffness of the spring corresponds to the size of the damage

Table 1. *Cont.*

S. No.	Reference Cited/Year	Structure Considered	Method of Damage Detection	Method of Crack Modeling in Finite Element/Analytical Approach
03	Krawczuk and Ostachowicz [25]/1995	Laminated Composite Beam	Change in Natural Frequencies and Mode Shapes	For analytical methods of beam like structures, the beam was separated to two halves and damage was considered to be a massless rotational spring where the stiffness of the spring corresponds to the size of the damage.
04	Ruotolo et al. [26]/1996	Isotropic Beam	Methods based on Frequencies Response Functions	Cracked element with the approach that the elements situated on one side is considered as external forces applied to the cracked element, while the elements on the other side is regarded as constraints.
05	Tsai and Wang [27]/1996	Isotropic Shaft	Change in Natural Frequencies and Mode Shapes	For analytical methods of beam like structures, the shaft was separated to two halves and damage was considered to be a massless rotational spring where the stiffness of the spring corresponds to the size of the damage.
06	Ratcliffe [28]/1997	Isotropic Beam	Change in Curvature Mode Shape	Cracks and other forms of localized damage in a structure can lead to reduction in the flexural stiffness (EI), but change in mass is minimal. For the uniform cross-section localized stiffness damage can be introduced by reducing the thickness for one element of the finite element model without altering the mass matrix.
07	ABdel Wahab et al. [19]/1999	Isotropic Concrete Beam	Change in Curvature Mode Shape	Approximate reduction in the elemental Young's modulus at the damaged location.
08	Fernandez-saez et al. [29]/1999	Isotropic Beam	Change in Natural Frequencies	For analytical methods of beam like structures, the beam was separated to two halves and damage was considered to be a massless rotational spring where the stiffness of the spring corresponds to the size of the damage.
09	Yang et al. [30]/2001	Isotropic Beam	Change in Strain Energy	Modeled cracked beam as a continuous system with varying moment of inertia over the length of the beam.
10	Roy Mahapatra and Gopalakrishnan [31]/2003	Laminated Composite Beam	Dynamic Stiffness Method	Modeled cracked beam as a continuous system with varying moment of inertia over the length of the beam.
11	Yong and Hong Hao [32]/2003	Isotropic Beam and Plate	Methods based on Change in Natural Frequencies	Approximate reduction in the elemental Young's modulus at the damaged location.
12	Ertugrul Cam et al. [33]/2005	Isotropic Beam	Change in Natural Frequencies	Modeled the crack as v-shaped groove by considering solid finite element available in ANSYS.

Table 1. *Cont.*

S. No.	Reference Cited/Year	Structure Considered	Method of Damage Detection	Method of Crack Modeling in Finite Element/Analytical Approach
13	Loya et al. [34]/2006	Isotropic Beam	Change in Natural Frequencies	For analytical methods of beam like structures, the beam was separated to two halves and damage was considered to be a massless rotational spring where the stiffness of the spring corresponds to the size of the damage.
14	Sadettin Orhan [35]/2007	Isotropic Beam	Change in Natural Frequencies	Modeled the crack as v-shaped groove by considering solid finite element available in ANSYS.
15	Yu et al. [36]/2007	Laminated Composite Shell	Change in Dynamic Response	Considered Piezoelectric patches as sensors and actuators to realize automatic damage detections in this finite element model of laminated composite shells partially filled with fluid.
16	Peng et al. [37]/2008	Isotropic Beam	Methods based on Frequencies Response Functions	Approximate reduction in second moment of area of the cross-section at the damaged location.
17	Lu et al. [20]/2013	Isotropic Plate	Change in Dynamic Response	Approximate reduction in the elemental Young's modulus at the damaged location.
18	Daniele Dessi and Gabriele Camerlengo [38]/2015	Isotropic Beam	Natural Frequencies and Modal strain Energy – Based Methods	Damage was modeled as a localized and uniform reduction of the bending stiffness distribution along the dimensional coordinate of the damage. Thus, the damaged beam of was considered as the union of three beam portions along its length.
19	Simon Laflamme et al. [39]/2016	Wind Turbine Blade as Cantilevered Composite Taper Plate.	Methods Based on Change in Strain	Damage is considered as a change in the stiffness at the damaged location of the laminate layer.
20	Zhi-BoYang et al. [40]/2017	Laminated Composite Plate	Two-Dimensional Chebyshev Pseudo Spectral Modal Curvature	In the damaged area of analytical model, the local thickness is reduced to 95% of the original thickness of the plate.
21	Mohammad-Reza Ashory et al. [41]/2018	Laminated Composite Plate	Modal Strain Energy-Based Damage Detection Methods	Assuming a spring model with six degree-of-freedom between adjacent layers corresponding to the six stiffness components of an orthotropic composite material, the elastic moduli in the damaged region was formulated by the stiffness reduction vector.
22	Jingwen Pan et al. [42]/2019	Laminated Composite Curved Plate	Natural Frequency – Based Methods	A "constrained mode" model was developed by adding pair of contact elements, TARGE170/CONTAC173, between the mating surfaces of the delaminated area.
28	Bao et al. [43]/2017	Smart Ultra-High-Performance Concrete (UHPC) overlays	Fully-Distributed Fiber Optic Sensor.	Developed delamination detection system for smart Ultra-High-Performance Concrete (UHPC) overlays using a fully-distributed fiber optic sensor.

Table 1. *Cont.*

S. No.	Reference Cited/Year	Structure Considered	Method of Damage Detection	Method of Crack Modeling in Finite Element/Analytical Approach
29	Saravana Kumar et al. [44]/2020	Glass/Epoxy Laminates Hybridized with Glass Fillers	Post Impact Flexural (FAI) test and Acoustic Emission (AE) monitoring	Investigated the low-velocity impact induced damage behavior and its influence on the residual flexural response of glass/epoxy composites improved with milled glass fillers.
30	Markus Linke et al. [45]/2020	Thin-Walled Composite Plates	Modified Compression-After-Impact Testing Device	Investigated Failure mechanisms of impact damage in composite structures based on the Compression After Impact (CAI) test procedure
31	Maurizio et al. [46]/2020	Composite Structures	Strain measurement	Provided a state-of-the-art review on strain detection techniques in composite structures.
32	Meng et al. [47]/2016	Composite Plate	Flexural Strength	Studied the flexural behavior of ultrahigh-performance concrete panels reinforced with embedded glass fiber-reinforced polymer grids.
33	Stamoulis et al. [48]/2019	Laminated Composites	Hashin criterion	In this paper, a finite element model based on explicit dynamics formulations is adopted. Hashin criterion is applied to predict the intralaminar damage initiation and evolution. The numerical analysis is performed using the ABAQUS programme.
34	Shweta et al. [49]/2020	CFRP Composites	Machine Learning (ML) algorithms	This work contributes to the improvement of intelligent damage classification algorithms that can be applied to health management strategies of composite materials, operating under complex working conditions.
35	Stelios et al. [50]/2020	Carbon Fiber-Graphene-Reinforced Hybrid Composite Plates	Finite Element Analysis	In this study, a computational procedure for the investigation of the vibration behavior of laminated composite structures, including graphene inclusions in the matrix, is developed. The material properties required to carryout the FEA are computed using the rule of mixtures.

From the above table it is clearly evident that most of the pioneering researches focus on damage detection in homogenous isotropic beam structures and a few in the recent past on laminated beam and plate like structures using the above-mentioned crack modelling techniques. However, there has not been much work on using the changes in curvature mode shapes for such detection purposes, especially using the proposed node releasing technique in laminated plates. In all the above research works the damage was considered as through the thickness perpendicular cracks and no research work has been carried out to detect damage in laminated structures by considering the damage running through partial thickness and also the slant cracks. Hence the present work concentrates on damage detection in laminated composite rectangular plates with variable sized damages (both perpendicular and slant cracks) using modal analysis through Finite Element Analysis (FEA). The FEA is carried out using the Ansys Parametric Design Language (APDL) code. A sample of the code developed to model the crack using the node releasing technique and do modal analysis of the laminated plate is given in the

Annexure. The proposed method will provide the base data for implementing online structural health monitoring of structures using the technologies such as Machine Learning and Artificial Intelligence etc with the support of higher efficiency sensors as suggested by Zengshun Chen et al. [51]. The modal behaviour of a cracked glass/epoxy composite plate is examined considering fundamental bending modes only. A numerical investigation is carried out to determine the presence of crack and its location through a linear approach based on changes in modal curvature. Furthermore, to validate the numerical model, the experimental modal analysis is carried out on intact and cracked composite plate specimens. The previous researchers mostly considered transverse through thickness cracks only and detection of partial thickness damage in laminated composite structures is scarce. In this paper, the single edge non-propagating open crack damage is modelled by the technique of releasing nodes at damage location for various crack lengths and depths. Extensive parametric studies are required to establish a reliable damage detection algorithm capable of detecting crack type damage in composite plate structures. Two different orientations viz. transverse and angular partial depth cracks are studied to investigate the effectiveness of three damage detection algorithms namely NCDF, Strain Energy Difference (SED) and DIM.

2. Materials and Methods

2.1. Theoretical Background

2.1.1. Vibration of Laminated Composite Plate

The geometrical view of the rectangular laminated composite plate structure is shown in the Figure 1.

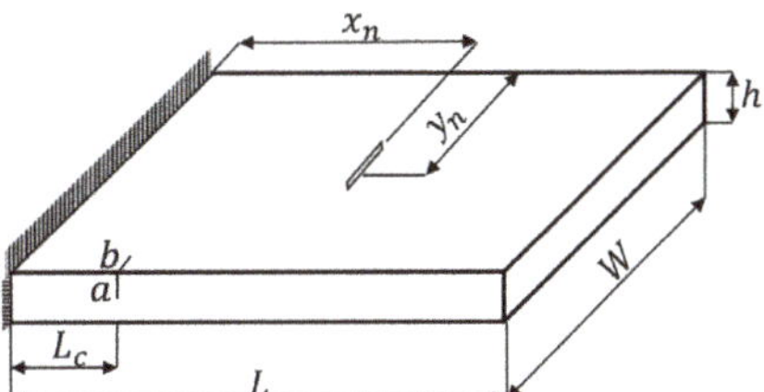

Figure 1. Geometrical sketch of the laminated composite plate with crack.

The notations of the length, width and height of the plate are L, W and h, L_c is the location of the crack from the fixed end, a and b are the crack depth and length respectively. The equilibrium differential equation for a free vibrating laminated composite plate is [48]

$$\frac{\partial^2 M_X}{\partial X^2} + \frac{\partial^2 M_Y}{\partial Y^2} + 2\frac{\partial^2 M_{XY}}{\partial X \partial Y} + N_X \frac{\partial^2 W}{\partial X^2} + N_Y \frac{\partial^2 W}{\partial Y^2} + 2N_{XY}\frac{\partial^2 W}{\partial X \partial Y} = \rho h \frac{\partial^2 W}{\partial t^2} \qquad (1)$$

where W is the displacement of the plate in the transverse direction (Z-axis), M_X, M_Y, M_{XY} and N_X, N_Y, N_{XY} are the moments and forces in the middle plane and ρ is the density of the plate.

2.1.2. Damage Detection Algorithms

The measured experimental or numerical dynamic response data such as natural frequencies and mode shapes are often accompanied with damage detection algorithms. This is because sometimes these data are not readily available for implementation in detecting damage. By the usage of algorithms, the useful data can be extracted thus neglecting external noise effects. The damage detection algorithms used in this study are highlighted below.

Curvature Damage Factor (CDF)

The second order derivative of mode shape known as the curvature mode shape was introduced by Pandey et al. [17] for detecting damage in beams which is defined by 1-d curvature. In the present work, the method is applied to a plate-like structure defined by 2-d curvature. The reduction in the bending stiffness of the structure is observed when a crack or other type of damage is introduced which will increase the magnitude of the curvature at the location of damage. By the usage of numerical approach known as the central difference approximation, the modal curvature can be obtained from the displacement mode shapes through finite element analysis as [20]

$$\vartheta'' = \frac{\vartheta_{i+1} - 2\vartheta_i + \vartheta_{i-1}}{l_e^2} \tag{2}$$

where ϑ'' is the curvature, ϑ is the normalized mode shape, i is the node number and l_e is the length of the element used in FE modelling. The absolute difference in the Curvature Mode Shape Difference (CMSD) is given by

$$\Delta\vartheta'' = \left| \vartheta''^{(d)} - \vartheta''^{(u)} \right| \tag{3}$$

where $\vartheta''^{(d)}$ and $\vartheta''^{(u)}$ are the curvature mode shapes of the damaged and undamaged plates respectively. The normalization of the above-mentioned parameter is carried out to obtain the Normalized Curvature Mode Shape Difference (NCMSD) as [10]

$$\Delta\vartheta'' = \left[1 + \frac{\Delta\vartheta''}{\max(\Delta\vartheta'') - \min(\Delta\vartheta'')} \right]^2 \tag{4}$$

The damage indicator in this method is simply a distinctive peak which appears in the plot of absolute difference in curvature. However, for higher modes, the peaks appeared at locations other than that of the damage. In order to minimize the misleading peaks, a parameter called Curvature Damage Factor (CDF) was proposed by [19], which is the average of the absolute differences in the modal curvature of the undamaged and damaged models. Mathematically, it can be expressed as

$$CDF_i = \frac{1}{n} \sum_{j=1}^{n} \left| \vartheta''^{(d)} - \vartheta''^{(u)} \right| \tag{5}$$

where j represents the mode number. Nevertheless, the above-mentioned parameter was still unsatisfactory in completely diminishing the effect of misleading peaks. Hence, to improve the accuracy of damage detection, an indicator called as NCDF (Normalized Curvature Damage Factor), which is the average of Normalized Curvature Mode Shape Difference (NCMSD) between intact and damaged structures of all modes of interest, has been used by [20]. It can be expressed as

$$NCDF_i = \frac{1}{n} \sum_{j=1}^{n} NCMSD_i \tag{6}$$

Modal Strain Energy-Damage Index Method (DIM)

The damage detection based on the changes in modal strain energy fraction was extended to plate-like structures by Cornwell et al. [21]. This is considered as an extension of curvature mode shape-based method since it exhibits a direct relation to modal curvature and it can be derived from the same for beam and plate structures [13]. It provides an essential early warning of damage in the structure even though the change in the response of the system caused by the damage is minimal. The total strain energy, U of the laminated composite plate during elastic deformation is given by [22]

$$U = \tfrac{1}{2} \int_0^b \int_0^a [D_{11}\left(\tfrac{\partial^2 w}{\partial x^2}\right)^2 + D_{22}\left(\tfrac{\partial^2 w}{\partial y^2}\right)^2 + 2D_{22}\left(\tfrac{\partial^2 w}{\partial x^2}\right)\left(\tfrac{\partial^2 w}{\partial y^2}\right)$$
$$+ 4\left(D_{16}\tfrac{\partial^2 w}{\partial x^2} + D_{26}\tfrac{\partial^2 w}{\partial y^2}\right)\tfrac{\partial^2 w}{\partial x \partial y} + 4D_{66}\left(\tfrac{\partial^2 w}{\partial x \partial y}\right)^2]\,dxdy \tag{7}$$

where w is the displacement in the transverse (Z-direction), a is length, b is width of the plate and D_{ij} is the flexural rigidity of the composite plate.

The theoretical basis of this method is that, if the primary location of the damage is at a sub-region, then the change in the fractional modal strain energy will be significant in that area and constant at the undamaged regions [49]. The strain energy associated with a sub-region in an undamaged plate for the kth mode is given by

$$U_{k,ij} = \tfrac{1}{2} \int_{y_j}^{y_{j+1}} \int_{x_i}^{x_{i+1}} \left[D_{11}\left(\tfrac{\partial^2 \phi_k}{\partial x^2}\right)^2 + D_{22}\left(\tfrac{\partial^2 \phi_k}{\partial y^2}\right)^2 + 2D_{22}\left(\tfrac{\partial^2 \phi_k}{\partial x^2}\right)\left(\tfrac{\partial^2 \phi_k}{\partial y^2}\right) \right.$$
$$\left. + 4\left(D_{16}\tfrac{\partial^2 \phi_k}{\partial x^2} + D_{26}\tfrac{\partial^2 \phi_k}{\partial y^2}\right)\tfrac{\partial^2 \phi_k}{\partial x \partial y} + 4D_{66}\left(\tfrac{\partial^2 \phi_k}{\partial x \partial y}\right)^2 \right]dxdy \tag{8}$$

where x_i, x_{i+1} and y_j, y_{j+1} represent the start and end points of the sub-region along the x and y axis respectively as shown in Figure 2.

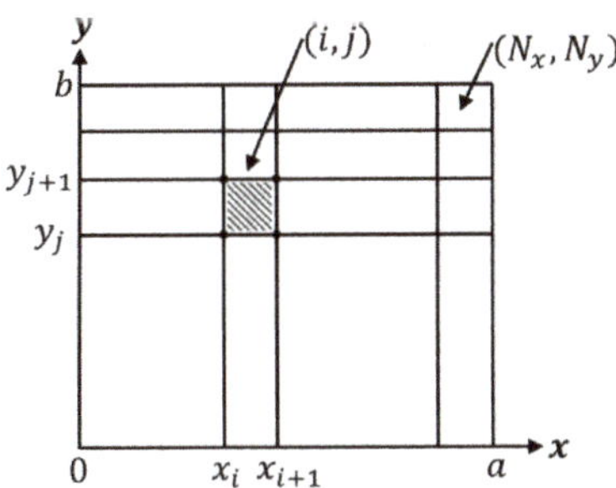

Figure 2. Schematic illustration of grid point arrangement in the plate.

Similar equations can be written for a damaged plate with displacement mode shapes, ϕ_k^*. The expression for fractal strain energy for an intact plate is given as

$$F_{k,ij} = \frac{U_{k,ij}}{U_k} \tag{9}$$

Similarly, the fractal strain energy for a damaged plate with the mode shape ϕ_k^* is given by

$$F_{k,ij}^* = \frac{U_{k,ij}^*}{U_k^*} \tag{10}$$

where U_k^* and $U_{k,ij}^*$ are the total and sub-regional strain energies of the damaged plate. By taking into consideration the measured fundamental mode shapes m, the damage index β_{ij} in the sub-region is given by

$$\beta_{ij} = \frac{\sum_{k=1}^m F_{k,ij}^*}{\sum_{k=1}^m F_{k,ij}} \tag{11}$$

The normalization of the damage index is given as

$$Z_{ij} = \frac{\beta_{ij} - \bar{\beta}_{ij}}{\sigma_{ij}} \tag{12}$$

where $\bar{\beta}_{ij}$ and σ_{ij} indicate mean and standard deviation of the damage index respectively.

Modal Strain Energy Difference (SED)

It is well known that a certain amount of strain energy is stored by a specific mode of vibration. The damage induced in a structure alters the stiffness, thus affecting the natural frequency and the displacement mode shape which are related to modal strain energy. The relation of strain energy with a particular mode shape is given as [18]

$$U_i = \frac{1}{2} \int_{x_k}^{x_{k+1}} EI \left(\frac{\partial^2 \phi_k}{\partial x^2} \right)^2 dx = \frac{1}{2} \int_{x_k}^{x_{k+1}} EI (\vartheta'')^2 dx \tag{13}$$

It is also known that, the flexural stiffness of the structure and the curvature parameter in the strain energy equation are interrelated. Thus, the stiffness reduction due to damage results in the increase of the curvature. Therefore, for indicating the location of damage, the square of the curvature mode shape found in the strain energy Equation (13) can be considered logically. The parameter for damage D_i by the concept of strain energy U_i given by the proportionality to the square of the modal curvature is expressed as

$$D_i \propto (\vartheta'')^2 = \left(\frac{\partial^2 \phi_k}{\partial x^2} \right)^2 \tag{14}$$

The damage indicator based on this method is defined as the absolute difference between the square of the modal curvature of undamaged and damaged structure given by

$$SED = \left| \vartheta''_{undamaged}(x)^2 - \vartheta''_{damaged}(x)^2 \right| \tag{15}$$

2.2. Numerical Modal Analysis

2.2.1. FE Modelling of the Laminated Composite Plate

The numerical modal analysis is carried out using commercial FE code ANSYS to formulate the finite element model of intact and cracked glass fiber reinforced laminated composite plates for investigating its free vibration behaviour. A laminated composite plate consisting of six plies $[0]_S$ and $[45]_S$ is considered. The glass fiber is used as the reinforcement and epoxy resin as the matrix with a unidirectional layup for the composite plates. The material properties calculated using rule of mixtures (Appendix A) and used for the FE analysis are listed in Tables 2 and 3. The composite plate has a length $L = 500$ mm, width $W = 250$ mm and depth $h = 5$ mm with each ply of thickness 0.833 mm, thus adding up to 6 layers. The FEA is carried out by employing a 3-D layered structural solid element (SOLID185) which has 8 nodes and 6 degrees of freedom per node. The translations and rotations about the X, Y and Z axes are the six degrees of freedom used in the analysis. The shell section is utilized to define the layer configuration as shown in the Figure 3.

Table 2. Properties of fiber and matrix.

Property	Glass Fiber	Epoxy Resin
Density, ρ (kg/m^3)	2600	1200
Elastic Modulus, E (GPa)	72	1.2
Shear Modulus, G (GPa)	30	0.807
Poisson's ratio, υ	0.25	0.3

Table 3. Properties of unidirectional laminated composite plate.

Property	Value
Density, ρ (kg/m^3)	1645
Elastic Modulus, E_x (GPa)	25.30
Elastic Modulus, $E_y = E_z$ (GPa)	4.01
Shear Modulus, $G_{xy} = G_{xz}$ (GPa)	1.55
Shear Modulus, G_{yz} (GPa)	1.49
Poisson's ratio, $\upsilon_{xy} = \upsilon_{xz}$	0.28
Poisson's ratio, υ_{yz}	0.35

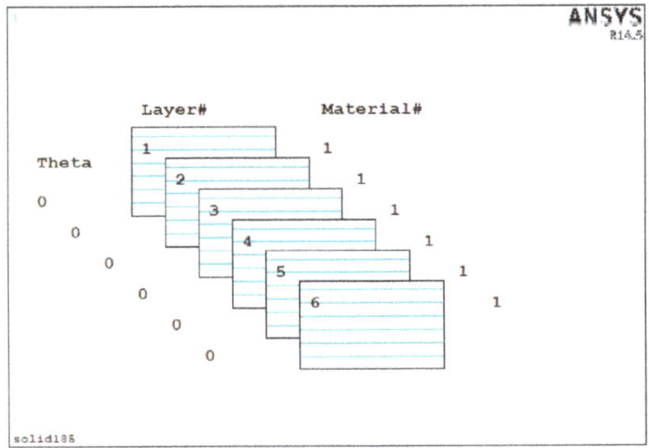

(a)

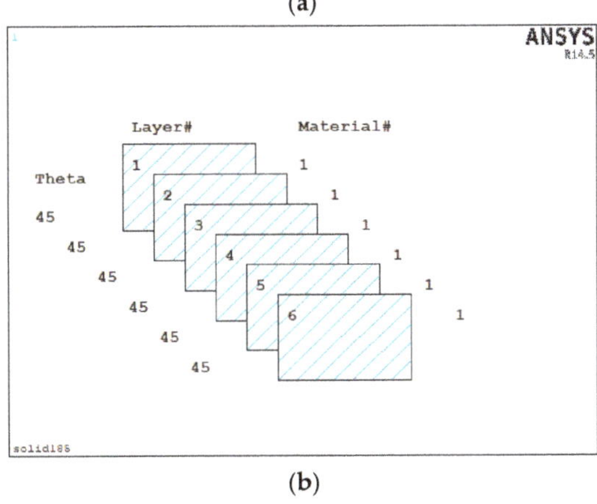

(b)

Figure 3. Layer stack-up in FE modelling of the Laminated Composite plate (**a**) 0 Degree Lay-up; (**b**) 45 Degree Lay-up.

The bottom layer is defined as Layer 1 and the stacking of the other layers is done from bottom to top along the positive Z-axis of the Cartesian coordinate system. The number of elements of the FE models was finalized based on the convergence test (Figure 4) resulting in a mesh size of 12,000 elements (100 along the length, 20 along the width and 6 along the depth). The natural frequencies and mode shapes of the laminated composite plates are determined theoretically by carrying out Eigen value modal analysis. The natural frequencies of the fundamental in-plane bending modes (Mode 1,2 and 3) are extracted using the Block Lanczos method. The result from the FEA i.e., displacements is utilized to determine the curvature mode shapes for damage detection. Validation of the FE model is done through comparison with the experimental results.

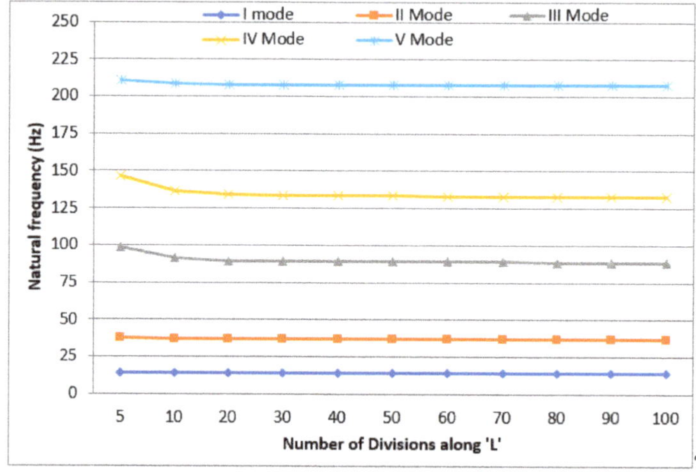

Figure 4. Convergence of the FE mesh of the plate model.

2.2.2. Modelling of the Crack

The crack is modelled in the structure by the technique of releasing of nodes at the location of the damage for the required damage depth 'a' and damage length 'b' as shown in Figures 1 and 5.

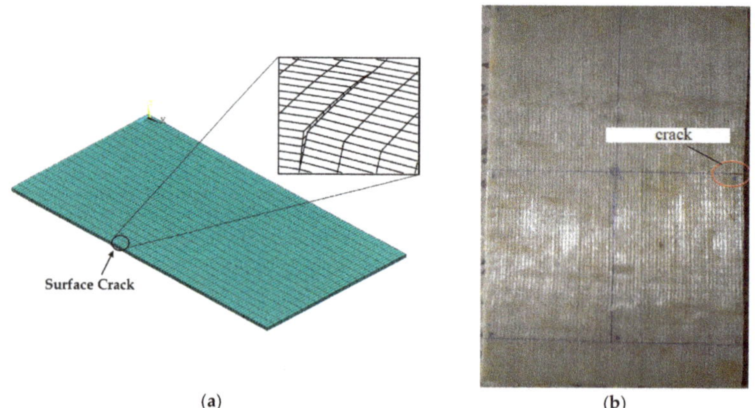

Figure 5. (a) FE model of cracked laminated composite plate, (b) Real plate with introduced crack.

This causes a discontinuity in the FE model of the composite plate structure thus simulating the scenario of damage. This technique allows modelling of damage through partial (in between layers) instead of through-thickness, where the latter has been considered in almost all the previous works. The size of the crack was considered as length (b/W) = 0.1, 0.3, and 0.5 for the depth (a/h) = 0.17, 0.33, 0.5, 0.67, 0.83 and 1 at 0.1 L, 0.2 L, 0.4 L and 0.6 L. The width of the crack is zero since it is considered as air crack. The APDL code developed for the modeling of the crack and FE analysis is given in the Appendix B.

2.3. Experimental Modal Analysis

The composite plate specimens were fabricated using unidirectional E-Glass fiber and Epoxy Resin (LY556) through manual lay-up method via open moulding technique. The pictorial view of the modal analysis setup with the fabricated composite plate specimen mounted on the fixture is shown in Figure 6.

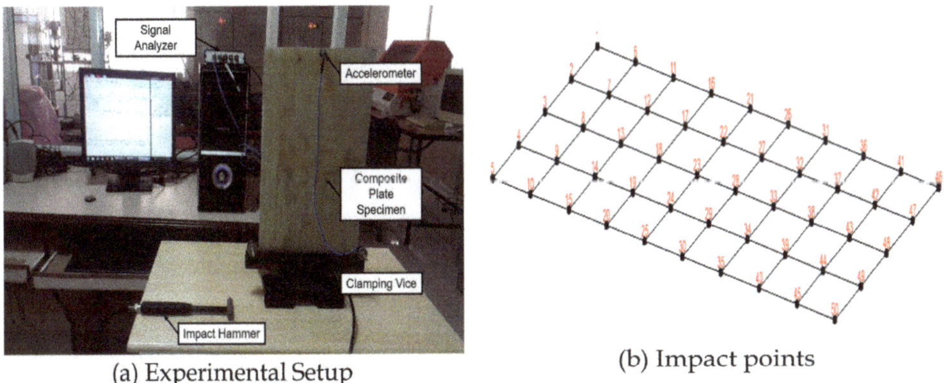

(a) Experimental Setup

(b) Impact points

Figure 6. Experimental modal analysis.

The elastic properties of the composite are evaluated using a simple rule of mixture [50] by knowing the properties of constituents (i.e., fiber and matrix) as obtained from the manufacturer and their corresponding weight fractions in the composite plate. The material and calculated properties of the composite are listed in Tables 2 and 3 respectively.

The cantilever edge fixity is achieved by clamping one end of the plate on a table using a bench-vice. The plate specimen is divided into 10 × 5 grids along longitudinal and transverse directions with 50 impact points. A PCB Impact hammer is used to excite all the grid points in the composite plate and the corresponding signal acquired from the force transducer is fed into NATIONAL INSTRUMENTS (NI) USB 4311, 4 channel dynamic signal analyser data acquisition system. There lies a difficulty in carrying out modal testing for the current thin-composite plate due to its lightweight and additional masses on the structure significantly affects the output response. Thus, the PCB accelerometer is fixed at the far end of the plate from the fixed end (at grid point 48) for measuring the acceleration outputs. It should be noted that the mass of the plate (1.028 kg) is far higher than that of the sensor (4.5 g).

In the final step of the analysis, LABVIEW-based modal analysis software MODALVIEW is used to extract the modal frequencies and displacement shapes by processing the Frequency Response Functions (FRFs) as shown in Figure 7.

This process involves the curve fitting using least square method by the MODAL VIEW software where each of the peaks of the FRFs indicates a single natural mode of vibration. The modal analysis was carried out in the frequency domain to extract the structural dynamic characteristics especially the natural frequencies and mode shapes which are the required characteristics for the present work.

The crack has been induced in the structure using a 1mm thick hacksaw blade cutter at 0.1 L from the fixed end of the plate.

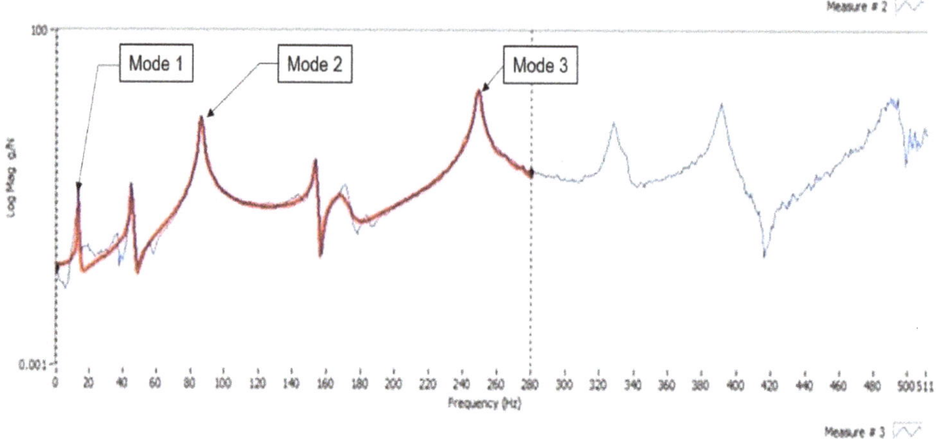

Figure 7. FRF curve of cantilevered composite plate (UDL-$[0]_S$) modal testing (curve fitting range—0 to 280 Hz).

3. Results and Discussion

The natural frequencies and corresponding mode shapes for the first three fundamental bending modes obtained through Finite Element Analysis (FEA) and Experimental Modal Analysis (EMA) are listed in Tables 4 and 5 respectively for undamaged plates and Tables 4 and 6 for damaged plates respectively.

Table 4. Natural frequencies of damaged and undamaged plates.

Mode	Before Damage			After Damage		
	EMA (Hz)	FEA (Hz)	Δ (%)	EMA (Hz)	FEA (Hz)	Δ (%)
1	13.14	14.19	7.4	12.85	14.04	0.08
2	88.38	88.67	0.39	85.31	88.59	0.04
3	248.91	248.64	0.11	245.05	247.41	0.95

The differences in the natural frequencies obtained through both analysis is less than 8% and 1% before and after damage. Also, the displacement shapes obtained through FEA and EMA are comparable with slight changes in the natural frequency since the fabricated plate is not exactly rectangular and error in measurement at certain impact points. It may be observed that for Mode-1, the difference in the values obtained through both analysis is more pronounced. The possible reasons may be attributed to the complexity introduced by non-homogeneity of the structure or defects in the fabricated specimens. Furthermore, the theoretical assumptions that are made in the FE modelling for simplification can cause the numerical models to behave slightly different from the real-life modal analysis tested specimens. The afore-mentioned comparison clearly shows that the FE model is validated and can be used for studying the vibration behaviour of cracked glass fibre reinforced fibre (GFRP) composite rectangular plates.

Table 5. Natural frequencies and mode shapes (undamaged).

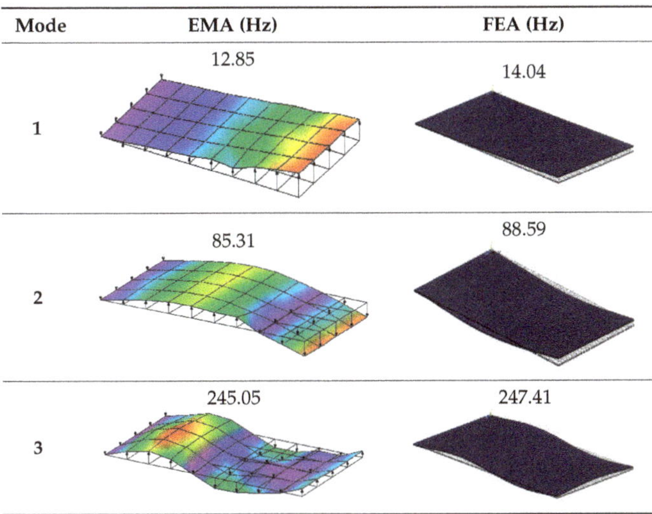

Mode	EMA (Hz)	FEA (Hz)
1	13.14	14.19
2	88.38	88.67
3	248.91	248.64

Table 6. Natural frequencies and mode shapes (damaged).

Mode	EMA (Hz)	FEA (Hz)
1	12.85	14.04
2	85.31	88.59
3	245.05	247.41

From the Table 7, it is evident that the percentage change in natural frequencies become more pronounced with the increase of damage depth and length. For example, the least value of percentage change (0.01 or even 0) occurs for the lowest damage depth ratio (a/h = 0.17) and damage length ratio (b/W = 0.1). On the other hand, the maximum change (31.05) occurs in the third mode for the highest a/h = 1.0 (through thickness damage) and b/W = 0.5. These variations in the frequency can be used to determine the presence but not for localization of damage. To realize the sensitivity of the mode-shape change for detection of damage, a test case is considered for crack size a/h = 0.1, b/W = 0.2 located at 0.2 L from the fixed end. The absolute difference in the mode shape is shown in Figure 8.

Table 7. Variation in Natural frequencies in the plate structure for different crack lengths and depths.

a/h	b/w	Change in Natural Frequencies (%)									
		Mode 1		Mode 2		Mode 3		Mode 4		Mode 5	
		$[0]_S$	$[45]_S$	$[0]_S$	$[45]_S$	$[0]_S$	$[45]_S$	$[0]_S$	$[45]_S$	$[0]_S$	$[45]_S$
	0.1	0.04	0.02	0.04	0.02	0.04	0.02	0.04	0.02	0.04	0.02
0.17	0.3	0.12	0.09	0.12	0.09	0.12	0.09	0.12	0.09	0.12	0.09
	0.5	0.42	0.22	0.42	0.22	0.42	0.22	0.42	0.22	0.42	0.22
	0.1	0.16	0.10	0.16	0.10	0.16	0.10	0.16	0.10	0.16	0.10
0.33	0.3	0.60	0.44	0.60	0.44	0.60	0.44	0.60	0.44	0.60	0.44
	0.5	1.06	0.75	1.06	0.75	1.06	0.75	1.06	0.75	1.06	0.75
	0.1	0.37	0.24	0.37	0.24	0.37	0.24	0.37	0.24	0.37	0.24
0.5	0.3	1.53	0.00	1.53	0.00	1.53	0.00	1.53	0.00	1.53	0.00
	0.5	2.80	0.00	2.80	0.00	2.80	0.00	2.80	0.00	2.80	0.00
	0.1	0.59	0.38	0.59	0.38	0.59	0.38	0.59	0.38	0.59	0.38
0.67	0.3	3.12	2.28	3.12	2.28	3.12	2.28	3.12	2.28	3.12	2.28
	0.5	6.16	4.46	6.16	4.46	6.16	4.46	6.16	4.46	6.16	4.46
	0.1	0.73	0.49	0.73	0.49	0.73	0.49	0.73	0.49	0.73	0.49
0.83	0.3	5.37	4.00	5.37	4.00	5.37	4.00	5.37	4.00	5.37	4.00
	0.5	12.58	9.36	12.58	9.36	12.58	9.36	12.58	9.36	12.58	9.36
	0.1	1.38	0.87	1.38	0.87	1.38	0.87	1.38	0.87	1.38	0.87
1.0	0.3	10.33	7.71	10.33	7.71	10.33	7.71	10.33	7.71	10.33	7.71
	0.5	23.73	18.5	23.73	18.5	23.73	18.5	23.73	18.5	23.73	18.5

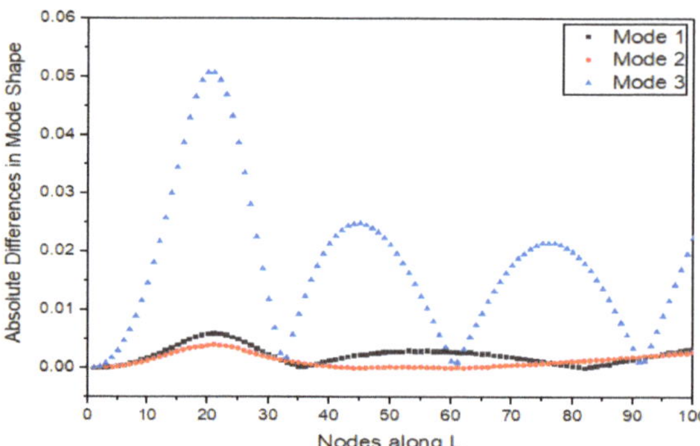

Figure 8. Absolute Difference between Displacement Mode Shapes for Intact and Damaged Composite Plate $[0]_S$ with damage at 0.2 L for a/h = 1.0 and b/W = 0.2.

From the graph, it is observed that the detection of damage using the change of mode shapes is very difficult as evidenced in the existing studies. Hence, the sensitivity of the second order derivative of the displacement mode shape (Modal Curvature) is examined in the following sections with a motivation from similar approaches in the literature.

Through the observation of natural frequencies and mode shapes obtained for undamaged and damaged structures, the presence of damage cannot be identified truthfully. It is hard to detect small sized crack damages based on the afore-mentioned modal parameters alone. The following section investigates the detection capability of modal curvature and its derivative parameters such as NCDF, DI and SED. The Figures 9 and 10 show the mapping of NCDF for various crack locations along the longitudinal edge of the plate i.e., 0.1 L, 0.2 L, 0.4 L, 0.6 L and for damage size, a/h = 0.33 and b/W = 0.5 for $[0]_S$ and $[45]_S$.

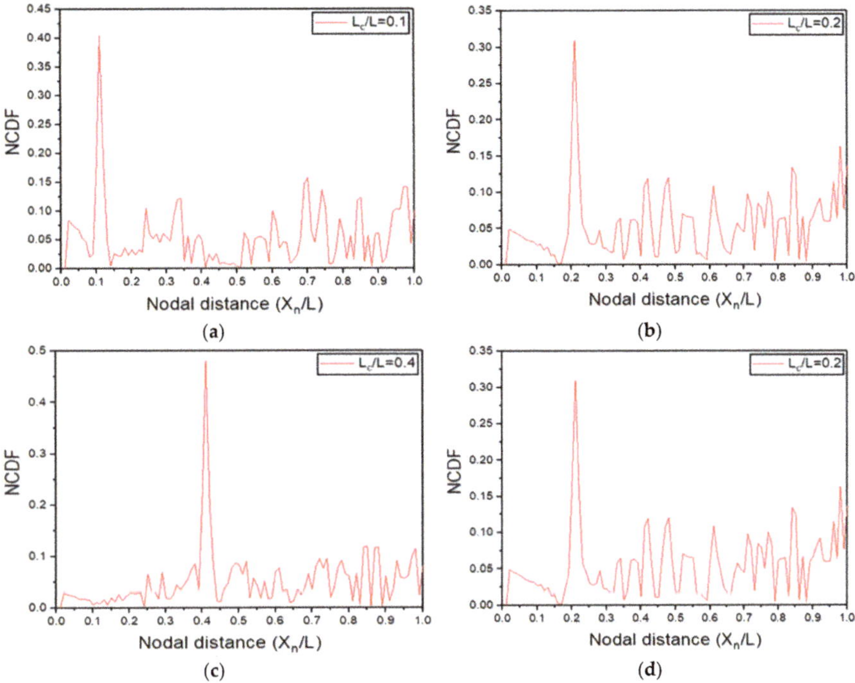

Figure 9. Normalized Curvature Damage Factor (NCDF) for cantilever composite plate $[0]_S$ with damage size for a/h = 0.33 and b/W = 0.5 at (**a**) 0.1 L (**b**) 0.2 L (**c**) 0.4 L (**d**) 0.6 L.

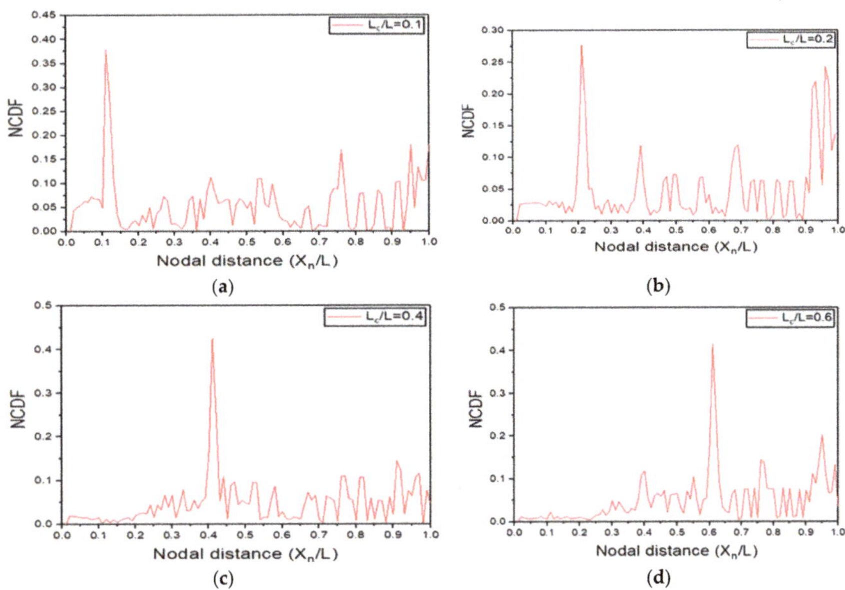

Figure 10. NCDF for cantilever composite plate $[45]_S$ with damage size for a/h = 0.33 and b/W = 0.5 at (**a**) 0.1 L (**b**) 0.2 L (**c**) 0.4 L (**d**) 0.6 L.

A distinctive peak is observed in all four plots, revealing the location of the crack. It can be noted that the normalization of the curvature damage factor has minimized the misleading peaks in regions away from the damage location. Irrespective of the location of the crack in the structure, NCDF clearly indicates the presence and location of the damage.

In all the above cases of damage detection, the NCDF value is computed at the central axis of the composite plate structure even though the damage is located far away from the central axis. This is considered for the sake of requirement of minimum number of sensors for mode shape measurement through experimental methods. The NCDF is unable to detect the damage in plate structures if b/W < 0.5 for any value of damage depth ratio as shown in Figure 11.

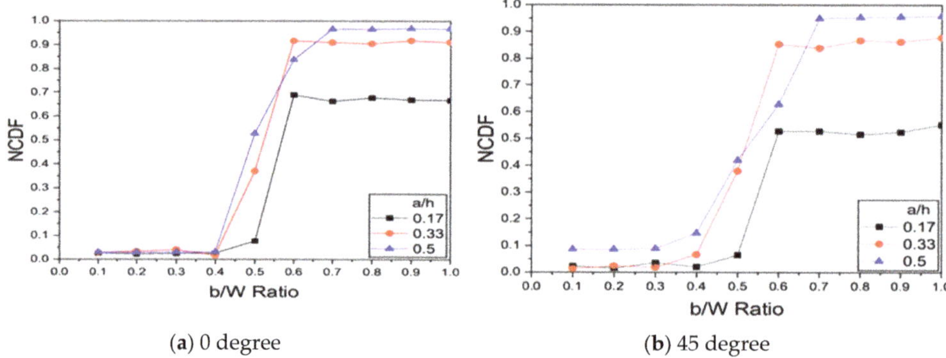

(a) 0 degree   (b) 45 degree

Figure 11. Variation of the NCDF over depth of layers at L_c/L = 0.1 in cantilever composite plate (**a**) $[0]_S$; (**b**) $[45]_S$.

Therefore, due to the limitation of the NCDF computed at central axis in detection of damage located at a distance from the central axis in plate-like structures, the damage indicator is computed at selected points on top surface of the plate. The following section examines the damage detection through NCDF and the damage detection algorithms based on modal strain energy i.e., SED and DI.

- Case 1: $[0]_S$ Layup

The NCDF, SED and DI for damage located at 0.1 L with a/h = 0.33, 0.5 and b/W = 0.2, 0.25 to 0.5 (crack located at a distance from the longitudinal edge) in the cantilever composite plate $[0]_S$ are shown in Figure 12.

From these plots, the location of damage is clearly detected through the appearance of characteristic peak at damage location. The results obtained are encouraging as it can be clearly seen that all the three algorithms perform fairly by indicating the presence of the damage as peaks, though SED shows a few extra peaks at other locations as well.

- Case 2: $[45]_S$ Layup

Figure 13 shows the plots for the three damage detection algorithms for a crack inclined at 45° to the longitudinal axis of $[45]_S$ for two sizes of damage, a/h = 0.17, 0.33 and b/W = 1 (through the width of the plate).

It can be inferred that NCDF clearly shows the peak and DI to some extent throughout the length of the crack across the width of the plate. Nevertheless, SED does not indicate the presence of crack as a few peaks are observed in both damage depth ratio cases and the damage is barely discernible. In NCDF, although the peak occurred at region surrounding the location of the surface crack, it can be observed that a few peaks emerged in locations away from the damage. It can also be noticed that the magnitude of the peak increases as the size of the damage increases which can be attributed to the corresponding loss in stiffness.

Since a plate structure is characterized by a two-dimensional curvature, it is interesting to investigate the effectiveness of the three damage detection algorithms mapped using the displacement response obtained at the lateral axis (Y-direction). It can be clearly seen from Figure 14 that NCDF shows the presence of damage throughout the width of the plate followed by Damage Index which shows the presence of damage up to some extent. The SED shows a bare minimum of a few peaks thus it cannot be applied as a potentially viable damage detection algorithm in comparison with the other two for slant cracks.

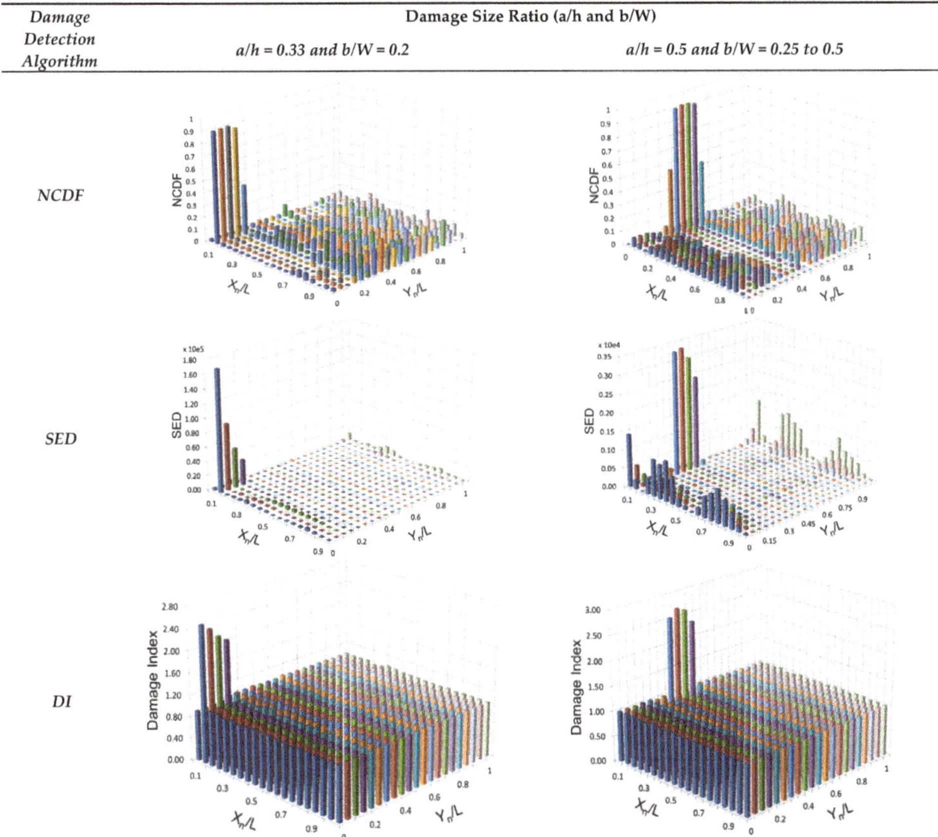

Figure 12. Cantilever composite Plate $[0]_S$ with damage at 0.1 L for a/h = 0.33 and b/W = 0.2, a/h = 0.5 and b/W = 0.25 to 0.5 (**a**) NCDF (**b**) SED (**c**) DI.

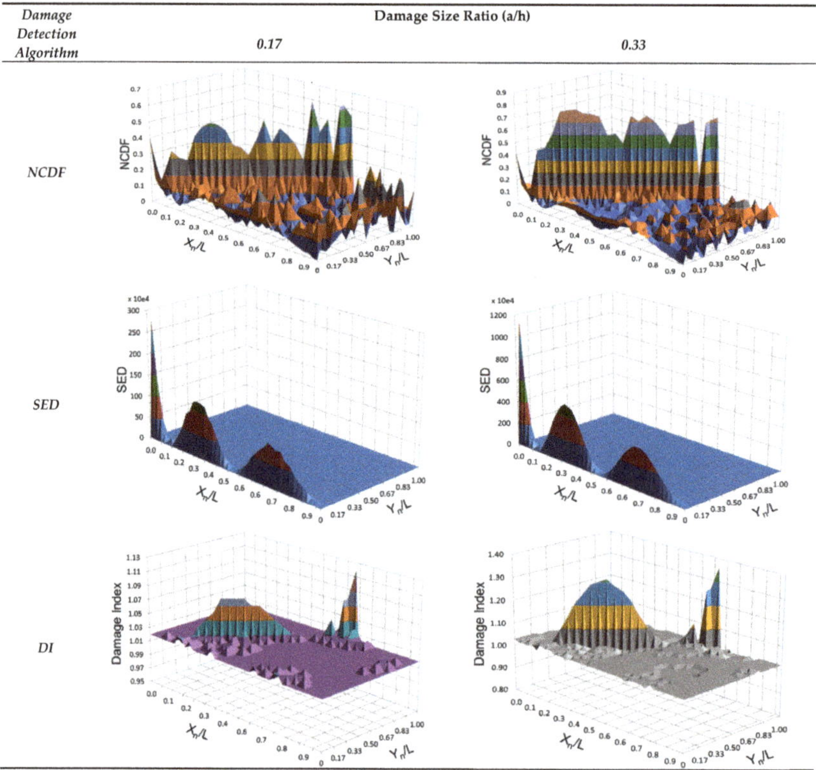

Figure 13. Composite Plate $[45]_S$ with damage at 0.1 L inclined at 45° for a/h = 0.17,0.33 and b/W = 1 (**a**) NCDF (**b**) SED (**c**) DI.

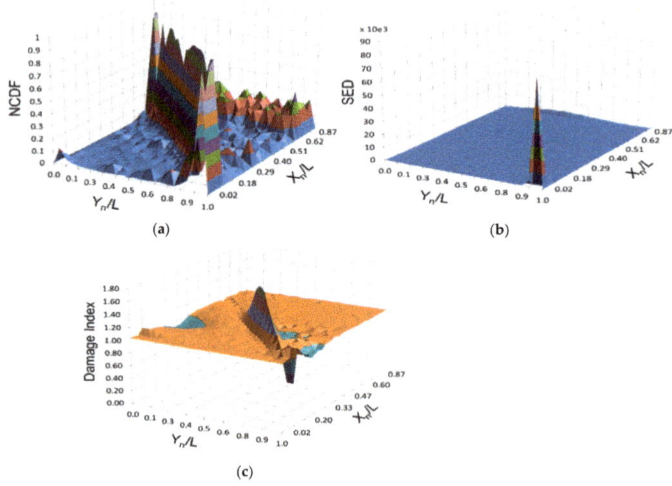

Figure 14. Cantilever Composite Plate $[45]_S$ with damage at 0.1 L inclined at 45° for a/h = 0.33 and b/W = 1 (**a**) NCDF (**b**) SED (**c**) DI plotted in Y-Direction.

4. Conclusions

The main objectives of this investigation are to ascertain the effectiveness of existing damage detection algorithms for laminated composite plate structures and observe their relative merits. Three damage detection algorithms are investigated to detect transverse and angular cracks. The damaged plate is successfully modelled using the novel method called as node releasing technique in a modal analysis by employing Finite element analysis. The reliability of the model is ensured by carrying out the convergence test and an experimental modal analysis which shows positive agreement to the numerical results. This study considers only the fundamental bending modes of the plate structure before and after the damage. The changes in natural frequency for various crack sizes are analysed and it could be concluded that the percentage change in the values increase with increasing size of the damage. The ineffectiveness of absolute mode shape difference as a tool for identifying the presence of damage led to the investigation of the damage indicators based on Curvature Mode shape (CMS) and modal strain energy.

The study based on NCDF shows that it is incapable of detecting damages for $b/W < 0.5$ for all damage depth ratios. This limitation is attributed to the fact that the computation of the damage indicator is carried out using the data recorded at the central axis of the plate and the location of the damage is far away from it. Thus, the algorithm is calculated at selected points at the top surface of the plate to examine the sensitivity of the three damage detection algorithms. The plots of NCDF, SED and DI clearly indicate the presence of the damage for transverse cracks in $[0]_S$ and $[45]_S$ in the form of distinguishing peaks at the damage location. In the case of damages located at a distance from the edge of the plate and in-between the layers, all the three algorithms display distinctive peaks. For the slant crack, NCDF detected the presence of damage better than the other two damage indices due to its incapability.

Further investigation can consider sensitivity analysis of the afore-mentioned damage indicators for multiple damages of various sizes in complex engineering applications such as non-circular cylindrical, conical and double curvature spherical shell structures. Also, exploration can be extended for upcoming complex fiber configurations such as 3-d pinned woven fabric and other types of damages in composites such as sub-surface defects.

The following are the future works may be implemented for damage detection in laminated composite structures based on the present work:

- In the present work, the damage detection is examined using the theoretical modal analysis of the composite plate structures with the help of the damage detection algorithms. The informative data of the damaged structures predicted by the theoretical method of damage detection can be useful for the purpose of implementing online structural health monitoring/condition monitoring via the emerging technologies like augmented reality, digital twin of products/structures in real life applications with the support of Internet of Things (IOT), neural networks and artificial intelligence like concepts.
- The findings of the theoretical damage detection procedure may be used as pre-requisites in the field of engineering fracture mechanics to predict the remaining life of the damaged structures in order to avoid sudden failure of the structures.
- Even though, the damage detection methods used in the present work are concentrated on the detection of damage which exists from top surface to partial thickness of a laminated composite structure, these methods may be extended to examine the effectiveness of damage detection methods based on curvature mode shape in detecting the sub-surface damage of a laminated composite structure.

Author Contributions: Conceptualization, C.K. and M.G.; methodology, M.G.; software, M.G.; validation, C.K., M.G. and G.K.; formal analysis, G.K.M.; investigation, M.G.; resources, C.K.; data curation, M.G.; writing—original draft preparation, C.K.; writing—review and editing, M.G.; visualization, G.K.M.; supervision, C.K.; project administration, M.G. All authors have read and agreed to the published version of the manuscript.

Funding: This research received no external funding.

Acknowledgments: The authors sincerely thank RMK Engineering College, Kavaraipettai, Tamil Nadu, India, for providing the testing facilities.

Conflicts of Interest: The authors declare no conflict of interest.

Appendix A

Micromechanical Analysis of Laminated Composite Plate:

$$\text{Weight of the Unidirectional fiber} = 950 \text{ g/m}^2 \text{ (GSM)} \tag{A1}$$

$$\text{Area of fiber} = 0.565*0.3 \text{ m}^2 \tag{A2}$$

$$\text{Area of six layers of fiber} = 0.565*0.3*6 \text{ m}^2 = 1.0509 \text{ m}^2 345 \tag{A3}$$

$$\text{Mass of fiber of the fabricated laminate, } m_f = 1.0509 \text{ m}^2*0.95 \text{ kg/m}^2 \tag{A4}$$

$$m_f = 0.965 \text{ kg} \tag{A5}$$

$$\text{Fiber weight fraction, } W_f = \frac{w_f}{w_c} = \frac{0.965}{1.883} \tag{A6}$$

$$W_f = 0.51248 \tag{A7}$$

Density of the laminated composite is given by

$$\frac{1}{\rho_c} = \frac{W_f}{\rho_f} + \frac{1 - W_f}{\rho_m} \tag{A8}$$

$$\frac{1}{\rho_c} = \frac{0.51248}{2.54} + \frac{1 - 0.51248}{1.2} \tag{A9}$$

$$\rho_c = 1645 \text{ kg/m}^3 \tag{A10}$$

$$\text{Fiber volume fraction, } v_f = \frac{\vartheta_f}{\vartheta_c} = 0.3318 \tag{A11}$$

$$v_f = 0.3318 \tag{A12}$$

Evaluation of Elastic Moduli

$$E_1 = E_f v_f + E_m (1 - v_f) \tag{A13}$$

$$E_1 = (72*0.3318) + (2.1*(1 - 0.3318)) \tag{A14}$$

$$E_1 = 25.295 \text{ Gpa} \tag{A15}$$

From the stress—strain graph obtained from UTM test,

$$E_{11} = 24.64 \text{ Gpa} \tag{A16}$$

$$E_2 = E_m \left[\frac{E_f + E_m + (E_f - E_m)v_f}{E_f + E_m - (E_f - E_m)v_f} \right] \tag{A17}$$

$$E_2 = 2.1 \left[\frac{72 + 2.1 + (72 - 2.1)0.3318}{72 + 2.1 - (72 - 2.1)0.3318} \right] \tag{A18}$$

$$E_2 = E_3 = 4.0137 \text{ Gpa} \tag{A19}$$

Evaluation of Poisson's Ratio

$$\upsilon_{12} = \upsilon_f v_f + \upsilon_m (1 - v_f) \text{ (3.15)} \tag{A20}$$

$$\upsilon_{12} = 0.25*0.3318 + 0.3(1 - 0.3318) \tag{A21}$$

$$\upsilon_{12} = \upsilon_{13} = 0.2834 \tag{A22}$$

$$\upsilon_{23} = \upsilon_f v_f + \upsilon_m(1 - v_f)\left[\frac{1 + \upsilon_m - \upsilon_{12}E_m/E_{11}}{1 - \upsilon_m^2 + \upsilon_m\upsilon_{12}E_m/E_{11}}\right] \tag{A23}$$

$$\upsilon_{23} = 0.25*0.3318 + 0.34(1 - 0.3318)\left[\frac{1 + 0.3 - 0.2834 * (2.1/25.295)}{1 - 0.3^2 + 0.2834 * 2.1 * (2.1/25.295)}\right] \tag{A24}$$

$$\upsilon_{23} = 0.3496 \tag{A25}$$

Evaluation of Rigidity Moduli

$$G_{12} = G_m\left[\frac{G_f + G_m + (G_f - G_m)v_f}{G_f + G_m - (G_f - G_m)v_f}\right] \tag{A26}$$

$$G_{12} = 0.807\left[\frac{30 + 0.807 + (30 - 0.807) * 0.3318}{30 + 0.807 - (30 - 0.807) * 0.3318}\right] \tag{A27}$$

$$G_{12} = G_{13} = 1.547 \text{ Gpa} \tag{A28}$$

$$G_{23} = \frac{E_{22}}{2(1 + \upsilon_{23})} \tag{A29}$$

$$G_{23} = \frac{4.0137}{2(1 + 0.3497)} \tag{A30}$$

$$G_{23} = 1.487 \text{ Gpa} \tag{A31}$$

Appendix B. A Sample ANSYS APDL Code Developed for Modal Analysis of the Cantilevered Composite Plate Structure with Damage

```
/CLEAR
/PREP7
/title,solid185
ET,1,solid185
KEYOPT,1,3,1
!KEYOPT,1,2,2
KEYOPT,1,8,1
SECTYPE,1,SHELL
SECDATA,.000833,1,0,9
MP,EX,1,39.805E9
,EY,1,9.98E9
,EZ,1,9.98E9
,PRXY,1,0.263
,PRYZ,1,0.318
,PRXZ,1,0.263
,GXY,1,3.98E9
,GYZ,1,3.78E9
,GXZ,1,3.98E9
,DENS,1,2082.18
N,1
,11,0.045
FILL
NGEN,21,11,1,11,1,,,0.012
```

```
ngen,7,231,1,231,1,,0.000833
e,12,13,2,1,243,244,233,232
EGEN,10,1,-1
EGEN,20,11,1,10,1
egen,6,231,1,200,1
N,1618,0.045
,1708,0.45
FILL
NGEN,21,91,1618,1708,1,,,0.012
ngen,7,1911,1618,3528,1,,0.000833
e,1709,1710,1619,1618,3620,3621,3530,3529
EGEN,90,1,-1
EGEN,20,91,1201,1290,1
egen,6,1911,1201,3000,1
NSEL,S,LOC,X,0.045
NSEL,R,LOC,Z,0,0.12
NSEL,A,LOC,Y,0,0.002499
NUMMRG,NODES
ALLSEL
!LSEL, S, LINE, ,1
!LSEL, A, LINE, ,3
!LSEL, A, LINE, ,6
!LSEL, A, LINE, ,8
!CM,TH,LINE
!CMSEL,,TH
!LESIZE,TH, , ,1, , , , ,1
!D,1,ALL,,,664,51
nsel,s,loc,x,0
d,all,all
allsel
FINISH
/SOLU
ANTYPE,MODAL
MODOPT,LANB,6
SOLVE
FINISH
/POST1
SET,LIST
SET,FIRST
PLDISP,1
nsel,s,loc,z,0.12
nsel,r,loc,y,0.002499
PRNSOL,U,Y
SET,NEXT
SET,NEXT
PLDISP,1
PRNSOL,U,Y
SET,LAST
PLDISP,1
PRNSOL,U,Y
```

References

1. Moreno-garcía, P.; Araújo, J.V.; Lopes, H. A new technique to optimize the use of mode shape derivatives to localize damage in laminated composite plates. *Compos. Struct.* **2014**, *108*, 548–554. [CrossRef]
2. Fan, W.; Qiao, P. A 2-D continuous wavelet transform of mode shape data for damage detection of plate structures. *Int. J. Solids Struct.* **2009**, *46*, 4379–4395. [CrossRef]
3. Schilling, P.J.; Karedla, B.P.R.; Tatiparthi, A.K.; Verges, M.A.; Herrington, P.D. X-ray computed microtomography of internal damage in fiber reinforced polymer matrix composites. *Compos. Sci. Technol.* **2005**, *65*, 2071–2078. [CrossRef]
4. Naito, K.; Kagawa, Y.; Kurihara, K. Dielectric properties and noncontact damage detection of plain-woven fabric glass fiber reinforced epoxy matrix composites using millimeter wavelength microwave. *Compos. Struct.* **2012**, *94*, 695–701. [CrossRef]
5. Bourchak, M.; Farrow, I.R.; Bond, I.P.; Rowland, C.W.; Menan, F. Acoustic emission energy as a fatigue damage parameter for CFRP composites. *Int. J. Fatigue.* **2007**, *29*, 457–470. [CrossRef]
6. Mian, A.; Han, X.; Islam, S.; Newaz, G. Fatigue damage detection in graphite/epoxy composites using sonic infrared imaging technique. *Compos. Sci. Technol.* **2004**, *64*, 657–666. [CrossRef]
7. Bastianini, F.; di Tommaso, A.; Pascale, G. Ultrasonic non-destructive assessment of bonding defects in composite structural strengthenings. *Compos. Struct.* **2001**, *53*, 463–467. [CrossRef]
8. Liang, T.; Ren, W.; Tian, G.Y.; Elradi, M.; Gao, Y. Low energy impact damage detection in CFRP using eddy current pulsed thermography. *Compos. Struct.* **2016**, *143*, 352–361. [CrossRef]
9. Aymerich, F.; Meili, S. Ultrasonic evaluation of matrix damage in impacted composite laminates. *Compos. Part B* **2000**, *31*, 1–6. [CrossRef]
10. Chandrashekhar, M.; Ganguli, R. Damage assessment of structures with uncertainty by using mode-shape curvatures and fuzzy logic. *J. Sound Vib.* **2009**, *326*, 939–957. [CrossRef]
11. Dimarogonas, A.D. Vibration of cracked structures: A state of the art review. *Eng. Fract. Mech.* **1996**, *55*, 831–857. [CrossRef]
12. Doebling, S.W.; Farrar, C.R.; Prime, M.B.; Shock, T.; Digest, V. Summary Review of Vibration-Based Damage Identification Methods. *Shock Vib. Dig. J.* **1998**, *30*, 91–105. [CrossRef]
13. Fan, W.; Qiao, P. Vibration-based Damage Identification Methods: A Review and Comparitative study. *Struct. Heal. Monit.* **2011**, *10*, 83–111. [CrossRef]
14. Chinchalkar, S. Determination of Crack Location in Beams Using Natural Frequencies. *J. Sound Vib.* **2001**, *247*, 417–429. [CrossRef]
15. Maia, N.M.; Silva, J.M.; Almas, E.A. Damage Detection in Structures: From Mode Shape to Frequency Response Function Methods. *Mech. Syst. Signal Process.* **2003**, *17*, 489–498. [CrossRef]
16. Abdo, M.A.B.; Hori, M. A Numerical Study of Structural Damage Detection Using Changes in the Rotation of Mode Shapes. *J. Sound Vib.* **2002**, *251*, 227–239. [CrossRef]
17. Pandey, A.; Biswas, M.; Samman, M. Damage Detection From Mode Changes in Curvature Mode Shapes. *J. Sound Vib.* **1991**, *145*, 321–332. [CrossRef]
18. Qiao, P.; Lu, K.; Lestari, W.; Wang, J. Curvature mode shape-based damage detection in composite laminated plates. *Compos. Struct.* **2007**, *80*, 409–428. [CrossRef]
19. Wahab, M.A.; Roeck, G.D.E. Damage detection in bridges using modal curvatures: Application to a real damage scenario. *J. Sound Vib.* **1999**, *226*, 217–235. [CrossRef]
20. Lu, X.B.; Liu, J.K.; Lu, Z.R. A two-step approach for crack identification in beam. *J. Sound Vib.* **2013**, *332*, 282–293. [CrossRef]
21. Cornwell, P.; Doebling, S.; Farrar, C. Application of the Strain Energy damage detection method to plate-like structures. *J. Sound Vib.* **1999**, *224*, 359–374. [CrossRef]
22. Hu, H.; Wang, B.; Lee, C.; Su, J. Damage detection of surface cracks in composite laminates using modal analysis and strain energy method. *Compos. Struct.* **2006**, *74*, 399–405. [CrossRef]
23. Wang, Z.X.; Qiao, P.; Xu, J. Vibration analysis of laminated composite plates with damage using the perturbation method. *Compos. Part B Eng.* **2015**, *72*, 160–174. [CrossRef]
24. Narkis, Y. Identification of crack location in vibrating simply supported beam. *J. Sound Vib.* **1992**, *174*, 549–558. [CrossRef]

25. Krawczuk, M.; Ostachowicz, W.M. Modelling and vibration analysis of a cantilever composite beam with a transverse open crack. *J. Sound Vib.* **1995**, *183*, 58–78. [CrossRef]
26. Ruotolo, R.; Surace, C.; Crespu, P.; Storer, D. Harmonic analysis of the vibrations of a cantilevered beam with a closing crack. *Compuars Swucrures* **1996**, *61*, 1057–1074. [CrossRef]
27. Tsai, T.C.; Wang, Y.Z. Vibration analysis and diagnosis of a cracked shaft. *J. Sound Vib.* **1996**, *192*, 607–620. [CrossRef]
28. Ratcliffe, C.P. Damage detection using a modified laplacian operator on mode shape data. *J. Sound Vib.* **1997**, *204*, 505–517. [CrossRef]
29. Fernadndez-saez, J.; Rubio, L.; Navarro, C. Approximate calculation of the fundamental frequency for bending vibrations of cracked beams. *J. Sound Vib.* **1999**, *225*, 345–352. [CrossRef]
30. Yang, X.F.; Swamidas, A.S.J.; Seshadri, R. Crack identification in vibrating beams using the energy method. *J. Sound Vib.* **2001**, *244*, 339–357. [CrossRef]
31. Roy Mahapatra, D.; Gopalakrishnan, S. A spectral finite element model for analysis of axial–flexural–shear coupled wave propagation in laminated composite beams. *Compos. Struct.* **2003**, *59*, 67–88. [CrossRef]
32. Xia, Y.; Hao, H. Statistical damage identification of structures with frequency changes. *J. Sound Vib.* **2003**, *263*, 853–870. [CrossRef]
33. Cam, E.; Orhan, S.; Luy, M. An analysis of cracked beam structure using impact echo method. *NDT E Int.* **2005**, *38*, 368–373. [CrossRef]
34. Loya, J.A.; Rubio, L.; Fernandez-Saez, J. Natural frequencies for bending vibrations of Timoshenko cracked beams. *J. Sound Vib.* **2006**, *290*, 640–653. [CrossRef]
35. Orhan, S. Analysis of free and forced vibration of a cracked cantilever beam. *NDT E Int.* **2007**, *40*, 443–450. [CrossRef]
36. Yu, L.; Cheng, L.; Yam, S.M.; Yan, Y.J.; Jiang, J.S. Online damage detection for laminated composite shells partially filled with fluid. *Compos. Struct.* **2007**, *80*, 334–342. [CrossRef]
37. Peng, Z.K.; Lang, Z.Q.; Chu, F.L. Numerical analysis of cracked beams using nonlinear output frequency response functions. *Comput. Struct.* **2008**, *86*, 1809–1818. [CrossRef]
38. Dessi, D.; Camerlengo, G. Damage identification techniques via modal curvature analysis: Overview and comparison. *Mech. Syst. Signal Process.* **2015**, *52*, 181–205. [CrossRef]
39. Laflamme, S.; Cao, L.; Chatzi, E.; Ubertini, F. Damage Detection and Localization from Dense Network of Strain Sensors. *Shock Vib.* **2016**, *2016*, 2562949. [CrossRef]
40. Yang, Z.-B.; Radzienski, M.; Kudela, P.; Ostachowicz, W. Two-dimensional Chebyshev pseudo spectral modal curvature and its application in damage detection for composite plates. *Compos. Struct.* **2017**, *168*, 372–383. [CrossRef]
41. Ashory, M.; Ghasemi-Ghalebahman, A.; Kokabi, M. An efficient modal strain energy-based damage detection for laminated composite plates. *Adv. Compos. Mater.* **2018**, *27*, 147–162. [CrossRef]
42. Pan, J.; Zhang, Z.; Wu, J.; RamRamakrishnan, K.; Singh, H.K. A novel method of vibration modes selection for improving accuracy of frequency-based damage detection. *Compos. Part B Eng.* **2019**, *159*, 437–446. [CrossRef]
43. Bao, Y.; Valipour, M.; Meng, W.; Khayat, K.H.; Chen, G. Distributed fiber optic sensor-enhanced detection and prediction of shrinkage-induced delamination of ultra-high-performance concrete overlay. *Smart Mater. Struct.* **2017**, *26*, 085009. [CrossRef]
44. Kannivel, S.k.; Subramanian, H.; Arumugam, V.; Dhakal, H.N. Low-Velocity Impact Induced Damage Evaluation and Its Influence on the Residual Flexural Behavior of Glass/Epoxy Laminates Hybridized with Glass Fillers. *J. Compos. Sci.* **2020**, *4*, 99. [CrossRef]
45. Linke, M.; Flügge, F.; Jose, A. Olivares–Ferrer, Design and Validation of a Modified Compression-After-Impact Testing Device for Thin-Walled Composite Plates. *J. Compos. Sci.* **2020**, *4*, 126. [CrossRef]
46. Arena, M.; Viscardi, M. Strain State Detection in Composite Structures: Review and New Challenges. *J. Compos. Sci.* **2020**, *4*, 60. [CrossRef]
47. Meng, W.; Khayat, K.H. Experimental and numerical studies on flexural behaviour of ultra high-performance concrete panels reinforced with embedded glass fiber-reinforced polymer grids. *Transp. Res. Rec.* **2016**, *2592*, 38–44. [CrossRef]
48. Stamoulis, K.; Georgantzinos, S.K.; Giannopoulos, G.I. Damage characteristics in laminated composite structures subjected to low-velocity impact. *Int. J. Struct. Integr.* **2019**, *11*, 670–685. [CrossRef]

49. Dabetwar, S.; Ekwaro-Osire, S.; Dia, J.P. Damage Classification of Composites Based on Analysis of Lamb Wave Signals Using Machine Learning. *ASCE-ASME J. Risk Uncert Engrg. Sys. Part B Mech. Engrg* **2020**. [CrossRef]
50. Georgantzinos, S.K.; Giannopoulos, G.I.; Markolefas, S.I. Vibration Analysis of Carbon Fiber-Graphene-Reinforced Hybrid Polymer Composites Using Finite Element Techniques. *Materials* **2020**, *13*, 4225. [CrossRef]
51. Chen, Z.; Zhou, X.; Wang, X.; Dong, L.; Qian, Y. Deployment of a smart structural health monitoring system for long-span arch bridges: A review and a case study. *Sensors* **2017**, *17*, 2151. [CrossRef] [PubMed]

Publisher's Note: MDPI stays neutral with regard to jurisdictional claims in published maps and institutional affiliations.

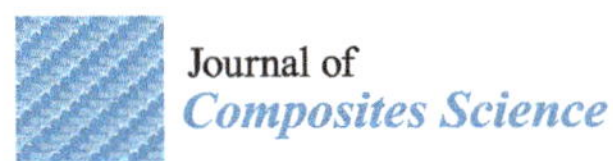

Journal of
Composites Science

Article

Mechanical Strength of Bamboo Filled PLA Composite Material in Fused Filament Fabrication

Scott Landes and Todd Letcher *

Mechanical Engineering Department, South Dakota State University, Brookings, SD 57007, USA;
scott.landes@jacks.sdstate.edu
* Correspondence: todd.letcher@sdstate.edu

Received: 22 September 2020; Accepted: 23 October 2020; Published: 26 October 2020

Abstract: Through the past two decades, there has been a continued push for renewable resources and future sustainability of materials and processes. This has prompted more developments of providing environmentally friendly practices and products, both in terms of higher recyclability and greater use of renewable resources. An important area of interest are materials for construction and manufacturing purposes, specifically "green" sustainable reinforcement materials for thermoplastic composite materials. During this time, there has also been an evolution in manufacturing methods. Additive manufacturing (AM) has continued to grow exponentially since its inception for its extensive benefits. This study aims to investigate an additive manufactured composite material that is a greener alternative to other composites that are not reinforced by natural fibers. A bamboo filled polylactic acid (PLA) composite manufactured by fused filament fabrication was evaluated in order to gather mechanical strength characteristics by means of tensile, flexure, compression, impact, and shear tests. In this material, the bamboo reinforcing material and the PLA matrix material can both be sourced from highly renewable resources. In this study, a variety of test samples were manufactured at different manufacturing parameters to be used for mechanical testing. The results were recorded with respect to varying manufacturing parameters (raster angle orientation). It was found that the 0° raster angle orientation performed the best in every category except tensile. Additively manufactured bamboo filled PLA was also seen to have comparable strength to certain traditionally manufactured bamboo fiber reinforced plastics.

Keywords: fused filament fabrication; PLA; bamboo; mechanical strength

1. Introduction

Growing demand for renewable resources and environmentally friendly practices have prompted research in many different areas [1–3]. One of which includes the mechanics (or strength) of materials. Plastics are especially notorious for negative environmental impact. Plastics are polymers, which consist of repeated monomers (simple molecules) that are covalently bonded to one another in a chain like structure. The benefits of such polymers are widely known and utilized, such as excellent strength to weight ratio as well as resistance to corrosion. In order to further increase the strength of the material while retaining the benefits of polymers, composite materials have been created to attempt to combine the advantageous properties of multiple materials while minimizing the disadvantages. One type of composite material incorporates continuous or chopped fibers of a strengthening material within a matrix polymer. Many polymer-matrix composites have been investigated and they have been widely used across many industrial applications. One of the most common is carbon fiber reinforced polymers (CFRP). However, with growing demand for sustainable and biodegradable materials, carbon fiber is not ideal, because much of it is manufactured using a petroleum-based product [4,5]. Additionally, it is manufactured using a variety of other costly materials and methods, making it not economical for many applications. Using natural fibers in composites provides a sustainable reinforcement material that could potentially address the growing demand for renewable environmentally friendly materials [6].

The cellulosic fibers found in bamboo is a natural fiber that can address this demand [7,8]. Bamboo is a perennial flowering plant in the grass family [9]. Bamboo is often referred to as a tree for its tree-like appearance and size. It is also known as one of the fastest growing plants and is easily harvested which make them especially attractive for sustainability and renewability [7]. Some stronger species of bamboo have structural properties that are similar to high grade hardwoods [10]. However, there are some difficulties that are seen when using bamboo fibers in composites. The extraction of consistent quality bamboo fibers due to brittleness, as well as variations of strength, depending on the well-being of the plant harvested, are just a few reasons for fiber strength variability [11,12]. Another challenge that has been seen is forming interphase between the polymer-matrix and bamboo fibers [5]. Adding an agent to promote bonding between the two phases has become standard for bamboo fiber reinforced polymers [5,11].

The manufacturing of composites can sometimes prove to be challenging, depending upon the reinforcement filler, matrix material, manufacturing structure, and others. Automating this task through additive manufacturing (AM) can relieve some of the challenges seen while also providing the additional benefits of AM. Additive manufacturing is the process in which a part is created by depositing material layer by layer. Some of the major and widely known benefits of additive manufacturing process are rapid prototyping, increased geometric complexity possibilities, low volume production affordability, etc. The process of fused filament fabrication (FFF) has been a major building block in AM polymer-matrix composites. FFF is the process in which a thermoplastic filament is heated and extruded through a nozzle that moves around a build plate to construct a layer of a part. Next, the nozzle moves vertically up by one layer thickness and the next layer is added on top of the previous layer. This continues until the part is fully constructed. Many composites have been investigated and successfully manufactured while using FFF. Some machines incorporate continuous reinforcement fibers placed in between layers of polymer [13–15], while other AM composites are produced using a filament in which already incorporates a chopped or ground reinforcement material [16–20]. A variety of types of natural reinforcement materials have been evaluated for FFF, such as cork and hemp [21,22]. One such FFF filament is a bamboo filled polylactic acid (PLA). Utilizing an AM process to produce a bamboo filled PLA component fulfills renewable and biodegradable requests for new materials by creating a material that is fully biodegradable and made from renewable materials.

Research has been completed on areas, such as mechanical and thermal properties of bamboo fiber reinforced epoxy and polymer composites manufactured by traditional means [23–27]. Research has also been conducted on the preparation of a bamboo filled PLA material intended for AM [28]. However, no research could be found on the mechanical properties of AM bamboo filled composite material. Presenting the mechanical strength of an AM bamboo filled PLA will result in readily available data, which could aid in the comparison and future advancement of modern manufacturing techniques of bamboo reinforced polymer composites. Anisotropy is regularly apparent in the parts produced due to the nature of the AM process. This requires the evaluation of material properties with respect to the AM build parameters. This paper aims to evaluate the mechanical properties of AM bamboo filler reinforced PLA properties accounting for varying raster orientation (orientation angle at which print infill lines are deposited). Testing utilized in the evaluation includes density, tensile, flexural, compression, impact, and shear.

2. Materials and Methods

2.1. Materials

All of the specimens used in this study were manufactured using a Flashforge Creator Pro (Zhejiang Flashforge3D Technology Co., Jinhua, China) with a 0.6mm nozzle. This nozzle size is slightly larger than usual nozzle sizes used for non-filled polymers, but is in line with common recommendations for composite materials that were used in FFF. Makerbot Desktop slicing software (Makerbot Desktop, Makerbot, Brooklyn, NY, USA) [29] was used for G-code compilation. Depending on the mechanical test, the following raster orientation were evaluated: 0°, 90°, alternating 0°/90°, and alternating −45°/45°. The bamboo filled PLA filament was manufactured by eSUN and is sold under the material name EBamboo [30]. Table 1 presents all of the relevant build parameters for the specimens.

Table 1. Print Parameters Used for All Specimen Creation.

Parameter	Value
Infill Style	Lines
Infill %	Solid (100%)
Nozzle	0.6 mm
Layer Height	0.2 mm
Contours/Shells	2
Nozzle Temperature	230 °C
Build Plate Temperature	50 °C
Print Speed	60 mm/s

Limited information pertaining to the bamboo filled PLA filament was available from the manufacturer. In order to gain a better understanding of the composition of the filament, the percentage of bamboo filler to PLA matrix was estimated by testing samples of the bamboo filament and recording volume measurements taken on Micromeritics AccuPyc II 1340 pycnometer (Micromeritics Instrument Corporation, Norcross, GA, USA). The mass of the samples was then recorded and used to calculate the density of the filament. This process was also completed with standard PLA filament available from eSUN, assumed to be the same as the EBamboo matrix material. An average density of 0.9 g/cm^3 was assumed for bamboo. The percentage of bamboo filler in the material was determined to be 11.8% using Equation (1).

$$\frac{\rho_{composite} - \rho_{matrix}}{\rho_{reinforcment} - \rho_{matrix}} = \% \, of \, Filler \, Material \tag{1}$$

FFF of the bamboo filled PLA composite produced satisfactory high-quality parts with limited defects and it was easily printed. It has a light brown color similar to bamboo with a woody-like texture and aroma, very similar to other wood-filled polymers available for FFF. Figure 1 displays magnified cross sectional views of the filament with the right side of the figure being powder size distribution measurements. Darker brown spots are the bamboo particles that are infused into the composite material. Through image processing, the bamboo could be isolated from the PLA matrix, and powder size measurements were taken. Assuming the measured areas to be circular, the powder sizes ranged from 5 μm to 35 μm in diameter. Figure 2 shows an average representation of print quality of Bamboo/PLA FFF printed parts. As can be seen, the parts printed well and were of high quality. Few defects were present and only in areas where the nozzle started/stopped. These problems can likely be reduced/removed by fine-tuning retraction settings and/or by using a Z-hop maneuver.

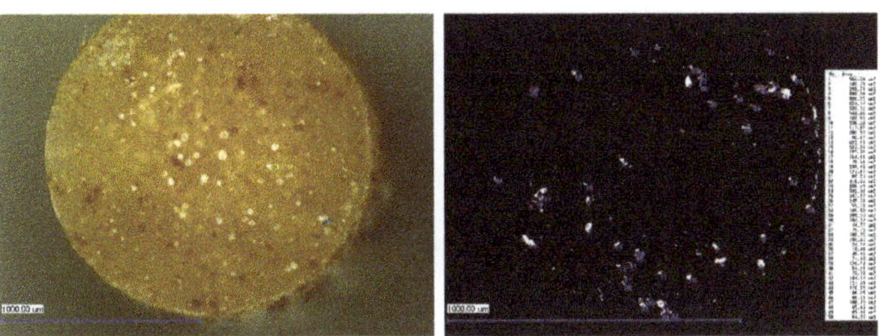

Figure 1. 150× Magnified Top View of Filament (**Left**) and Process Image for Bamboo Powder Size Evaluation (**Right**).

Figure 2. Example of representative printed part quality.

One potential cause for concern was the degradation of bamboo during the AM process. The filament was extruded at a temperature of 230 °C in this study. Prior research has evaluated thermal degradation of bamboo reinforced epoxy composites through Thermogravimetric Analysis (TGA). It was found that degradation occurred at 250 °C independent upon filler percentage. Contrary, specifically, Bambusa vulgaris fibers showed low thermal stability and degradation occurring at 220 °C [31]. It should be noted that, during TGA, the temperature was increased at a rate of 10 °C/min. Being that the TGA process slowly heated the material (10 °C/min), it cannot be directly compared to the AM extrusion process where the filament is heated to 230 °C and then immediately starts cooling. However, we could expect very little bamboo degradation during the FFF process that was based on the previous research [31].

2.2. Mechanical Testing

Each specimen's cross-sectional area was measured at multiple locations along the test section prior to conducting testing, as per the American Society for Testing and Materials (ASTM) testing standard for each test type. These cross-sectional area measurements were then averaged and used to calculate stress/strain for each test. Tensile, flexural and shear tests were all conducted using an MTS Insight load frame with a 5 kN MTS load cell. Compression testing was completed on an MTS Landmark load frame with a 100 kN load cell.

For each type of mechanical test conducted, samples from a variety of raster angles were tested, similar to [32]. In all cases, raster angles of 0° are the printing orientation that aligns the toolpath in the lengthwise direction (strongest orientation). The orientation angle increases in the CCW direction that was measured from the specimen's longitudinal axis. The different raster angles employed in this study are shown in Figure 3, where the x-axes represent the specimen's longitudinal axis.

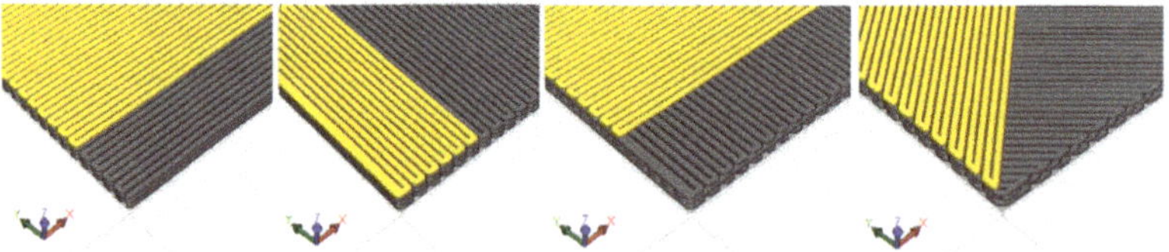

Figure 3. Raster Orientations 0°, 90°, 0°/90°, −45°/45°, respectively from left to right.

2.2.1. Tensile

Tensile tests were conducted in accordance to ASTM D638 Standard Test Methods for Tensile Properties of Plastics [33]. Tensile tests were displacement controlled at a rate of 5 mm/min. while data (force, grip displacement, and strain) was collected at a rate of 100 Hz. Strain was measured using an MTS Model 634.31F-24 extensometer (MTS, Eden Prairie, MN, USA). Six specimens of each of the four raster orientations were tested. The dimensions of the tensile specimens follow Type IV in ASTM D638, which means cross sectional areas of approximately 6 mm × 4 mm.

2.2.2. Flexural

Flexural specimens were tested in accordance to ASTM D790 Standard Test Methods for Flexural Properties of Unreinforced and Reinforced Plastics and Electrical Insulating Materials [34]. Flexural tests were conducted using an MTS 3-point flexure fixture with a support span of 5 cm, while displacement controlled at a rate of 1.3 mm/min. Following the recommendations for specimen size in ASTM D790 leads to samples of cross-sectional area of 4 mm × 12.7 mm.

2.2.3. Compression

Compression tests were completed in accordance to ASTM D695 Standard Test Method for Compressive Properties of Rigid Plastics [35]. Two platens that were attached to each grip were used to apply the compressive load while the displacement of each specimen was measured with the built-in linear variable differential transformer (LVDT). The tests were displacement controlled at a rate of 1.3 mm/min. Specimens were round cylinders measuring approximately 12.75 mm in diameter and 50 mm in length. Due to the simplicity/symmetry of the geometry, the 0° and 90° as well as the 0°/90° and −45°/45° raster orientations were effectively the same. Therefore, six specimens of just two orientations (0° and −45°/45°) were tested (see Figure 4).

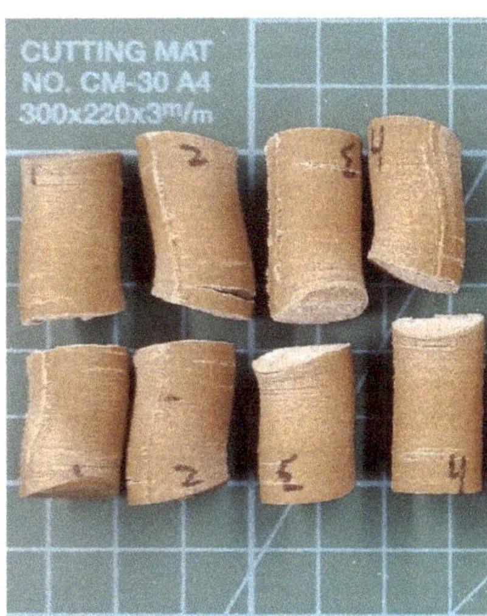

Figure 4. Example Compression Specimens Before (**Left**) and After (**Right**) Testing.

2.2.4. Impact

The impact tests were conducted in accordance with ASTM D256 Standard Test Methods for Determining the Izod Pendulum Impact Resistance of Plastics while using a calibrated custom build Izod Impact tester [36]. Figure 5 shows a picture of the Impact testing setup along with an example of impact specimens. Six specimens of 0°, 90°, and −45°/45° raster orientations were tested.

Figure 5. Impact Testing Setup (**Left**) and Examples of Impact Testing Specimens (**Right**).

2.2.5. Shear

The shear tests were conducted in accordance to ASTM D5379-19 Standard Test Method for Shear Properties of Composite Materials by the V-Notched Beam Method using a v-notched beam test fixture as shown in Figure 6 [37]. Prior to testing, the specimens were prepped for digital image correlation (DIC) strain measurement by spray painting a fully white background and then black speckles were applied by spray paint can. The speckles would be tracked by DIC software. Three specimens of 0°, 90°, and −45°/45° raster orientations were tested.

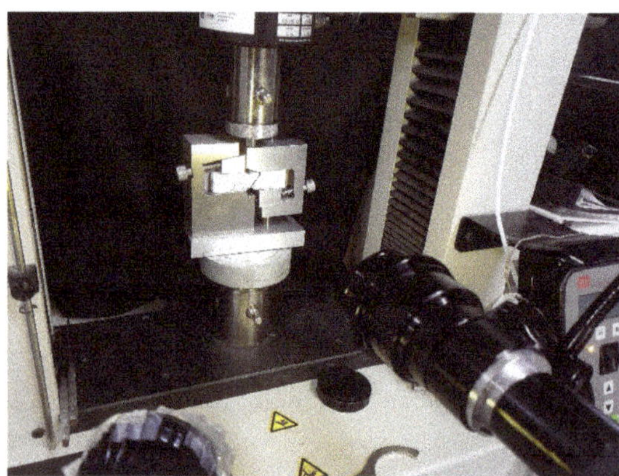

Figure 6. Shear Test Setup after Specimen Failure (**Left**) and Prepped Specimens for Shear Test (**Right**).

Strain data were collected while using a custom created DIC capture system. This system uses a LabView program that captures and image and a load cell measurement simultaneously. The pictures are captured using an Allied Vision Technologies Marlin F-046 camera (Allied Vision Technologies, Stadtroda, Thuringia, Germany). Pictures and corresponding load values are then converted into stress/strain values using GOM Correlate [38].

3. Results

3.1. Tensile

The average tensile testing results with standard deviations are shown for each raster orientation in Table 2. Figure 7 shows representative stress-strain curves of each raster orientation.

Table 2. Tensile Testing Results.

Raster Orientation	Ultimate Tensile Strength (MPa)	Modulus of Elasticity (GPa)	Yield Stress (MPa)	Maximum Elongation (%)
−45°/45°	32.50 ± 0.92	2.63 ± 0.16	30.33 ± 1.22	3.161 ± 0.64
0°	29.21 ± 0.98	2.36 ± 0.09	27.64 ± 0.75	2.428 ± 0.49
90°	31.10 ± 1.53	2.49 ± 0.15	29.38 ± 1.63	3.502 ± 0.96
0°/90°	30.00 ± 1.23	2.49 ± 0.14	27.50 ± 1.52	5.074 ± 1.23

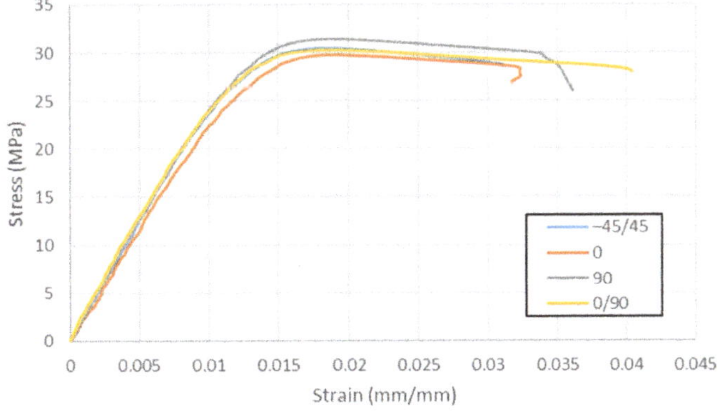

Figure 7. Representative Tensile Stress-Strain Curves.

All of the raster orientations resulted in comparable ultimate tensile strengths with variations between the four orientations. However, there was a notable difference seen in the ductility of the material, depending on raster orientation. The 0°/90° orientation was found to be the most ductile.

3.2. Flexural

Table 3 shows the average flexural testing results and standard deviations for each raster orientation. Figure 8 displays representative stress-strain curves from the flexural testing for each raster orientation.

Table 3. Flexural Testing Results.

Raster Orientation	Ultimate Flexural Strength (MPa)	Modulus of Flexure (GPa)	Yield Stress (MPa)	Maximum Elongation (%)
−45°/45°	56.76 ± 3.00	2.40 ± 0.14	48.50 ± 2.77	4.224 ± 0.22
0°	54.57 ± 1.44	2.26 ± 0.05	46.33 ± 1.96	4.581 ± 0.49
90°	49.35 ± 1.65	2.15 ± 0.12	41.50 ± 2.55	4.001 ± 0.36
0°/90°	52.38 ± 0.82	2.22 ± 0.04	44.45 ± 2.26	4.481 ± 0.46

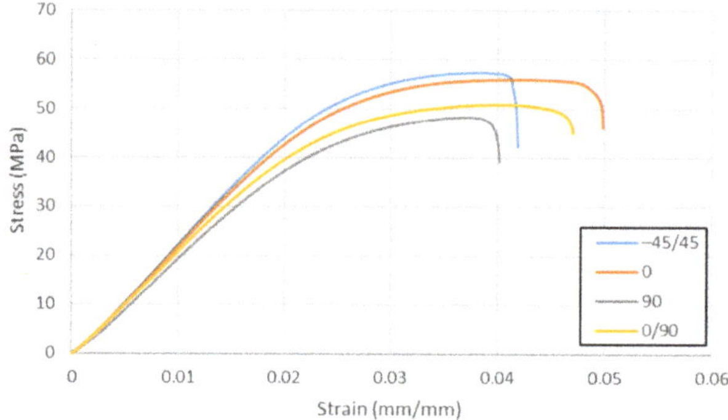

Figure 8. Representative Flexural Stress-Strain Curves.

The 90° orientation resulted in the lowest ultimate flexural strength as well as the least ductility. This result matches intuition. Because it should be expected that side by side bonding would be not as strong as continuous length of material extruded (as in the 0° direction). The 0° and −45°/45° orientation produced higher ultimate flexural strengths. Alternating orientation of 0°/90° had lower strength at the benefit of better ductility. Overall, the 0° orientation had the best fracture toughness by having comparable strength to the best results, while also being the most ductile.

3.3. Compression

The average results and standard deviations from compression testing is shown in Table 4. No significant difference was seen in the compressive properties between the different orientations. Figure 9 illustrates the representative compressive stress–strain curves.

Table 4. Compression Testing Results.

Raster Orientation	Ultimate Compressive Strength (MPa)	Compressive Modulus (GPa)
−45°/45°	48.55 ± 4.31	1.89 ± 0.05
0°	48.07 ± 2.43	1.95 ± 0.09

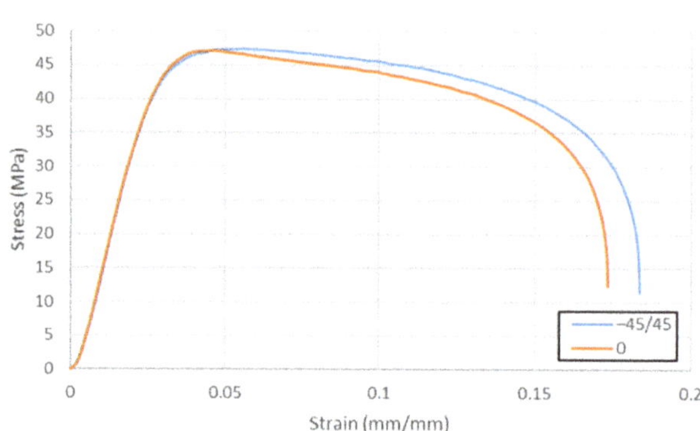

Figure 9. Representative Compressive Stress-Strain Curves.

3.4. Impact

Izod impact tests were conducted in order to record the energy absorption of the material before fracturing in a fast impact fracture. Table 5 shows the average results and standard deviation for various raster orientations. Significant differences were seen between the varying orientations. The 0°

orientation clearly performed the best with an impact energy absorption double the −45°/45° and triple the 90°.

Table 5. Izod Impact Testing Results.

Raster Orientation	Impact Energy (J/m)
−45°/45°	21.01 ± 0.84
0°	42.94 ± 1.22
90°	13.94 ± 2.68

3.5. Shear

The ultimate shear strength and shear chord modulus of elasticity were recorded from shear tests. Table 6 shows the results. Figure 10 shows shear stress–strain curves. It can be noted that all of the shear tests show brittle behavior. The 0° orientation had the highest ultimate shear strength (29 MPa) when compared to the other orientations, as expected. The −45°/45° orientations had the lowest ultimate shear strength at 25 MPa. Interestingly, the 0° orientation also had the lowest shear modulus of elasticity.

Table 6. Shear Testing Results.

Raster Orientation	Ultimate Shear Strength (MPa)	Shear Chord Modulus of Elasticity (GPa)
−45°/45°	25.11 ± 4.371	1.32 ± 0.25
0°	29.13 ± 0.97	1.13 ± 0.34
90°	26.84 ± 2.06	2.07 ± 0.68

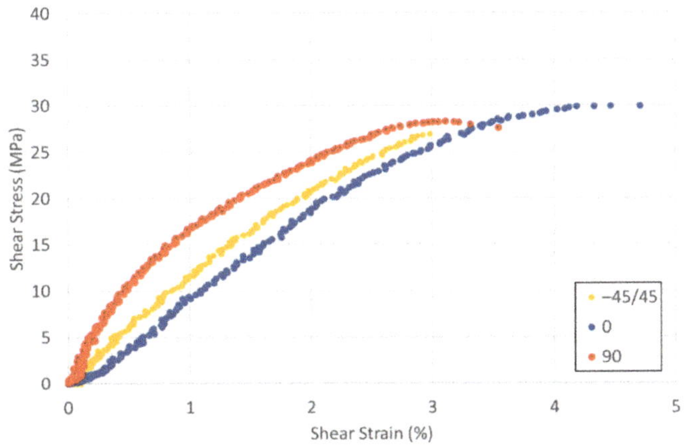

Figure 10. Representative Shear Stress-Strain Curves.

4. Discussion

The composite material that was evaluated in this study used bamboo as a reinforcement material, not fibers, but a bamboo powder filler. Bamboo's sustainable natural reinforcement fibers provides an environmentally friendly option to compete with other composites. Specifically, bamboo infused PLA for AM is a topic that has limited data available.

From the results of the testing in this study, orientation had little effect on ultimate tensile strength with all orientations being within 3 MPa of each other. However, a more notable change in ductility was seen in different orientations. The 0° orientation was seen to be the least ductile at a maximum elongation of 2.428 ± 0.49%, while the 0°/90° orientation was the most ductile at 5.074 ± 1.23%. Flexural testing showed slight changes that were based on orientation. The 0° orientation was determined in order to perform the best under flexure, for it had nearly the highest ultimate flexure strength while also being the must ductile, providing the highest fracture toughness. Compression testing concluded that there were no notable differences in mechanical strength between orientations and similar results can be expected independent upon orientation. Impact testing found that nearly triple the amount of impact energy can be absorbed in the 0° orientation when compared to the 90° and double the amount of impact energy when compared to the −45°/45°. Shear testing found that the composite material was extremely brittle in shear and no notable differences were seen in the ultimate shear strength, while, again, the 0° orientation still had a slight advantage over the other orientations.

When comparing the AM bamboo filled PLA to other bamboo fiber reinforced polymer composites made from tradition methods, it was found to have comparable if not better (assuming 11.8% bamboo filler as calculated in Equation (1)) tensile and flexural strength [25]. The AM of the material was also found to produce satisfactory parts of limited defects with relative ease. Reviewing neat PLA, the AM bamboo/PLA composite produced strength that was nearly half that of neat PLA in tensile and flexural testing [32]. However, the bamboo/PLA looked to be less mechanically susceptible to print orientations differences, which could be of use when used in applications of varying loading directions. Overall, the results of this study found that the AM of bamboo filled PLA composite is an effective means of producing an environmentally friendly, recyclable, and sustainable bamboo reinforced polymer-matrix composite.

5. Conclusions

Bamboo filled PLA composite material was AM using FFF. Four raster orientation angles (0°, 90°, 0°/90°, and −45°/45°) were evaluated for anisotropic effects. Tensile, flexure, compression, impact, and shear tests were all conducted in order to provide mechanical strength characteristics. It was found that the raster orientation had the largest effect on ductility and impact energy absorption. The 0° orientation was found to perform the best in all categories, except tensile testing, where it was found to have the least strength and ductility of the four orientations.

AM bamboo filled PLA was found to produce satisfactory parts with limited defects. The AM bamboo composites were also seen to have similar mechanical properties to certain bamboo fiber reinforced plastics, of which were traditionally manufactured and have nearly half the strength of neat PLA. Future work could include a mechanistic analysis of bamboo percentage affects. The material is also an environmentally friendly option when compared to other polymer-matrix composites due to the natural and sustainable reinforcement fibers that bamboo offers.

Author Contributions: Conceptualization, T.L.; methodology, T.L.; analysis, T.L. and S.L.; data curation, T.L. and S.L.; writing—original draft preparation, S.L.; writing—review and editing, T.L. and S.L.; visualization, T.L. and S.L. Both authors have read and agreed to the published version of the manuscript.

Funding: Funding to purchase Esun Ebamboo filament was provided by the SD Space Grant Consortium. The Mechanical Engineering Department at South Dakota State University provided the use of material testing equipment.

Acknowledgments: The authors would like to thank the Mechanical Engineering Department and the Materials and Evaluation Testing Laboratory at South Dakota State University for the use of the material testing equipment.

Conflicts of Interest: The authors declare no conflict of interest.

References

1. Jiménez-González, C.; Poechlauer, P.; Broxterman, Q.B.; Yang, B.-S.; Ende, D.A.; Baird, J.; Bertsch, C.; Hannah, R.E.; Dell'Orco, P.; Noorman, H.; et al. Key Green Engineering Research Areas for Sustainable Manufacturing: A Perspective from Pharmaceutical and Fine Chemicals Manufacturers. *Org. Process. Res. Dev.* **2011**, *15*, 900–911. [CrossRef]
2. Jang, Y.-S.; Lee, J.; Malaviya, A.; Seung, D.Y.; Cho, J.H.; Lee, S.Y. Butanol production from renewable biomass: Rediscovery of metabolic pathways and metabolic engineering. *Biotechnol. J.* **2012**, *7*, 186–198. [CrossRef]

3. Mohanty, A.K.; Wibowo, A.; Misra, M.; Drzal, L.T. Development of renewable resource-based cellulose acetate bioplastic: Effect of process engineering on the performance of cellulosic plastics. *Polym. Eng. Sci.* **2003**, *43*, 1151–1161. [CrossRef]

4. Johnson, T. All about Carbon Fiber and How It's Made. Available online: https://www.thoughtco.com/how-is-carbon-fiber-made-820391 (accessed on 27 August 2020).

5. Kannan, R.; Ahmad, M.M.H.M. A review on mechanical properties of bamboo fiber reinforced polymer composite. *Aust. J. Basic Appl. Sci.* **2013**, *7*, 247–253.

6. Faruk, O.; Bledzki, A.K.; Fink, H.-P.; Sain, M. Progress Report on Natural Fiber Reinforced Composites. *Macromol. Mater. Eng.* **2014**, *299*, 9–26. [CrossRef]

7. Delgado, P.S.; Lana, S.L.B.; Ayres, E.; Patrício, P.O.S.; Oréfice, R.L. The potential of bamboo in the design of polymer composites. *Mater. Res.* **2012**, *15*, 639–644. [CrossRef]

8. Verma, D.; Jain, S.; Zhang, X.; Gope, P.C. (Eds.) *Green Approaches to Biocomposite Materials Science and Engineering*; IGI Global: Hershey, PA, USA, 2016; p. 322.

9. Chand, N.; Fahim, M. Bamboo Reinforced Polymer Composites. In *Tribology of Natural Fiber Polymer Composites*; Elsevier: Amsterdam, The Netherlands, 2008; pp. 162–179.

10. Kaminski, S.; Lawrence, A.; Trujillo, D. Structural use of bamboo: Part 1: Introduction to bamboo. *Struct. Eng.* **2016**, *94*, 40–43.

11. Song, K.; Ren, X.; Zhang, L. Bamboo Fiber-Polymer Composites: Overview of Fabrications, Mechanical Characterizations and Applications. *Green Energy Technol.* **2017**, 209–246. [CrossRef]

12. Kaur, N.; Saxena, S.; Gaur, H.; Goyal, P. A review on bamboo fiber composites and its applications. In Proceedings of the 2017 International Conference on Infocom Technologies and Unmanned Systems (ICTUS 2017), Dubai, UAE, 18–20 December 2017; pp. 843–849.

13. Metal and Carbon Fiber 3D Printers for Manufacturing|Markforged. Available online: https://markforged.com/ (accessed on 15 October 2020).

14. Sauer, M. Evaluation of the Mechanical Properties of 3D Printed Carbon Fiber Composites. Master's Thesis, South Dakota State University, Brookings, SD, USA, 2018.

15. Caminero, M.; Chacón, J.; García-Moreno, I.; Rodríguez, G. Impact damage resistance of 3D printed continuous fibre reinforced thermoplastic composites using fused deposition modelling. *Compos. Part B Eng.* **2018**, *148*, 93–103. [CrossRef]

16. Hu, Z.; Thiyagarajan, K.; Bhusal, A.; Letcher, T.; Fan, Q.H.; Liu, Q.; Salem, D.; Kaushik, T. Design of ultra-lightweight and high-strength cellular structural composites inspired by biomimetics. *Compos. Part B Eng.* **2017**, *121*, 108–121. [CrossRef]

17. Coughlin, N.; Drake, B.; Fjerstad, M.; Schuster, E.; Waege, T.; Weerakkody, A.; Letcher, T. Development and Mechanical Properties of Basalt Fiber-Reinforced Acrylonitrile Butadiene Styrene for In-Space Manufacturing Applications. *J. Compos. Sci.* **2019**, *3*, 89. [CrossRef]

18. Wang, X.; Jiang, M.; Zhou, Z.; Gou, J.; Hui, D. *3D Printing of Polymer Matrix Composites: A Review and Prospective*; Elsevier Ltd.: Amsterdam, The Netherlands, 2017; Volume 110, pp. 442–458.

19. Compton, B.G.; Lewis, J.A. 3D-Printing of Lightweight Cellular Composites. *Adv. Mater.* **2014**, *26*, 5930–5935. [CrossRef]

20. 3DXTECH. CarbonX™ CF-PLA 3D Printing Filament. Available online: https://www.3dxtech.com/tech-data-sheets-safety-data-sheets/ (accessed on 15 October 2020).

21. Daver, F.; Lee, K.P.M.; Brandt, M.; Shanks, R. Cork–PLA composite filaments for fused deposition modelling. *Compos. Sci. Technol.* **2018**, *168*, 230–237. [CrossRef]

22. Xiao, X.; Chevali, V.S.; Song, P.; He, D.; Wang, H. Polylactide/hemp hurd biocoposites as sustaainable 3D printing feedstock. *Compos. Sci. Technol.* **2019**, *184*, 107887. [CrossRef]

23. Prabhu, R.; Joel, C.; Bhat, T. Development and Characterization of Low Cost Bamboo Fibre Reinforced Polymer Composites. *Am. J. Mater. Sci.* **2017**, *2017*, 130–134.

24. Okubo, K.; Fujii, T.; Yamamoto, Y. Development of bamboo-based polymer composites and their mechanical properties. *Compos. Part A Appl. Sci. Manuf.* **2004**, *35*, 377–383. [CrossRef]

25. Xian, Y.; Ma, D.; Wang, C.; Wang, G.; Smith, L.; Cheng, H. Characterization and Research on Mechanical Properties of Bamboo Plastic Composites. *Polymers* **2018**, *10*, 814. [CrossRef] [PubMed]

26. Zhang, K.; Wang, F.; Liang, W.; Wang, Z.; Duan, Z.; Yang, B. Thermal and Mechanical Properties of Bamboo Fiber Reinforced Epoxy Composites. *Polymers* **2018**, *10*, 608. [CrossRef]

27. Shah, A.U.M.; Sultan, M.T.; Jawaid, M.; Cardona, F.; Abu Talib, A.R. A Review on the Tensile Properties of Bamboo Fiber Reinforced Polymer Composites. *Bioressources* **2016**, *11*. [CrossRef]
28. Zhao, D.X.; Cai, X.; Shou, G.Z.; Gu, Y.Q.; Wang, P.X. Study on the preparation of bamboo plastic composite intend for additive manufacturing. *Key Eng. Mater.* **2016**, *667*, 250–258. [CrossRef]
29. MakerBot. MakerBot—Sofware and Apps. Available online: https://www.makerbot.com/3d-printers/apps/ (accessed on 15 October 2020).
30. eSUN. eSUN eBamboo 3D Filament. Available online: http://www.esun3d.net/products/208.htmll (accessed on 15 October 2020).
31. Shah, A.U.M.; Sultan, M.T.; Cardona, F.; Jawaid, M.; Abu Talib, A.R.; Yidris, N. Thermal Analysis of Bamboo Fibre and Its Composites. *Bioresources* **2017**, *12*, 2394–2406. [CrossRef]
32. Letcher, T.; Waytashek, M. Material Property Testing of 3D-Printed Specimen in PLA on an Entry-Level 3D Printer. In *ASME International Mechanical Engineering Congress and Exposition*; American Society of Mechanical Engineers: New York, NY, USA, 2014.
33. ASTM. *ASTM D638-14 Standard Test Method for Tensile Properties of Plastics*; ASTM International: West Conshohocken, PA, USA, 2014.
34. ASTM. *ASTM D790-17 Standard Test Methods for Flexural Properties of Unreinforced and Reinforced Plastics and Electrical Insultating Materials*; ASTM International: West Conshohocken, PA, USA, 2017.
35. ASTM. *ASTM D695-15 Standard Test Method for Compressive Properties of Rigid Plastics*; ASTM International: West Conshohocken, PA, USA, 2015.
36. ASTM. *ASTM D256-10 Standard Test Methods for Determining the Izod Pendulum Impact Resistance of Plastics*; ASTM International: West Conshohocken, PA, USA, 2010.
37. ASTM. *ASTM D5379/D5379M—19 Standard Test Method for Shear Properties of Composite Materials by the V-Notched Beam Method*; ASTM International: West Conshohocken, PA, USA, 2019.
38. GOM. GOM Correlate—Software for 3D Testing Data. Available online: https://www.gom.com/3d-software/gom-correlate.html (accessed on 15 October 2020).

Publisher's Note: MDPI stays neutral with regard to jurisdictional claims in published maps and institutional affiliations.

Journal of
Composites Science

Article

Modelling and Experimental Investigation of Hexagonal Nacre-Like Structure Stiffness

Rami Rouhana [1],* and Markus Stommel [2]

[1] Chair of Plastics Technology, TU Dortmund University, Leonhard-Euler-Str. 5, 44227 Dortmund, Germany
[2] Leibniz Institute of Polymer Research, Hohe Str. 6, 01069 Dresden, Germany; stommel@ipfdd.de
* Correspondence: rami.rouhana@tu-dortmund.de

Received: 26 June 2020; Accepted: 11 July 2020; Published: 15 July 2020

Abstract: A highly ordered, hexagonal, nacre-like composite stiffness is investigated using experiments, simulations, and analytical models. Polystyrene and polyurethane are selected as materials for the manufactured specimens using laser cutting and hand lamination. A simulation geometry is made by digital microscope measurements of the specimens, and a simulation is conducted using material data based on component material characterization. Available analytical models are compared to the experimental results, and a more accurate model is derived specifically for highly ordered hexagonal tablets with relatively large in-plane gaps. The influence of hexagonal width, cut width, and interface thickness are analyzed using the hexagonal nacre-like composite stiffness model. The proposed analytical model converges within 1% with the simulation and experimental results.

Keywords: nacre; stiffness; hexagonal tablets; analytical model; finite element simulations; composite; Abaqus

1. Introduction

Nature presents several structures optimized through evolution. Nacre is currently a central inspiration for biomimetic structure design and manufacturing methods for toughened ceramics, as showcased by Bouville [1]. Nacre consists of microscale tablets of brittle ceramic material assembled in brick mortar form and connected by a relatively tough binding organic material. Natural nacre consists of calcium carbonate bonded by protein molecules and polysaccharides [2]. This construction of overlapping tablets increases the toughness of the material by two orders of magnitude or more than bulk brittle material, as described by Bathelat et al. [3]. The most attractive feature in this construction is that it maintains considerable stiffness and strength, as measured by Wang et al. [4] and Currey [5]. This feature is becoming attractive to composite engineers for the development of novel composite types, which has been an active research area for the last two decades. Budarapu [6] further adds that the nacre structure resists crack propagation running across the tablets and main deformation of the structure, which is due to tablet sliding until pull-out.

Several methods at different scales have replicated the structure. At the macro-scale, constructed sheets are described in the work of Yin et al. 2019 [7] and Valashani and Barthelat [8], where millimeter-scale tablets were produced from laser cut glass sheets and were assembled using polymeric ethylene-vinyl acetate or polyurethane as matrix material, and then tested for side impact loads. Pre-preg fiber-reinforced sheets are cut into tiles and carefully assembled to overlap by half of their length in the work of Narducci and Pinho [9].

At the micro-scale, layers of reinforcing material are randomly assembled, or oriented/positioned with limited control. In the work of Grossman et al. [10], alumina tablets were magnetically oriented and bonded using polyetherimide (PEI) polymer as a matrix. Launey et al. [11] presented a method using ice-templating to produce a ceramic-metal composite with good mechanical properties. Another method demonstrated at the micro-scale was described by Munoz et al. [12], which allows the manufacturing of high-toughness composite using a freeze-granulation method, in which 7-μm-diameter Alumina tablets were oriented by sedimentation and later sintered with silica as the matrix interface.

At the nanoscale, De Luca et al. [13] demonstrated the deposition of well-aligned sub-micron inorganic tablets of 90% weight percentage on the surface of glass fibers, improving the interfacial strength and energy dissipation by up to 30%. The processes developed in this study have different limitations, as they are complicated, time-consuming, non-scalable, have limited accuracy, and are difficult, as supported by the review of Mirkhalaf and Zreiqat [14]. Finally, the work of Wegst et al. [15] summarized the manufacturing methods and added the use of additive manufacturing methods for generating nacre-like structures as a potential path for future research.

The finite element method (FEM) has been used to estimate structure stiffness, energy absorption, and other mechanical properties with acceptable accuracy. In the work of Grujicic et al. [16], FEM was used to assess the impact damage of a rigid projectile on a nacre-like body armor structure consisting of Boron Carbide; at the millimeter scale, hexagonal tablets were bonded by polyuria matrix. The study concluded that a nacre-like structure had superior penetration resistance when compared to a monolithic structure. Raj et al. [17] used FEM to predict the toughness of a highly organized hierarchical structure of nacre consisting of calcium carbonate tablets bonded by an organic matrix with the work being in good agreement with experimental measurements. Li et al. [18] also made use of FEM to compare the predicted stiffness and strength of nacre-like tablets and compared them to experimentally produced specimens using an additive manufacturing method.

Analytical models have been developed for composite structures since the proposal of the rule of mixture model by Voigt [19] in 1889. This model, which depends on the volume quantities of the composite components, is accurate enough to estimate the stiffness of long fiber composites. Nacre structures are discontinuous reinforcement composites and thus have a non-uniform cross-section. However, if nacre structures are accurately assembled with a high length-to-thickness ratio, the structures behave as a continuous body and revert to the rule of mixture, while at the same time possessing in-plane isotropy, as explained by Kim et al. [20]. Models for nacre structures have been developed, but with several conditions and assumptions to accept their validity, as in the work of Kotha et al. [21]. Such models have several limitations and do not apply to certain geometries. Liu et al. [22] presented a model for analyzing the stress fields between the tablets at the nanoscale, concluding that a smaller aspect ratio of tablet size to thickness and a larger elastic modulus ratio between reinforcement and matrix led to a more uniform shear stress distribution in the matrix. Most studies presented here conclude that further analysis is still required to predict the mechanical properties of nacre-like structures accurately. Further improvement is needed for experimental methods to produce nacre-like composites and measure the mechanical properties of such structures; these statements are repeated in the work of Bonderer et al. [23], Miao et al. [24], Mirkhalaf et al. [25], and Luz and Mao [26].

The method presented in this work details the selection of two readily available materials that fit the criteria of nacre brittle/tough combination and can be processed by easily accessible tools. This simplification should allow for fast geometry development, leading to direct experimental validation, more accurate analytical models, and reliable simulation models. Polystyrene (PS) sheets are selected as the reinforcing component, with brittle behavior and relatively high stiffness. Polyurethane (PU) is chosen as the nacre matrix component due to its higher toughness and the fact that, as a liquid solution, it can be cast similarly to epoxy. The nacre geometry is laser cut from thin polystyrene sheets, and then several alternating layers are assembled to ensure optimum overlap. The polystyrene sheets are held together by interlayers of polyurethane solution. These assembled sheets have each component characterized mechanically, and then each assembled sheet is tested in tensile and has its stiffness measured. These composite sheets are then modeled in computer-aided design (CAD), following digital microscope measurements and simulated by Simulia/Abaqus (version 2017, Dassault Systemes, Vélizy-Villacoublay, France) [27] FEM software to determine the stiffness.

The measured and simulated values can then be compared to analytical models developed for nacre mechanical properties prediction and, in this specific case, stiffness. The analytical model presented by Padawer and Beecher [28] is to specifically predict the stiffness of nacre like structures with reinforcing tablets overlapping. The overlap of tablets causes increases in shear stresses of the structure. In the case of nacre-like assemblies, the shear modulus of the matrix material is an important variable in the mechanical performance of the composite. A third analytical model is presented by Lusis et al. [29] with a correction to the model of Padawer and Beecher, as it also considers the effect of the interaction of adjacent flakes. The fourth model, further improved for ordered nacre structure, is the shear tension model described by Kotha el Al [21] and adapted by Barthelat [30]. This "Shear Tension Model" considers the gap distance between tablets, the overlap ratio, and several geometrical variables and material property variables. Several assumptions are necessary to consider the model accurate. For this work, not all assumptions are valid, which is discussed further in the model section of this paper. Another model for the tablet–matrix composite was developed by Sakhvan et al. [31–33], to be a universal map for nacre-like structure design. However, with similar assumption requirements as the previous model, it has limited accuracy for the prediction of hexagonal-based tablet structures. We present a model that is suitable for hexagonal-shaped tablets inspired from the work of Barthelat [30] with the use of stiffness of volumes in series, and the work of Bar-On and Wagner [34] and Jaeger and Fratzl [35] with the use of Voigt model to calculate the stiffness of partial volumes of a representative volume element (RVE). As previously presented models assume that gaps between tablets are negligible and do not accurately predict the stiffness of the structure, the model and method for experimental validation presented in this work could allow for the acceleration of the design and development of highly ordered hexagonal nacre-like structures.

Carbon foils referred to as nanocrystalline diamond foils have measured stiffness of 600 GPa with a 50 μm thickness, as presented by Lodes et al. [36]. If the laser-cutting method is used to produce carbon-based hexagonal connected tablet, it can be combined with epoxy matrix material with stiffness of 3 GPa and maximum strain of 8.3%, as measured in the work of Shi et al. [37]. The manufacturing process and combination method could produce a composite with exceptional stiffness, strength, and toughness.

2. Materials and Methods

Nacre is a combination of brittle high stiffness reinforcing component and a tough low-stiffness binder matrix. The soft matrix stiffness in natural nacre is three orders of magnitude lower than the ceramic stiffness, as stated by Kim et al. [38], but analytical and simulation work has been done with the interface to reinforcement stiffness ratio ranging between 10 and 0.001. Ni et al. [39] summarize several nacreous composites with reinforcement elastic modulus to matrix shear modulus ratio ranging from 27 to 3800. Stiffness combination is not enough to provide toughness; the matrix material must have a relatively high plastic strain so that the composite can absorb energy [30]. The almost incompressibility of the matrix (high Poisson's ratio) is favorable for increased stiffness and strength [40]. An additional requirement to obtain a high-strength composite is bonding strength and the shear strength of the matrix component. However, as the focus of this investigation is stiffness validation, the material selection did not take strength requirement as an essential criterion [41]. Among polymers, polystyrene has a very high brittleness with a brittleness index B [42,43], nine times larger than other polymers on average. The brittleness is calculated using Equation (1), where E' is the storage modulus determined by dynamic mechanical analysis at 1 Hz and room temperature, and ε_b is the elongation at break.

$$B = 1/(\varepsilon_b E')$$
(1)

Polystyrene is available in sheet form of micron range thickness. Polystyrene can be cut-processed into shape with laser energy methods, without toxicity risks, and with relatively low energy values compared to laser cut glass [25].

Glass would not be a suitable candidate as it is challenging to process with laser and is energy-intensive. Glass is also difficult and expensive to obtain in thin sheet form [30,44]. The thinnest currently available glass sheet is called D263 at 0.03 mm (SCHOTT North America Inc., Duryea, PA, USA). Glass is also so brittle that several layers would be difficult to laminate by hand, and would tend to crack when applying pressure in order to squeeze the matrix thin between the layers. Other ceramics would also not be suitable for laser processing and costly to obtain as thin sheets.

Polystyrene sheets are available at 135 μm thickness (Evergreen Scale Models, Des Plaines, IL, USA). The sheets are perfectly transparent and with a mirror polish surface. They are cleaned in an ultrasonic bath then cut with a laser to the desired shape. Laser cutting involves high temperatures that cause thermal degradation of polymers. The molecular weight of polystyrene near the cut areas is expected to drop. However, the work of McCormick et al. [45] showed that the stiffness of polystyrene does not change considerably with a reduction in molecular weight. The cutting geometry is similar to the hexagonal shape of nacre but leaves small bridges connecting adjacent plates. Such bridges are also described in the work of Ghazlan et al. [46] but with a different purpose. The bridges in this work allow the tablets to stay in position when assembling the structure, while, in previous work, the bridges have considerably larger cross-section and are used to increase stiffness in the structure.

Polystyrene has an acceptable bond strength to polyurethane, as stated by Hall et al. [47]; it can be further enhanced through surface plasma treatment or surface sulfonation [48]. Polyurethane is available as a relatively low viscosity mixture with reaction times longer than ten minutes, which is enough time to make a composite assembly. Polyurethane has considerable plasticity and stiffness, which is expected to be five to ten times less than polystyrene. One additional advantage of this selected polyurethane is that the final solid form is translucent, which allows for microscopic imagining of the assembly and the detection of trapped air cavitation.

Polyurethane resin SK6812 with a hardener of code 7802 (S U. K Hock GmbH, Regen, Germany) is mixed in an equal weight ratio as recommended by the manufacturer to obtain the expected properties. Twelve minutes is the curing time during which the viscosity increases until the solution turns solid, and it is no longer possible to laminate polystyrene layers. The resin has a viscosity of 130–150 mPas, and the hardener has a viscosity of 70 to 80 mPas. According to the material data sheet of the manufacturer, cured material is expected to have an ultimate tensile strength of 23 MPa and a maximum strain of 20%, as well as a flexural modulus (assumed equal to tensile elastic-modulus) by three point-bending of 0.56 GPa and impact strength of 10 KJ/m^2 based on DIN 53457 and DIN 53454 respectively.

Before proceeding with the composite assemblies, each material is characterized mechanically. Polystyrene sheet samples are tested following standard DIN EN ISO 527-3:2019-02 – ASTM D882, resulting in tensile stiffness of 2.66 GPa along the extrusion direction of the sheets. A stiffness of 2.64 GPa is measured perpendicular to the extrusion direction. As the difference in stiffness is not large, the average stiffness of 2.65 ± 0.04 GPa can be assumed. A more significant difference is measured in the strength of the film, with 50.06 ± 3.7 MPa in the extrusion direction and 43.76 ± 1.6 MPa perpendicular to extrusion. Such a difference does not affect the stiffness measurement of the composite assembly. PU samples were cast in additively manufactured molds. The molds are made using PolyVinyl Alcohol (PVA) polymer filament, which can be dissolved in water. PU mix is cast in the dissolvable mold, left for seven days to cure, then dissolved in water to release the specimens, the specimens are then dried and tested following the standard (DIN EN ISO 527-2) which resulted in a stiffness value of 0.35 ± 0.02 GPa. Delamination tests are also conducted following the standard ASTM D3163 with PU bonded to PS sheets. The delamination tests resulted in a shear strength value of 0.39 MPa.

The tensile test data, including strain measurements obtained by digital image correlation (DIC) method, also allowed the calculation of the Poisson's ratios of PS and PU, which were, respectively, 0.35 and 0.41. A high Poisson ratio confirms higher incompressibility of PU. The software used for (DIC) is Istra4D (Dantec Dynamics A/S, Skovlunde, Denmark), which uses a full-field image analysis. Square quadratic elements of 15 pixels were used to divide the region of interest (ROI), which consists of 1936

$\times$ 1216 pixels. The ROI is divided into 10660 (130 $\times$ 82) quadratic elements. A displacement uncertainty of 0.4 μm was recorded for a displacement of 100 μm. With the same cameras and software used, the work of Hack et al. [49] reported a strain measurement error of less than 3% in-plane. Using the Poisson's ratio and stiffness value of PU, the shear modulus G can be calculated using Equation (2) [50], resulting in G = 0.124 GPa.

$$G = \frac{E}{2\left(1+\upsilon\right)} \tag{2}$$

The layers are laminated on a designed plate with alignment pins to ensure the best possible overlap between the tablets. Three specimens were produced by combining six alternative layers for each. The manufactured composite laminates are measured by digital microscope Olympus DXS500 (Olympus Deutschland GmbH, Hamburg, Germany) to determine matrix layer thickness, overlap, and alignment accuracy. The measurements were used to create an accurate CAD model of the assembly. The model is meshed with quadratic tetrahedral elements. Material properties of the bi-components are assigned to the model, and then periodic boundary conditions are used to simulate an infinite laminate of also infinite layers. The samples in the simulation are under a uniaxial strain of 1% in full elastic behavior. With a small strain value and elastic material behavior, the element size used was half of the smallest geometrical feature, which is the reinforcement thickness of 0.135 μm.

3. Results

3.1. Experimental

Thin polystyrene films of 135 μm thickness with a measured stiffness of 2.65 GPa and polyurethane liquid resin mixture with resulting stiffness measured to be 0.35 GPa are then combined composite materials to investigate nacre-like structure stiffness. The PS sheets are processed with a laser-cutting machine Rayjet 50 (Trotec Laser GmbH, Wels, Austria) to obtain the final geometry shown in Figure 1a. Several laser-cut tests are run to obtain the nacre geometry with minimal cut line width. The optimal cut settings are 1% power of a 30 Watts CO_2 laser head with two to four passes, which resulted in a cut width (kerf) of 280 μm. The generated cut sheet is shown in Figure 1b with final tablet width of 4.22 mm.

The cut nacre sheets are cleaned in an ultrasound water bath and then accurately assembled on an assembly board using the cut guide holes shown in Figure 1c to align the tablet overlap. At the end of the assembly, a deadweight is added above the laminates to apply 20 kPa pressure, so that the matrix layers flow and become thinner, while the polystyrene is not damaged. The assembled sheets are left for seven days to cure properly before releasing them from the assembly board. Three nacre-like specimens are produced and then tested in tensile load using a universal high-precision tensile-testing machine Shimadzu AGS-X (Shimadzu, Kyoto, Japan) at 0.5 mm/min displacement. The strain is measured using the DIC system with high precision cameras Limess Q400 (LIMESS Messtechnik u. Software GmbH, Krefeld, Germany). The measured stiffness values for three specimens are 1.221 $\pm$ 0.012 GPa. The specimens failed at a stress of 6.7 $\pm$ 0.3 MPa and a maximum strain of 0.217 $\pm$ 0.0134.

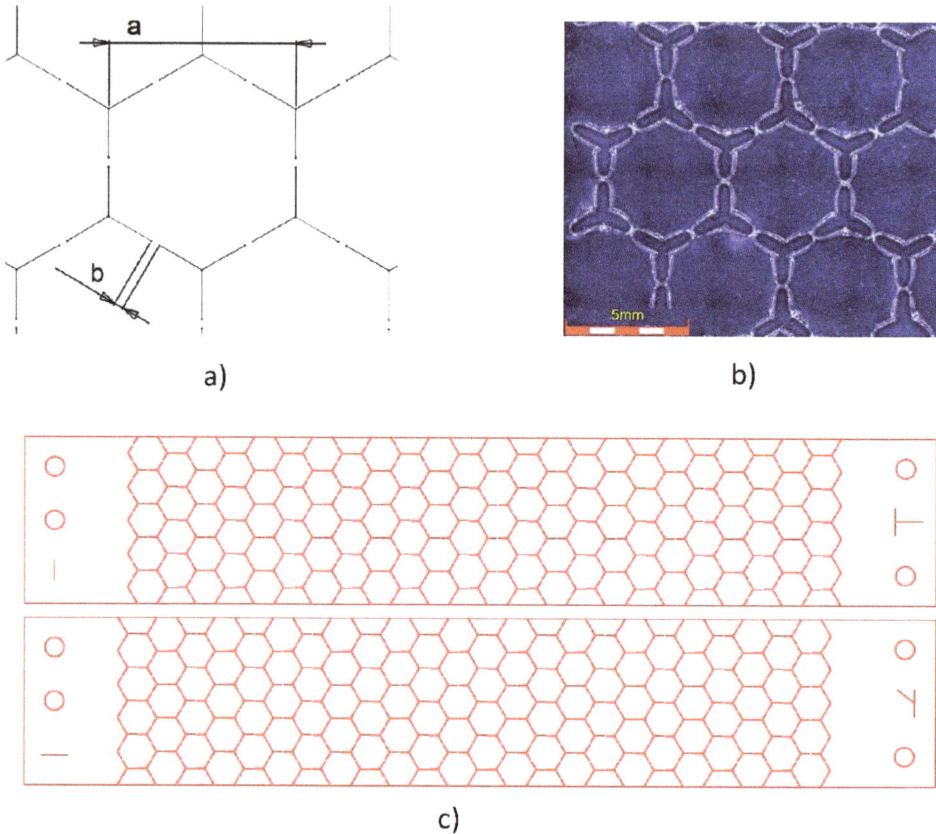

Figure 1. Polystyrene Laser cutting. (**a**) Designed cut lines (**b**) Laser cut view under a microscope (**c**) Laminate cut design with guide holes and orientation labels.

3.2. FEM

The manufactured experimental nacre laminates are measured with the Opto-digital microscope before being destroyed in the tensile tests. All measured features are modeled in CAD, and then the specimens are meshed in FEM software Simulia/Abaqus with a mesh size of 0.065 μm of quadratic tetrahedral elements. A smaller mesh element size was tested but resulted in no significant change in calculated stiffness. The materials are defined as linear elastic, with properties described in Table 1. The Poisson's ratios are calculated based on the strain values measured in the DIC of each composite component.

Table 1. Material mechanical properties.

Material	Stiffness (GPa)	Poisson's Ratio
Polystyrene	2.65	0.35
Polyurethane	0.35	0.41

The components of this FEM are partitioned from the same part, so that no interaction conditions where necessary, and thus debonding was not possible in the model. As the materials are linear, and the goal is to calculate composite stiffness, a total strain of 1% is applied to the representative volume element shown in Figure 2c. The measured hexagonal width was 4.22 mm, the side length was

2.442 mm, the cut gap was 280 μm, and the bridge width was 0.4 mm, shown in Figure 2a. Figure 2b shows the increase in the thickness of the polystyrene sheets due to the laser energy melting the polymer along the cut line. The increase in width that we refer to as a bulge has a total thickness of 250 μm and a width of 160 μm. The bulge is added to the CAD model of the RVE to represent the reinforcement tablets more accurately. Due to the bulge and limited pressure used when laminating the layers, the matrix layer thickness remained relatively large, with a value of 168 μm.

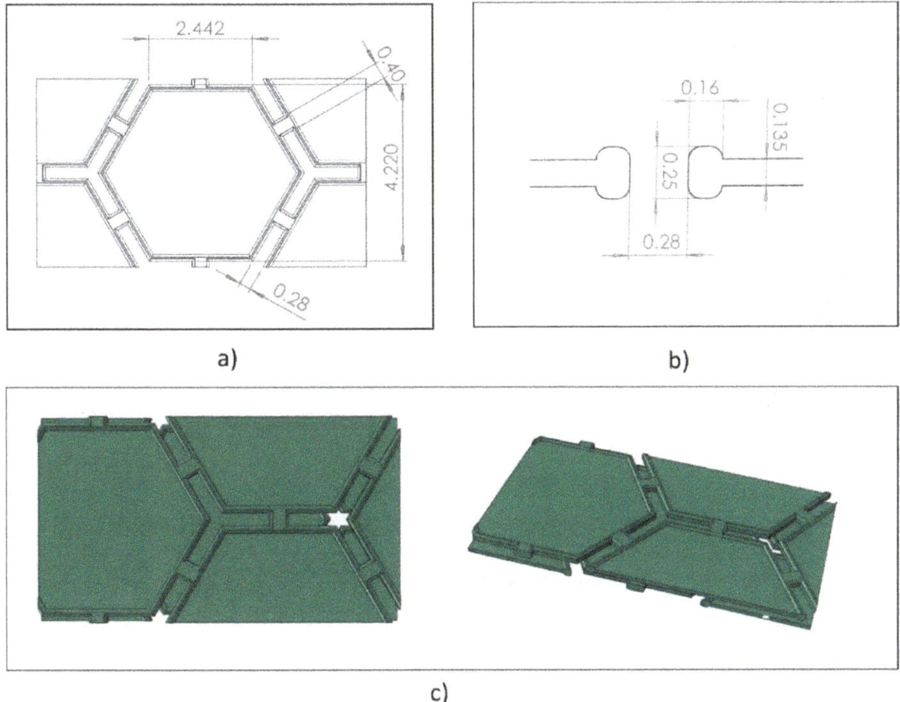

Figure 2. Measured and drawn geometry of the laminated polystyrene plates. (**a**) Tablet dimensions, (**b**) Bulge section view with dimensions. (**c**) Representative volume element (RVE) CAD model representing polystyrene cut tablets.

Periodic boundary conditions are used so that the stress calculations are, for an infinite repetitive body, in all three directions: X, Y, and Z. The resultant stiffness value was 1.21 GPa, which has a ~1% difference compared to the experimental value. It is important to note that a nacre RVE is not symmetric but only periodic, which makes some analytical models not suitable for a hexagonal-based tablet when compared to simplified rectangular tablets with half-length overlap. Periodic models such as RVE have been previously shown to more accurately estimate the composite structure stiffness than analytical models if they are correctly modeled [51]. This RVE of nacre is the minimum required volume to represent the composite, as the structure is periodic and very ordered, with known reinforcement dimensions and orientation governed by the lamination assembly process.

3.3. Analytical Stiffness

Once the experimental value and simulation value of stiffness was obtained, calculations were made using several analytical models to compare the analytical values and identify the error in every model.

3.3.1. Voigt Model

The Voigt model, represented by Equation (3), is used to calculate the axial loading stiffness of endless fiber composites [19]. It is generally known as the rule of mixture and is only dependent on the volume of the reinforcing material relative to the matrix material. This is the simplest model to make a prediction of stiffness, but as a nacre structure is not fiber-based with endless length, and has no specific orientation, this model cannot be considered as valid for nacre. Nevertheless, it is used to compare the nacre stiffness to a fiber composite of the same volume ratios and materials

$$E_c = f\,E_f + (1 - f)\,E_m \tag{3}$$

where E_f is the stiffness of the fibers, E_m is the stiffness of the matrix, and f is the volume fraction of fibers relative to total volume or, in this case, polystyrene hexagonal plates relative to polyurethane matrix. Based on the CAD model generated from microscope measured values, the volume of reinforcement element relative to total composite volume is 42.59%, which results in a stiffness of 1.33 GPa, which is 8.9% higher than the simulation value. Tablet composite stiffness is expected to approach the Voigt model when the length to thickness ratio exceeds 100 and tends to infinity [28,52]. With a ratio averaged at ~34 (4.55/0.135) for a hexagonal plate, this composite has a stiffness value lower than the Voigt model predicts.

3.3.2. Padawer and Beecher

Padawer and Beecher [28] made an early attempt to develop a predictive stiffness model for nacre-like structures. As the rule of mixture is only valid for continuous sheet, a model for discontinuous planar reinforcement was needed. Assuming that the reinforcing flakes have uniform width and uniform thickness, the aspect ratio α is the first variable used to estimate the stiffness of the flake-reinforced composites. Important assumptions are also necessary, as the reinforcing element needs to be uniformly spaced and aligned to the plane. Such conditions are only valid when the composite is accurately designed and built, which is the case for the PS/PU composite. Additional assumptions are necessary for strength prediction, which do not fall in the scope of this work.

The Padawer–Beecher model depends on a constant "u" which is itself dependent on geometry, and material properties are shown in Equation (4). G_M is the shear modulus of the matrix material, in this case, polyurethane. vf is the volume fraction of reinforcement and E_R is the stiffness modulus of reinforcing material, in this case, the polystyrene films.

$$u = \alpha\left[\frac{(G_M)(vf)}{(E_R)(1 - vf)}\right]^{1/2} \tag{4}$$

The model makes use of a term called Modulus Reduction Factor (MRF). It is used as a stiffness reduction constant dependent on "u", as shown in Equation (5).

$$\mathrm{MRF} = \left[1 - \frac{\tanh(u)}{u}\right], \tag{5}$$

The stiffness model equation is very similar to the rule of mixture model but with the added multiplication with the MRF value, as shown in Equation (6).

$$E_C = (E_R)(vf)(\mathrm{MRF}) + (E_M)(1 - vf), \tag{6}$$

As we have experimentally obtained all material properties required for the calculation and the flakes are designed to have the same constant width of 4.22 mm and thickness of 135 μm, we can use the model and calculate a stiffness of 1.15 GPa where the MRF value is 0.841. It is important to note that a higher flake volume ratio leads to an increase in MRF, and thus stiffness, but the increase is not significant above $\alpha = 100$, as explained by Padawer and Beecher [28].

3.3.3. Lusis et al.

The model developed by Padawer and Beecher does not take into consideration the interaction between the adjacent flakes. An adjustment was presented by Lusis et al. [29], with the MRF calculated following a logarithmic equation shown in Equation (7).

$$\text{MRF} = \left[1 - \frac{\ln(u+1)}{u}\right], \tag{7}$$

With this adjustment, the MRF value is expected to be lower, and thus the calculated material stiffness is 0.973 GPa. Based on the paper by Jackson et al. [52], generally, the prediction by Lusis tends to fall lower than experimental values, which is also the case in this work.

3.3.4. Shear Tension Model

The shear tension model is first described by Kotha et al. [21], and then adapted by Barthelat [30], like in most analytical models; there are important assumptions to make so that the model is considered valid. Kotha et al. [21] made four assumptions. The first is that the tablets are of uniform width, rectangular shape, and isotropic. The geometry of the tablets in the experimental work is hexagonal but has a defined width, and the material is quasi-isotropic, with little variation in stiffness with the change of direction. There is less than 0.7% difference in stiffness for PS along extrusion relative to lateral. In the second assumption, the tablets need to be uniformly arranged, which is valid in this work. With an overlap close to 30% and tablets accurately positioned, we can assume that the majority of the stresses are transferred from tablet to tablet by shear, which is the third assumption required. With the matrix having a low stiffness compared to the reinforcement, the fourth assumption is also valid, and the end of the tablets contribute little to stress transfer.

The model is dependent on several parameters shown in Table 2 with the relative equations to calculate the respective values of which volume fraction, aspect ratio for the tables, aspect ratio of tablet length to thickness and overlap ratio of tablet length. For a composite system dependent on shear transfer of stress, an elastic shear transfer number β_0 is derived and used to estimate the composite stiffness. It is important to note that the volume fraction value ϕ is dependent on tablet and interface thickness and assumes gaps between tablets to be negligible.

Table 2. Shear lag model equations and variables.

Parameter	Equation	Variables
Volume fraction	$\phi = \frac{t_t}{t_t + t_i}$	t_t tablet thickness t_i interface thickness
Aspect ratio of overlap region	$\rho_0 = \frac{L_0}{t_t}$	L_0 overlap length t_t tablet thickness
Overlap ratio	$k = \frac{L_0}{L}$	L tablet length $0 < k \leq 0.5$
Elastic shear transfer number	$\beta_0 = \rho_0 \sqrt{\frac{G_i \, t_t}{E_t \, t_i}}$	G_i matrix shear modulus E_t tablet stiffness modulus
Non-Symmetric RVE Composite stiffness	$E = \dfrac{\phi E_t}{1 + \frac{k}{\beta_0}[\coth(\beta_0) + \coth(\frac{1-k}{k}\beta_0)]}$	-

The hexagonal tablet overlap has symmetry, as shown in Figure 3, which can be used to calculate the required variables. If the gaps between hexagons are neglected, we should obtain a stiffer model. Due to a non-rectangular geometry, half a hexagon with an edge length "a" of 2.44 mm, and a half area of 7.72 mm^2 with a half-width of 2.115 mm, can be approximated to a rectangle of 3.656 mm length.

Thus, an area overlap of 0.29% would be equivalent to an overlap length of 1.0627 mm. The volume fraction ϕ would be 0.4455 and the aspect ratio of the overlap region would be 7.87. The resultant shear transfer number would be $\beta_0 = 1.527$. The calculated stiffness with the simplified model is 0.845 GPa, which is considerably lower than the value calculated by the Lusis model [29].

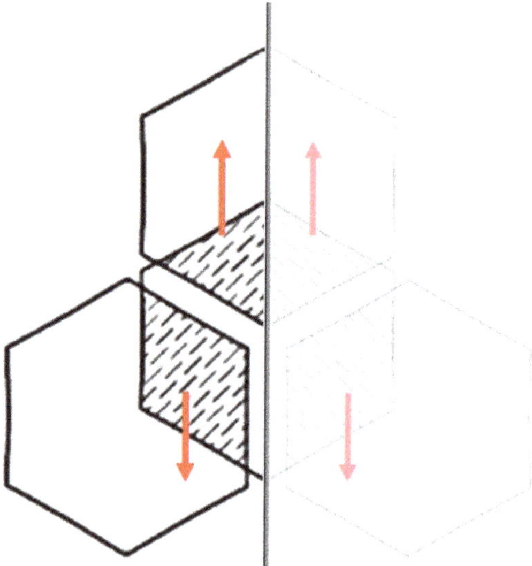

Figure 3. Volume symmetry of hexagonal reinforcement.

3.3.5. The Proposed Model for Highly Ordered Hexagonal Tablets

Using the well-defined geometry of the hexagonal tablets and assuming the complete linear elastic behavior of the material, we can simplify the composite representative volume as a rectangle divided into several zones of either one layer of reinforcement or two layers of reinforcement. Using the rule of mixture for every continuous and homogeneous area of overlap, we can calculate the stiffness of the areas. What other models do not take into consideration is the size of the gap between tablets, which is calculated in this model. Another assumption for the shear-lag model [30] is that the end of the tablets does not contribute to the stiffness of the model. This is taken into consideration in this model.

The RVE geometry of hexagonal nacre-like composite can be simplified, as in the half hexagonal volume shown in Figure 3; however, to be able to calculate using homogeneous volumes, the RVE is divided into several zones of known stiffness, shown in Figure 4.

Known parameters that are required would be, most importantly, the stiffness of the materials, the thickness of the reinforcing tablets, and matrix layers between them. In this specific case of hexagonal tablets, it is important to know at least one defining dimension of the hexagon, which is the short diagonal (The width of a hexagon). From the short diagonal "d", all other dimensions of a hexagon can be calculated. The edge length "a" is calculated by dividing the short diagonal by $\sqrt{3}$. Another important feature that is also relevant to the manufacturing method of these connected tablets would be the width of cut "c" (kerf). The width of the cut dictates the gap dimensions between the tablets in the same layers.

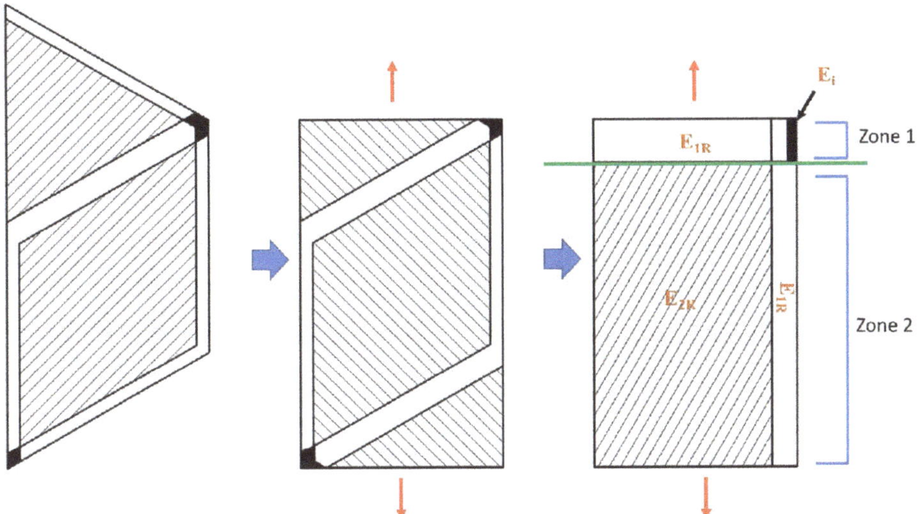

Figure 4. Volume simplification of hexagonal reinforcement.

The rectangular volume shown in Figure 4 would have a width W_r of half the short diagonal d_1. The rectangular length L_r is 1.5 times the edge length a. The double reinforced zone has a width W_1, which is equal to the width of the rectangle without a kerf c width. The height of the double reinforcement zone (hatch-filled zone) is the length of the rectangle with two projected cut width removed $L_r - 2c/\cos30$. The height of the single-layer reinforcement (zone 1) is the width of two cuts projected at 30-degree angle $2c/\cos30$. The remaining zone highlighted in black is a non-reinforced area that can also be seen in Figure 2c; it has an equal area to a hexagon with short diagonal equal to the cut width c. The non-reinforced zone has a width of 0.375c and the same height of Zone 1.

In Figure 4, there are zones of a single layer of reinforcement with a stiffness we shall call E_{1R}. The remaining zones are of double layers of reinforcement with a stiffness we shall call E_{2R}. These stiffness values are calculated following the equations shown in Table 3, where t_t is the tablet thickness, t_i is the interface thickness, E_p is the tablet stiffness, and E_i is the interface stiffness.

Table 3. Hexagonal Composite Stiffness Model.

Parameter	Equation	Variables
Single reinforcing layer zone stiffness	$E_{1R} = \dfrac{t_t E_t + (t_t + 2t_i)E_i}{2(t_t + t_i)}$	t_t tablet thickness t_i Interface thickness
Double reinforcing layer zone stiffness	$E_{2R} = \dfrac{t_t E_t + t_i E_i}{t_t + t_i}$	E_t Tablet stiffness modulus E_i interface stiffness modulus
Hybrid zone stiffness 1	$E_{\text{Hyb1}} = \dfrac{0.75c}{d} E_i + \dfrac{d-0.75c}{d} E_{1R}$	-
Hybrid zone stiffness 2	$E_{\text{Hyb2}} = \dfrac{2c}{d} E_{1R} + \dfrac{d-2c}{d} E_{2R}$	c cut width d short diagonal
Stiffness ratio 1	$r_1 = \dfrac{2c}{L \cos 30 \, E_{\text{Hyb1}}}$	a edge length L = 1.5a rectangle height
Stiffness ratio 2	$r_2 = \dfrac{1}{E_{\text{Hyb2}}} - \dfrac{2c}{L \cos 30 \, E_{\text{Hyb2}}}$	-
Hexagonal Composite stiffness	$E_c = \dfrac{1}{r_1 + r_2}$	-

Along the load axis, the stiffness of Zone 1 is calculated following the Voigt model, with a dependence on width similar to the work of Jäger and Fratzl [35], resulting in hybrid zone stiffness

E_{Hyb1}. The stiffness of Zone 2 is calculated similarly, resulting in hybrid zone stiffness E_{Hyb2}. To calculate resultant composite stiffness of Zone 1 and Zone 2, the equation for zones acting in series is used, as suggested by Barthelat [30].

With the dimensions produced experimentally known, and the material properties measured, we can calculate E_{1R} to be 0.862 GPa, E_{2R} is 1.375 GPa, E_{Hyb1} is 0.838 and E_{Hyb2} is 1.358 GPa. The stiffness ratio values are as follows: r_1 is 0.198 GPa^{-1}, r_2 is 0.614 GPa^{-1}, resulting with a composite stiffness of 1.231 GPa.

The same method can be used with volume simplification to calculate stiffness for rectangular tablets with cut width and perfect positioning. The resulting equations are similar to the hexagonal equations but with cos30 angle projection removed, and the unreinforced, black-colored zone in Figure 4 has a width of c/2 in a rectangular composite when compared to a hexagonal composite with a non-reinforced zone of width 0.75c. The equations for rectangular tablet composite with cut width are shown in Table 4, with w being the rectangle width and L being the rectangle height. The advantage of hexagonal tablets remains, with quasi-isotropic stiffness in the plane.

Table 4. Rectangular Composite Stiffness Model.

Parameter	Equation	Variables
Single reinforcing layer zone stiffness	$E_{1R} = \dfrac{t_t E_t + (t_t + 2t_i)E_i}{2(t_t + t_i)}$	t_t tablet thickness t_i Interface thickness
Double reinforcing layer zone stiffness	$E_{2R} = \dfrac{t_t E_t + t_i E_i}{t_t + t_i}$	E_t Tablet stiffness modulus E_i interface stiffness modulus
Hybrid zone stiffness 1	$E_{Hyb1} = \dfrac{0.5c}{d}E_i + \dfrac{d-0.5c}{d}E_{1R}$	-
Hybrid zone stiffness 2	$E_{Hyb2} = \dfrac{2c}{d}E_{1R} + \dfrac{d-2c}{d}E_{2R}$	c cut width d rectangle width
Stiffness ratio 1	$r_1 = \dfrac{2c}{L\,E_{Hyb1}}$	L rectangle height
Stiffness ratio 2	$r_2 = \dfrac{1}{E_{Hyb2}} - \dfrac{2c}{L\,E_{Hyb2}}$	-
Hexagonal Composite stiffness	$E_c = \dfrac{1}{r_1 + r_2}$	-

4. Discussion

The calculated, simulated, and measured stiffness values are summarized in Table 5, with the percentage difference relative to the experimental value. The simulated value has a difference of 1% relative to the experimental value. The small difference can be explained by an accurate representation of the geometry by optical measurements and the accurate material characterization data used in the FEM model. However, the FEM is a long process and cannot allow quick iterations in design unless a parametric model is developed for specifically hexagonal tablets connected by thin bridges. The rule of mixture results in a considerably higher value because it assumes a continuous reinforcement with constant cross-section. At the same time, a hexagonal nacre-like structure is discontinuous and has a varying reinforcement cross-section, which leads to stiffness reduction. However, tablet-reinforced structures converge to the rule of mixture, as the aspect ratio tends towards infinity, as stated by Jackson et al. [52] and Bekah et al. [53]. In the case of the "Padawer & Beecher" and "Lusis" models, they both predict a lower stiffness than what is measured. This could be due to the shear modulus of the matrix material and the thickness of the interface layer. Jackson et al. [52] have shown that the value of shear modulus influences the prediction error in both models relative to the experimental. The model's accuracy is highly dependent on the interface layer thickness and mechanical properties, as it assumes that all loads are conveyed by shear stresses and ignores stresses in tension between the tablets. However, it is important to note that both models are developed for single-lap joint rectangular flakes [29].

Table 5. Stiffness of PS-PU Hexagonal Nacre-like Composite.

Method	Stiffness (GPa)	Difference from Experimental
Experimental	1.221	0%
Finite Element Method (FEM)	1.21	−1%
Rule of Mixture (Voigt Model)	1.33	+9%
Padawer and Beecher Model	1.15	−6%
Lusis et al. Model	0.973	−20%
Shear-lag Model	0.845	−31%
Hexagonal Model	1.231	+0.8%

Calculating following the shear lag model presented by Bathelat [30] resulted in the most significant difference of 31% from the measured value. While it was necessary to make important assumptions and approximations from hexagonal geometry to rectangular, and the model neglects any loads being conveyed between the tablets except by shear. The model assumes the spacing between the tablets to be negligible. In this example, the spacing is 6% of the tablet width. The model again appears to diverge from experimental results when the interface layer is comparable to the reinforcement layer in thickness. The hexagonal model presented here is derived from the exact geometry of the hexagonal nacre-like composite, assuming perfect positioning and consistent cut width. A 0.8% difference from the experimental value is as good as the FEM result. As the model is strictly dependent on material property, hexagon size, cut width, and reinforcement and interface thickness, it is easier to run fast iterations to identify more efficient composites with higher stiffness. It is also easy to predict the influence of important variables on the resultant stiffness, as shown in Figure 5. The matrix thickness reduces the stiffness of the hexagonal composite strongly as it nears the reinforcement tablet thickness, and as it becomes thicker, the stiffness no longer drops sharply. As shown in Figure 5a, with a matrix thickness of 0 mm, the maximum stiffness would be 2.392 GPa, which is 10% lower than the stiffness of the reinforcing material (while the cut width remains 0.28 mm), which indicates a considerable reduction of stiffness due to cut width. The PS-PU stiffness, with an interface of thickness 0.168 mm, in this case, is 52% of the maximum possible stiffness. Above a tablet width that is a hundred times the thickness (aspect ratio 100), the stiffness tends to the value calculated by the rule of mixture. With the PS-PU samples produced having a width to thickness ratio of 33, the stiffness is 91% of the rule of mixture value, as shown in Figure 5b. While maintaining an interface thickness of 0.168 mm and Hexagonal dimension, the reduction in cut width tends to increase the stiffness closer to the rule of mixture value without changing the tablet aspect ratio. Again, the PS-PU composite has a stiffness that is 91% of the rule of mixture value, as shown in Figure 5c. Thus, we could summarize that the reduction in matrix layer thickness could yield the highest stiffness value, close to the reinforcing material stiffness. Cut width reduction and hexagonal size increase brings the stiffness value closer to what the rule of mixture suggests.

In Figure 5a, the plotted stiffness calculated by the Padawer and Beecher model is consistently lower than the simulation and experimental work. If cut width is considered negligible in the model, thus adjusting aspect ratio and reinforcement volume, the plotted stiffness curve referred to with "NC" (no cut width) is brought closer to simulation values. The hexagonal model is closest to simulation values. Figure 5b shows both Padawer and Beecher models mentioned and the Hexagonal model Ec. The adjusted Padawer and Beecher model (NC) is brought closer to simulated values and is more accurate when the tablets are smaller relative to the gaps. The significant advantage of the Hexagonal Model is shown in Figure 5c, where the Ec plot aligns better with simulated values when the Kerf increases.

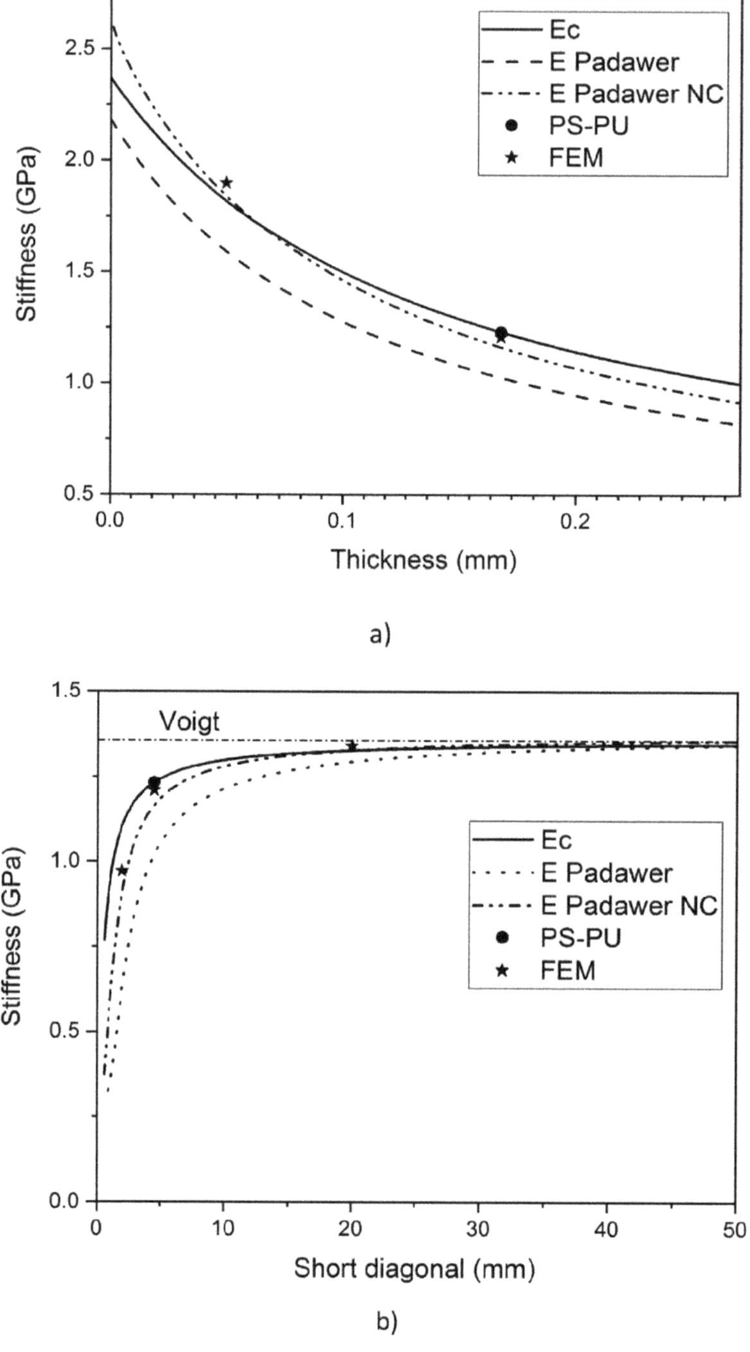

a)

b)

Figure 5. *Cont.*

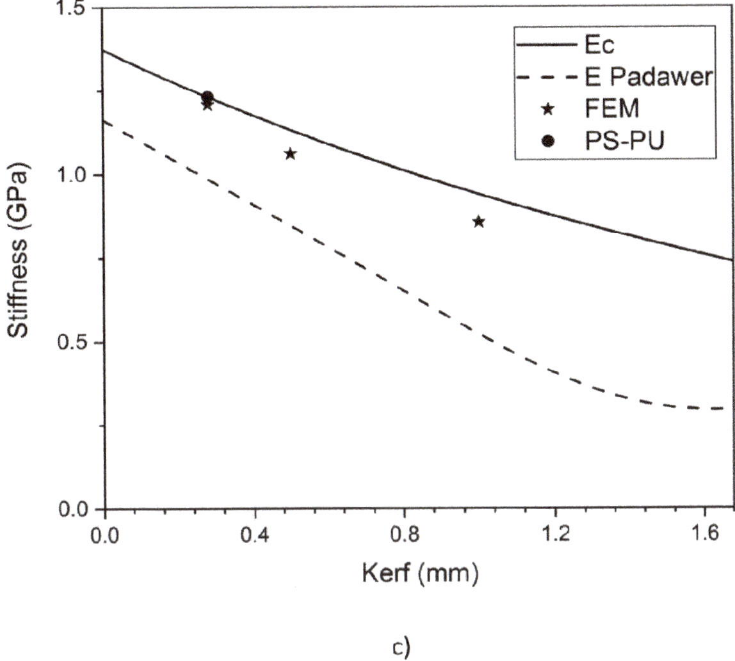

c)

Figure 5. Processing influence on hexagonal composite stiffness, (**a**) Matrix thickness, (**b**) Hexagonal short diagonal, (**c**) cut width.

5. Example

As polystyrene is not the right candidate for high strength composites and is only used for experimental validation of the model, a carbon film with epoxy matrix composite has its stiffness predicted using the hexagonal model using the carbon films described earlier with a stiffness of 600 GPa and a thickness of 50 μm [36] and epoxy resins matrix of stiffness 3 GPa [37], assuming that interface thickness is equal to reinforcing layer thickness and the cut width is 35 μm, based on the laser cut notches using YAG solid-state laser as described by Lodes et al. [36]. The determined values can then be used to plot the change in composite stiffness depending on the size of the hexagonal tablet shown in Figure 6.

With a hexagon diagonal width of 1.53 mm, the stiffness of the composite is at 95% of the Voigt model stiffness value. Increasing the hexagonal width to 5.25 mm brings the stiffness to 99% of the Voigt model. The size of the hexagon, in this case, should also be connected to the energy spent to do the cutting, as smaller hexagons require more cutting travel perimeter in a unit area. For later development, the hexagonal size must be optimized to minimize energy, maintain stiffness and strength, and maximize toughness.

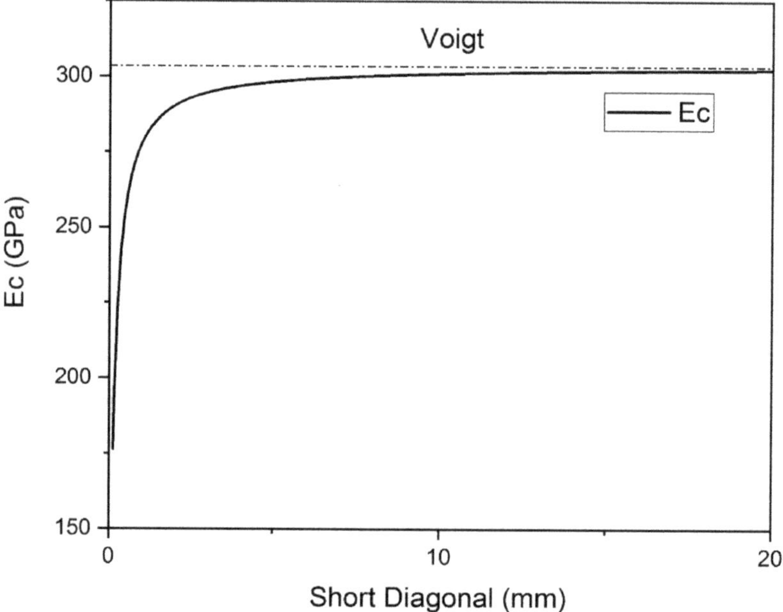

Figure 6. Carbon-Epoxy hexagonal composite stiffness.

6. Conclusions

The hexagonal stiffness prediction model could allow fast design iterations for hexagonal nacre-like composites. The resultant geometry can then be validated using the experimental method described using polystyrene thin film and polyurethane as a matrix material. The FEM simulation method would serve as an additional validation method for the design. It is important to note that the presented manufacturing methods of a hexagonal nacre-like composite results in unavoidable cut width from an endless sheet of reinforcing material. Such a cut width can be reduced depending on the cutting method used and can be optimized for each method. In the example of laser cutting, the process involves parameters such as laser power and cutting speed. The thickness of the reinforcing tablets depends on the material selected and the method used to produce the material film. The original film thickness is the base dimension on which a hexagonal composite can be designed. It leads to the selection of hexagonal size to maintain an optimal aspect ratio for maximizing stiffness.

For practical and industrial applications, the Hexagonal Model could be used to predict the stiffness of mass-produced composites, using cutting methods and with controlled interface thickness. As the mechanical performance of a composite is also dependent on strength and toughness, specific prediction models are still required for those properties, and this stiffness model could serve as a guide.

Author Contributions: Conceptualization, R.R.; methodology, R.R.; software, M.S.; validation, M.S., and R.R.; formal analysis, R.R.; investigation, R.R.; resources, M.S.; data curation, R.R.; writing—original draft preparation, R.R.; writing—review and editing, R.R.; visualization, R.R.; supervision, M.S.; project administration, R.R.; funding acquisition, M.S. All authors have read and agreed to the published version of the manuscript.

Funding: This research received no external funding.

Acknowledgments: The authors would like to Thank Kevin Breuer and Patrick Gahlen for providing the python script to run the FEM simulations. The authors acknowledge the financial support by the German Research Foundation (DFG)and TU Dortmund University within the funding program "Open Access Publishing".

Conflicts of Interest: The authors declare no conflict of interest.

References

1. Bouville, F. Strong and tough nacre-like aluminas: Process–structure–performance relationships and position within the nacre-inspired composite landscape. *J. Mater. Res.* **2020**, *9*, 1–19. [CrossRef]
2. Al-Maskari, N.S.; McAdams, D.A.; Reddy, J.N. Modeling of a biological material nacre: Waviness stiffness model. *Mater. Sci. Eng. C Mater. Biol. Appl.* **2017**, *70*, 772–776. [CrossRef] [PubMed]
3. Barthelat, F.; Yin, Z.; Buehler, M.J. Structure and mechanics of interfaces in biological materials. *Nat. Rev. Mater.* **2016**, *1*, 1144. [CrossRef]
4. Wang, R.Z.; Suo, Z.; Evans, A.G.; Yao, N.; Aksay, I.A. Deformation mechanisms in nacre. *J. Mater. Res.* **2001**, *16*, 2485–2493. [CrossRef]
5. Currey, J.D. Mechanical properties of mother of pearl in tension. *Proc. R. Soc. Lond. B* **1977**, *196*, 443–463. [CrossRef]
6. Budarapu, P.R.; Thakur, S.; Kumar, S.; Paggi, M. Micromechanics of engineered interphases in nacre-like composite structures. *Mech. Adv. Mater. Struct.* **2020**, *6*, 1–16. [CrossRef]
7. Yin, Z.; Hannard, F.; Barthelat, F. Impact-resistant nacre-like transparent materials. *Science* **2019**, *364*, 1260–1263. [CrossRef]
8. Valashani, S.M.M.; Barthelat, F. A laser-engraved glass duplicating the structure, mechanics and performance of natural nacre. *Bioinspir. Biomim.* **2015**, *10*, 26005. [CrossRef]
9. Narducci, F.; Pinho, S.T. Exploiting nacre-inspired crack deflection mechanisms in CFRP via micro-structural design. *Compos. Sci. Technol.* **2017**, *153*, 178–189. [CrossRef]
10. Grossman, M.; Pivovarov, D.; Bouville, F.; Dransfeld, C.; Masania, K.; Studart, A.R. Hierarchical Toughening of Nacre-Like Composites. *Adv. Funct. Mater.* **2019**, *29*, 1806800. [CrossRef]
11. Launey, M.E.; Munch, E.; Alsem, D.H.; Saiz, E.; Tomsia, A.P.; Ritchie, R.O. A novel biomimetic approach to the design of high-performance ceramic-metal composites. *J. R. Soc. Interface* **2010**, *7*, 741–753. [CrossRef] [PubMed]
12. Muñoz, M.; Cerbelaud, M.; Videcoq, A.; Saad, H.; Boulle, A.; Meille, S.; Deville, S.; Rossignol, F. Nacre-like Alumina Composites Based on Heteroaggregation. *J. Eur. Ceram. Soc.* **2020**. [CrossRef]
13. de Luca, F.; Sernicola, G.; Shaffer, M.S.P.; Bismarck, A. "Brick-and-Mortar" Nanostructured Interphase for Glass-Fiber-Reinforced Polymer Composites. *ACS Appl. Mater. Interfaces* **2018**, *10*, 7352–7361. [CrossRef] [PubMed]
14. Mirkhalaf, M.; Zreiqat, H. Fabrication and Mechanics of Bioinspired Materials with Dense Architectures: Current Status and Future Perspectives. *JOM* **2020**, 1458–1476. [CrossRef]
15. Wegst, U.G.K.; Bai, H.; Saiz, E.; Tomsia, A.P.; Ritchie, R.O. Bioinspired structural materials. *Nat. Mater.* **2015**, *14*, 23–36. [CrossRef]
16. Grujicic, M.; Ramaswami, S.; Snipes, J. Nacre-like ceramic/polymer laminated composite for use in body-armor applications. *AIMS Mater. Sci.* **2016**, *3*, 83–113. [CrossRef]
17. Raj, M.; Patil, S.P.; Markert, B. Mechanical Properties of Nacre-Like Composites: A Bottom-Up Approach. *J. Compos. Sci.* **2020**, *4*, 35. [CrossRef]
18. Liu, F.; Li, T.; Jia, Z.; Wang, L. Combination of stiffness, strength, and toughness in 3D printed interlocking nacre-like composites. *Extrem. Mech. Lett.* **2020**, *35*. [CrossRef]
19. Voigt, W. Ueber die Beziehung zwischen den beiden Elasticitätsconstanten isotroper Körper. *Ann. Phys.* **1889**, *274*, 573–587. [CrossRef]
20. Kim, Y.; Kim, Y.; Libonati, F.; Ryu, S. Designing tough isotropic structural composite using computation, 3D printing and testing. *Compos. Part B Eng.* **2019**, *167*, 736–745. [CrossRef]
21. Kotha, S.P.; Li, Y.; Guzelsu, N. Micromechanical model of nacre tested in tension. *J. Mater. Sci.* **2001**, 2001–2007. [CrossRef]
22. Liu, G.; Ji, B.; Hwang, K.-C.; Khoo, B.C. Analytical solutions of the displacement and stress fields of the nanocomposite structure of biological materials. *Compos. Sci. Technol.* **2011**, *71*, 1190–1195. [CrossRef]
23. Bonderer, L.J.; Chen, P.W.; Kocher, P.; Gauckler, L.J. Free-Standing Ultrathin Ceramic Foils. *J. Am. Ceram. Soc.* **2010**, *93*, 3624–3631. [CrossRef]
24. Miao, T.; Shen, L.; Xu, Q.; Flores-Johnson, E.A.; Zhang, J.; Lu, G. Ballistic performance of bioinspired nacre-like aluminium composite plates. *Compos. Part B Eng.* **2019**, *177*, 107382. [CrossRef]

25. Mirkhalaf, M.; Dastjerdi, A.K.; Barthelat, F. Overcoming the brittleness of glass through bio-inspiration and micro-architecture. *Nat. Commun.* **2014**, *5*, 3166. [CrossRef] [PubMed]
26. Luz, G.M.; Mano, J.F. Biomimetic design of materials and biomaterials inspired by the structure of nacre. *Philos. Trans. A Math. Phys. Eng. Sci.* **2009**, *367*, 1587–1605. [CrossRef]
27. *Abaqus 2017*; Dassault Systèmes Simulia: Vélizy-Villacoublay, France, 2020.
28. Padawer, G.E.; Beecher, N. On the strength and stiffness of planar reinforced plastic resins. *Polym. Eng. Sci.* **1970**, *10*, 185–192. [CrossRef]
29. Lusis, J.; Woodhams, R.T.; Xanthos, M. The effect of flake aspect ratio on the flexural properties of mica reinforced plastics. *Polym. Eng. Sci.* **1973**, *13*, 139–145. [CrossRef]
30. Barthelat, F. Designing nacre-like materials for simultaneous stiffness, strength and toughness: Optimum materials, composition, microstructure and size. *J. Mech. Phys. Solids* **2014**, *73*, 22–37. [CrossRef]
31. Sakhavand, N. Mechanics of Platelet-Matrix Composites across Scales: Theory, Multiscale Modeling, and 3D Fabrication. *Mater. Sci.* **2015**, *79–05*, 292.
32. Sakhavand, N.; Muthuramalingam, P.; Shahsavari, R. Toughness governs the rupture of the interfacial H-bond assemblies at a critical length scale in hybrid materials. *Langmuir* **2013**, *29*, 8154–8163. [CrossRef] [PubMed]
33. Sakhavand, N.; Shahsavari, R. Universal composition-structure-property maps for natural and biomimetic platelet-matrix composites and stacked heterostructures. *Nat. Commun.* **2015**, *6*, 6523. [CrossRef] [PubMed]
34. Bar-On, B.; Wagner, H.D. New insights into the Young's modulus of staggered biological composites. *Mater. Sci. Eng. C Mater. Biol. Appl.* **2013**, *33*, 603–607. [CrossRef] [PubMed]
35. Jäger, I.; Fratzl, P. Mineralized Collagen Fibrils: A Mechanical Model with a Staggered Arrangement of Mineral Particles. *Biophys. J.* **2000**, *79*, 1737–1746. [CrossRef]
36. Lodes, M.A.; Kachold, F.S.; Rosiwal, S.M. Mechanical properties of micro- and nanocrystalline diamond foils. *Philos. Trans. A Math. Phys. Eng. Sci.* **2015**, *373*. [CrossRef]
37. Shi, H.-Q.; Sun, B.-G.; Liu, Q.; Yang, Z.-Y.; Yi, K.; Zhang, Y.; Fu, S.-Y. A high ductility RTM epoxy resin with relatively high modulus and Tg. *J. Polym. Res.* **2015**, *22*, 435. [CrossRef]
38. Kim, Y.; Kim, Y.; Lee, T.-I.; Kim, T.-S.; Ryu, S. An extended analytic model for the elastic properties of platelet-staggered composites and its application to 3D printed structures. *Compos. Struct.* **2018**, *189*, 27–36. [CrossRef]
39. Ni, Y.; Song, Z.; Jiang, H.; Yu, S.-H.; He, L. Optimization design of strong and tough nacreous nanocomposites through tuning characteristic lengths. *J. Mech. Phys. Solids* **2015**, *81*, 41–57. [CrossRef]
40. Tushtev, K.; Murck, M.; Grathwohl, G. On the nature of the stiffness of nacre. *Mater. Sci. Eng. C* **2008**, *28*, 1164–1172. [CrossRef]
41. Wei, X.; Naraghi, M.; Espinosa, H.D. Optimal length scales emerging from shear load transfer in natural materials: Application to carbon-based nanocomposite design. *ACS Nano* **2012**, *6*, 2333–2344. [CrossRef]
42. Brostow, W.; Hagg Lobland, H.E.; Narkis, M. Sliding wear, viscoelasticity, and brittleness of polymers. *J. Mater. Res.* **2006**, *21*, 2422–2428. [CrossRef]
43. Brostow, W.; Hagg Lobland, H.E.; Khoja, S. Brittleness and toughness of polymers and other materials. *Mater. Lett.* **2015**, *159*, 478–480. [CrossRef]
44. Chintapalli, R.K.; Breton, S.; Dastjerdi, A.K.; Barthelat, F. Strain rate hardening: A hidden but critical mechanism for biological composites? *Acta Biomater.* **2014**, *10*, 5064–5073. [CrossRef] [PubMed]
45. McCormick, H.W.; Brower, F.M.; Kin, L. The effect of molecular weight distribution on the physical properties of polystyrene. *J. Polym. Sci.* **1959**, *39*, 87–100. [CrossRef]
46. Ghazlan, A.; Ngo, T.; van Le, T.; Nguyen, T.; Remennikov, A. Blast performance of a bio-mimetic panel based on the structure of nacre—A numerical study. *Compos. Struct.* **2020**, *234*, 111691. [CrossRef]
47. Hall, J.R.; Westerdahl, C.A.L.; Devine, A.T.; Bodnar, M.J. Activated gas plasma surface treatment of polymers for adhesive bonding. *J. Appl. Polym. Sci.* **1969**, *13*, 2085–2096. [CrossRef]
48. Erickson, B.L.; Asthana, H.; Drzal, L.T. Sulfonation of polymer surfaces—I. Improving adhesion of polypropylene and polystyrene to epoxy adhesives via gas phase sulfonation. *J. Adhes. Sci. Technol.* **1997**, *11*, 1249–1267. [CrossRef]
49. Hack, E.; Lin, X.; Patterson, E.A.; Sebastian, C.M. A reference material for establishing uncertainties in full-field displacement measurements. *Meas. Sci. Technol.* **2015**, *26*, 75004. [CrossRef]

50. Landau, L.D.; Lifshitz, E.M.; Sykes, J.B.; Reid, W.H.; Dill, E.H. Theory of Elasticity: Vol. 7 of Course of Theoretical Physics. *Phys. Today* **1960**, *13*, 44–46. [CrossRef]
51. Breuer, K.; Stommel, M. RVE modelling of short fiber reinforced thermoplastics with discrete fiber orientation and fiber length distribution. *SN Appl. Sci.* **2020**, *2*, 140. [CrossRef]
52. Jackson, A.P.; Vincent, J.; Turner, R.M. The mechanical design of nacre. *Proc. R. Soc. Lond. B* **1988**, *234*, 415–440. [CrossRef]
53. Bekah, S.; Rabiei, R.; Barthelat, F. Structure, Scaling, and Performance of Natural Micro- and Nanocomposites. *BioNanoScience* **2011**, *1*, 53–61. [CrossRef]

 Journal of
Composites Science

Article

Developing an Equivalent Solid Material Model for BCC Lattice Cell Structures Involving Vertical and Horizontal Struts

Tahseen A. Alwattar and Ahsan Mian *

Department of Mechanical and Material Engineering, Wright State University, 3640 Colonel Glenn Hwy, Dayton, OH 45435, USA; al-wattar.2@wright.edu
* Correspondence: ahsan.mian@wright.edu

Received: 4 May 2020; Accepted: 15 June 2020; Published: 17 June 2020

Abstract: In this study, a body-centered cubic (BCC) lattice unit cell occupied inside a frame structure to create a so-called "InsideBCC" is considered. The equivalent quasi-isotropic properties required to describe the material behavior of the InsideBCC unit cell are equivalent Young's modulus (E_e), equivalent shear modulus (G_e), and equivalent Poisson's ratio (v_e). The finite element analysis (FEA) based computational approach is used to simulate and calculate the mechanical responses of InsideBCC unit cell, which are the mechanical responses of the equivalent solid. Two separates finite element models are then developed for samples under compression: one with a $6 \times 6 \times 4$ cell InsideBCC lattice cell structure (LCS) and one completely solid with equivalent solid properties obtained from a unit cell model. In addition, $6 \times 6 \times 4$ cell specimens are fabricated on a fused deposition modeling (FDM) uPrint SEplus 3D printer using acrylonitrile butadiene styrene (ABS) material and tested experimentally under quasi-static compression load. Then, the results extracted from the finite element simulation of both the entire lattice and the equivalent solid models are compared with the experimental data. A good agreement between the experimental stress–strain behavior and that obtained from the FEA models is observed within the linear elastic limit.

Keywords: lattice cell structures; InsideBCC; equivalent solid properties; three-dimensional printing

---

1. Introduction

Lattice cell structures (LCS) are the engineered porous structures that are composed of periodic unit cells in three dimensions. Such structures have many scientific and engineering applications, such as in thermal systems, gas technology, mechanical and aerospace structures, etc., for which lightweight, high strength, and energy absorption capabilities are essential properties [1–8]. Additionally, combining different unit cell configurations, such as a frame structure and a body-centered cubic (BCC) structure (Figure 1), will provide higher strength and higher stiffness LCS. Such a combined LCS is termed as InsideBCC in this paper. It has been shown that the combining unit cells such as lattice structures inside the tube, BCC with vertical struts at each node, BCC with vertical struts in alternate layers, and BCC with gradient distributed vertical struts, contribute to the buckling and bending resistance and enhancement in energy absorption performance as well as the high specific strength and stiffness [9–12].

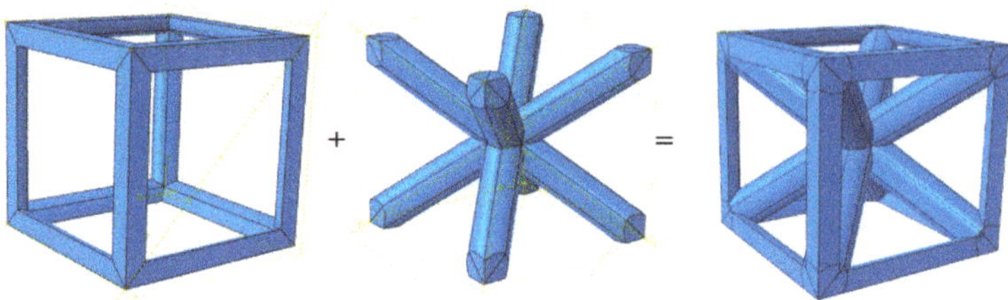

Figure 1. Combining frame and body-centered cubic (BCC) unit cells to develop an InsideBCC unit cell.

The finite element analysis (FEA) technique is usually used for designing complex structures at the system and subsystem levels to reduce cost and time. While modeling large scale structures with lattice cells, the numerical simulations require a large number of degrees of freedom to get accurate analysis, which requires more computation time [13,14]. For example, the BCC lattice structure with a strut diameter of 1 mm and the unit cell dimensions of 5 mm × 5 mm × 5 mm having the overall LCS dimensions of 25 mm × 25 mm × 20 mm under the compression was modeled using FEA [9]. In this case, the time needed to complete the solution was 48 h for mass scaling of 1E-7 times the time increment. Additionally, by using the FEA simulation, a BCC unit cell with a strut diameter of 0.7 mm and dimensions of a single unit cell are 5 mm × 5 mm × 5 mm in *x*, *y*, and *z* direction respectively under tension the computational time of solution was 10 h [15]. To model large scale structures involving LCS, an equivalent solid material can be used to replace and represent the whole lattice structure model to reduce computational time. The equivalent solid section will not have any struts. Displacement results from the equivalent solid model can be used as an input to different sub-models. Here, the sub-modeling approach can be utilized to find localized stress/strain within lattices in critical regions for large scale structures. Thus, the main objective of this study is to develop and employ a computational methodology to find equivalent solid material properties of InsideBCC LCS to conduct large scale finite element simulation easier and computationally less expensive.

This paper is organized as follows. First, the load–displacement and stress–strain results is first obtained from the FEA of InsideBCC unit cell under compression and shear in all three orthogonal directions. The stress–strain plots are then used to calculate the equivalent mechanical properties such as the elastic modulus (E_e), shear modulus (G_e), and Poisson's ratio (ν_e) for a unit cell. In this case, the unit cell size is 5 mm × 5 mm × 5 mm with a strut diameter of 1 mm and the bulk material is considered to be acrylonitrile butadiene styrene (ABS). Next, two separate finite element models are developed for samples under compression: one with 6 × 6 × 4 cell InsideBCC LCS and one completely solid with equivalent solid properties obtained from the unit cell model. In addition, 6 × 6 × 4 cell specimens are fabricated on a 3D printer using ABS material and tested experimentally under quasi-static compression load. Then, the results extracted from the finite element simulation of both the entire lattice and the equivalent solid models were compared with the experimental data to validate the FEA modeling scheme. This modeling scheme can be used to obtain equivalent solid properties for InsideBCC LCSs with other unit cell dimensions and strut diameters.

2. General Methodology

The main goal of this research was to develop an elastic material to predict the mechanical response of combining a lattice cell structure, which is called InsideBCC. Additionally, the current study could be used to find the effect of vertical and horizontal struts on the mechanical response of lattice cells within the elastic limit when compared with its BCC counterpart [16]. An InsideBCC unit cell is a representative volume element (RVE) that is used to generate equivalent solid mechanical responses. These mechanical properties would be used to create the equivalent solid material model, which is identical to the mechanical response of a whole lattice structure. Only periodic boundary conditions provide the correct equivalent properties of a unit cell since it was used as an RVE.

The general methodology strategy is schematically showed in Figure 2 in which the InsideBCC unit cell was used as a model for predicting the mechanical responses of the full-scale lattice structure. The following were the steps process flow of this research:

a. A model of the InsideBCC unit cell was simulated using FEA to calculate the elastic modulus for x, and y direction ($E_x = E_y = E_z = E_e$). For this case, the bulk ABS mechanical properties were used.

b. Calculate the Poisson's ratio for different orientations ($v_{zx} = v_{zy} = v_{xz} = v_{yz} = v_{xy} = v_{yx} = v_e$), obtained in part (a) models.

c. A model of the InsideBCC unit cell was simulated using FEA to predict the shear modulus ($G_{xy} = G_{xz} = G_{yz} = G_e$). For this case, the bulk ABS mechanical properties were used.

d. A FEA model of $6 \times 6 \times 4$ cell InsideBCC was generated and run to find its load–displacement response under compression. For this case, the bulk ABS mechanical properties were used.

e. A separate FEA model was generated for a complete solid material with overall dimension to be same as a $6 \times 6 \times 4$ cell InsideBCC. In this case, the equivalent solid properties obtained from steps (a–c) above were used.

f. Three samples of $6 \times 6 \times 4$ cell InsideBCC were fabricated on a fused deposition modeling (FDM) uPrint SEplus 3D printer using acrylonitrile butadiene styrene (ABS). The samples are then tested under quasi-static compression test to compare their load–displacement behavior with that simulated from both solid material (step d) and LCS models (part c).

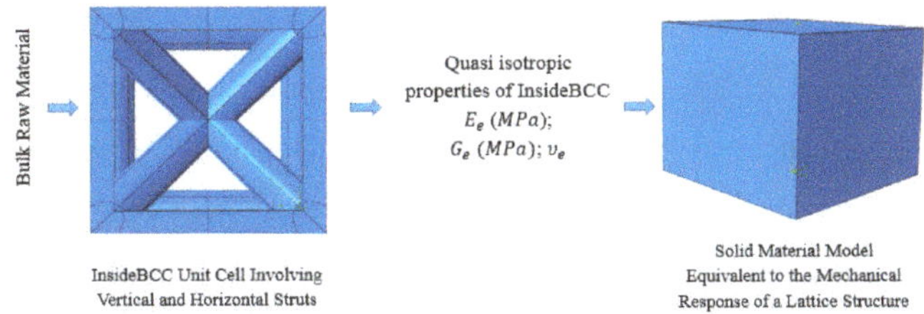

Figure 2. A methodology strategy from unit cell finite elemental analysis (FEA) model to equivalent solid model of InsideBCC configuration.

3. Experimental Procedure

3.1. InsideBCC LCS Layout, Fabrication and Material Used

The micromechanics ABAQUS software 6.17 was used to design all models for finite element (FE) simulation [17], printing purposes, and implementing experimental tests. The micromechanics technique has more flexibility for creating hexahedral mesh elements to increase the accuracy of results. InsideBCC unit cell was designed by micromechanics ABAQUS, as illustrated in Figure 3a. The dimensions of a single InsideBCC unit cell are 5 mm × 5 mm × 5 mm with a truss diameter of 1 mm. The overall dimensions to create the lattice structure of InsideBCC were 30 mm × 30 mm × 20 mm in x, y, and z-directions respectively as shown in Figure 3b. The configurations of InsideBCC geometry that were designed by micromechanics technique software are saved in stereolithography (STL), which is a format used to define the sample geometry of a 3D printer software Stratasys Catalyst [18]. These samples were fabricated by a fused deposition modeling (FDM) based uPrint SE plus 3D printer [19] using acrylonitrile butadiene styrene (ABS) material. Furthermore, the printer nozzle temperature of 300 °C and the chamber temperature of 77 °C were maintained, which were used as default temperature settings for all the specimens that were printed. Three samples were fabricated to conduct the experimental quasi-static compression test for validation of the results obtained from both the FEA models of equivalent solid material and whole lattice structure. Those specimens were fabricated with support material, which was removed using a Stratasys cleaning apparatus, SCA, 1200HT [20]. The printing parameters selected for this research were based on the references [9,10,16,21].

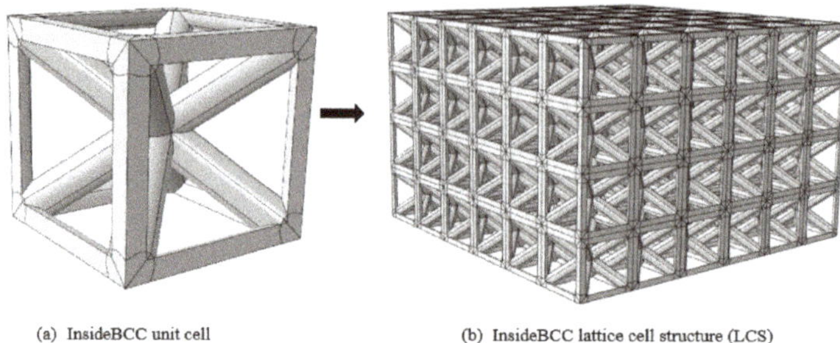

Figure 3. Micromechanics technique design (MTD). (**a**) InsideBCC unit cell. (**b**) InsideBCC lattice structure.

3.2. Quasi-Static Compression Test

The quasi-static compression test was performed to calculate material behaviors such as modulus of elasticity, and yield point of LCS. A universal testing machine, which is Instron 5500R, was used to conduct the compression test on the fabricated specimens of InsideBCC CLS with dimensions of 30 mm × 30 mm × 20 mm, consisting of 6 × 6 × 4 InsideBCC unit cells in x, y, and z direction respectively. All the samples of polymer cellular structure were compressed by 10 mm displacement with a constant displacement rate of 0.4 mm/min. A software Bluehill2$^{(r)}$ connected with the Instron machine 5500R was used to collect the load–displacement data, which was saved in Excel. The completed final specimen of LCS for compression testing is illustrated in Figure 4.

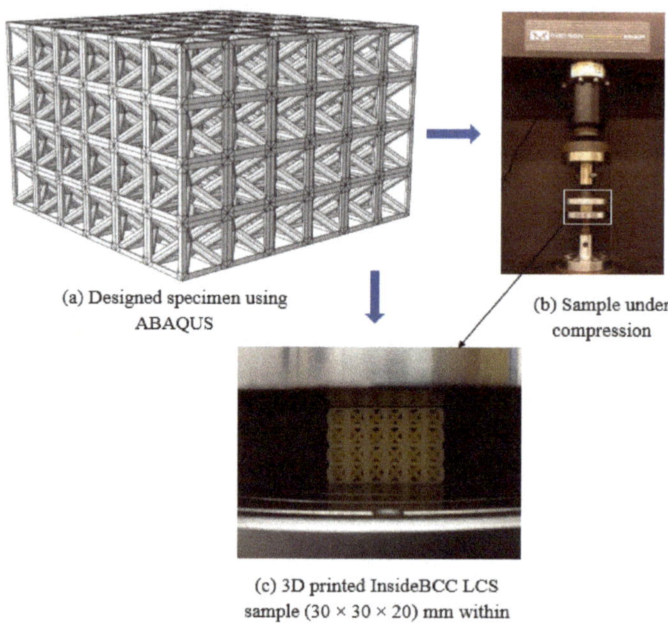

Figure 4. Experimental procedures, (**a**) specimen design in MTD; (**b**) fabricated sample under compression; and (**c**) zoomed-in view of 3D printed sample within compression test fixture.

4. Finite Element Modeling

Both the InsideBCC unit cell (for steps a and c in Section 2) and LCS (for step d in Section 2) were designed and meshed using the intervention technique, which was developed in the micromechanics technique of ABAQUS/CAE 6.17 software. The developing micromechanics technique (DMT) has drastically simplified the selection of element types during mesh generation for a combining cellular structure. The DMT method addresses the challenges associated with the complexity of the InsideBCC unit cell and LCS to generate hexahedral mesh. The general process flow of the DMT method is schematically illustrated in Figure 5 in which the InsideBCC unit cell and LCS were designed and meshed. Seven essential steps are to: (1) create the single strut by using five points in one plane as a deformable coordinates point; (2) create four struts as a basis of a frame; (3) generate the frame by a mirror and then by using cell partition, which is defined as a cutting plane in order to complete the whole frame; (4) as a separate process, generate a single BCC unit lattice cell; (5) join the BCC unit cell inside the frame to create the InsideBCC unit cell; (6) using the linear pattern in order to generate the number of unit cells in all direction x, y, and z; and (7) finally, generate mesh using mesh controls by selecting the Hex option.

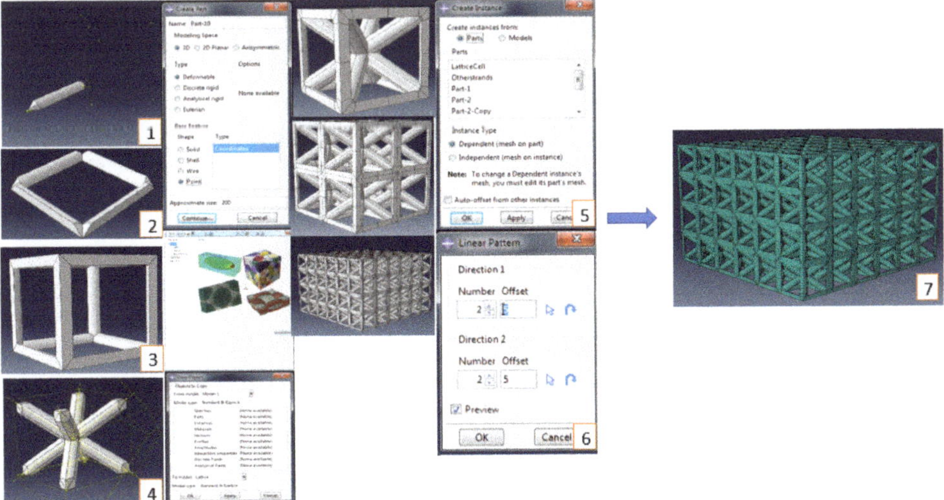

Figure 5. Framework for developing micromechanics technique using ABAQUS.

4.1. Unit Cell Modeling for Equivalent Solid Properties

In this step, the FEA was employed to model the compression and shear test of an InsideBCC unit cell geometry for different directions within the elastic limit to predict equivalent solid properties of the whole cellular structure. The unit cell has dimensions 5 mm × 5 mm × 5 mm and the diameter struts of 1 mm (Figure 3a). ABS polymer material was employed for all finite element FE models.

4.1.1. Mesh Generation

In general, two element types are usually used for mesh generation, which are tetrahedron and hexahedron. In the field of mesh generation, researchers have used tetrahedral elements that can be easily created automatically in most finite element models [22–24]. However, user intervention is required to create a hexahedral mesh element, which provides high performance and it gives accurate results as compared with the tetrahedral mesh elements [20,25,26]. Therefore, in this research to generate the mesh for all FE simulation, a hexahedral element type is used by the MTD technique, which in turn gives the capability to use a first-order hexahedron continuum solid element with linear brick reduced integration (C3D8R).

Since meshing is significantly better at providing accurate results, both mesh sensitivity analysis and the type of mesh generation were considered. To achieve FEA with high performance, in this work, a mesh sensitivity analysis was performed on a unit cell. Accordingly, mesh sensitivity was performed by observing the stiffness versus the total number of the elements under compression, which is illustrated in Figure 6. Additionally, this curve illustrated the mesh convergence that occurred when the mesh size decreased from 1.25 (coarse) to 0.19 (fine) when the percentage variation of stiffness was within 2%. Based on this percentage, the acceptable mesh size chosen from the mesh convergence study was 0.35 mm, the total number of elements was 2400 elements, and the amount of stiffness was 230.39 N/mm. A discretized model shows the acceptable mesh size employed for the InsideBCC model is illustrated in Figure 6 as an insert. In this research, the same procedure was followed to perform the mesh sensitivity analysis for all FE models.

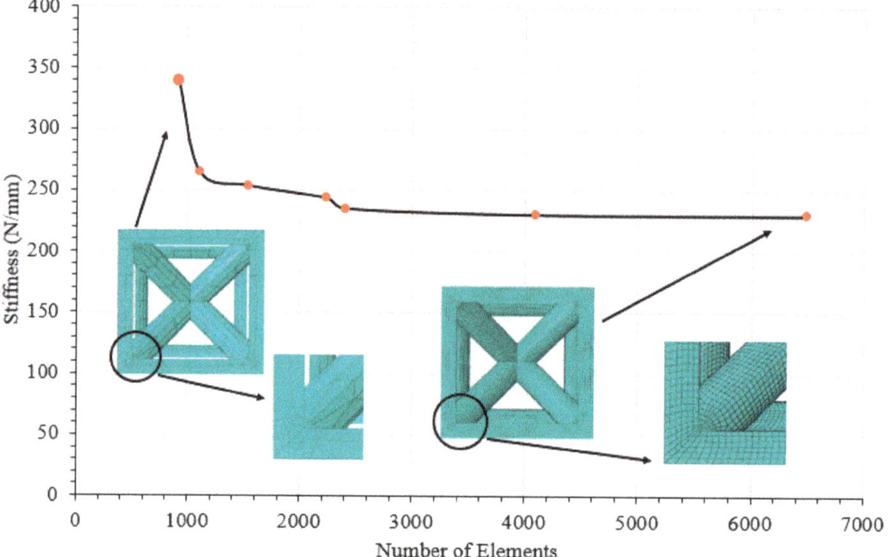

Figure 6. Mesh sensitivity study of the InsideBCC unit cell (5 × 5 × 5) mm and d = 1 mm for mesh size from 1.25 mm (coarse) to 0.19 mm (fine).

4.1.2. Boundary Conditions

In the current research, for the aim of capturing the behavior of the entire heterogeneous cellular structures based on the analyses of the InsideBCC unit cell, only the periodic boundaries would suffice. Regarding both the elastic modulus and Poisson's ratio, the lattice unit cell faces are free to move, including all translational and rotational degrees of freedom with respect to rigid plates as illustrated in Figure 7a. Typically, the rigid plate is a non-deformable plate when subjected to any certain load. Consequently, the applied displacement load with a specific amount on the reference point of the rigid plate will be distributed with the same value for the entire plate. To determine the elastic modulus and Poisson's ratio, a compressive displacement was applied on one face (here, the top plate in Figure 7a toward the opposite face (here, the bottom plate in Figure 7b in the downward direction)). The reference plate on the bottom plate was to be fixed in all three directions as illustrated in Figure 7a. For the shear modulus simulation, the model was placed between two rigid plates, where each of the opposite unit cell faces were clamped with the corresponding plates as illustrated in Figure 7b. Shear displacement was applied on the top plate while the other face was kept fixed for all degrees of freedom as shown in Figure 7b. The displacement loading was applied using the dynamic explicit FEA simulation.

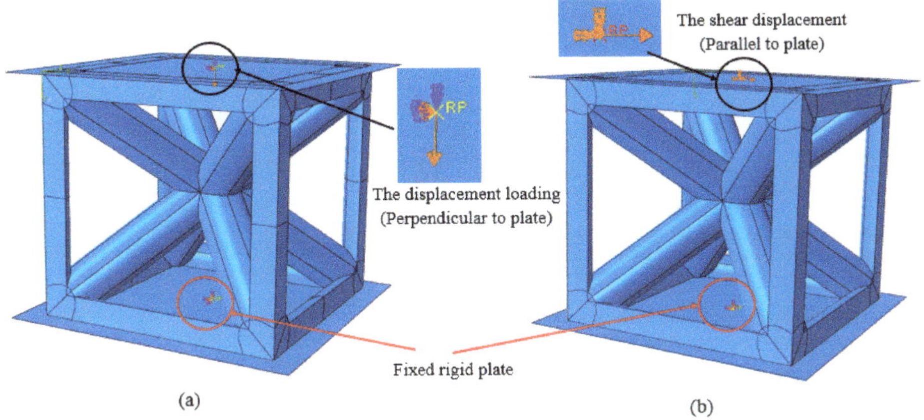

Figure 7. (a) Boundary conditions of the unit cell FEA model for the elastic modulus and Poisson's ratio in all three directions x, y, and z. (b) Boundary conditions of the unit cell FEA model for the shear modulus in all three directions x, y, and z.

4.1.3. Material Properties

The bulk material properties of ABS used in the finite element models of the InsideBCC unit cell are illustrated in Table 1. Experimentally, three specimens were investigated to measure the material properties of the bulk material according to ASTM-D695, ISO 604 for standard compression and ASTM-D882 for tension tests [9–16].

Table 1. Bulk material properties of ABS material.

Young's Modulus (MPa)	Poisson's Ratio	Density (g/mm^3)	Yield Strength (MPa)	Ultimate Tensile Strength (MPa)	Plastic Strain (mm/mm)
861.55	0.35	7.92×10^{-4}	25.75	33.33	0.045

4.2. InsideBCC LCS and Equivalent Solid Material Model

To demonstrate the performance of development of an elastic material model for InsideBCC lattice cell structures, two separate finite element models were developed under the compression load: one with a $6 \times 6 \times 4$ cell InsideBCC lattice structure (for step d in Section 2) and one entirely solid (for step e in Section 2). The optimized discretized simulations of a 30 mm × 30 mm × 20 mm combining a lattice structure with a 5 mm × 5 mm × 5 mm unit cell having a strut diameter of 1 mm (Figure 8a) and 30 mm × 30 mm × 20 mm equivalent solid material (Figure 8b) are illustrated in Figure 8. The heterogeneous LCS in Figure 8a and the equivalent solid material in Figure 8b were simulated with hexagonal elements. The LCS model was generated and meshed using the DMT micromechanics technique as described in the previous section. Interestingly, the study of the mesh optimum sensitivity analysis shows the number of elements for the heterogeneous lattice structure model was 345,600 while the number of elements for the solid material model was 144,000, illustrated in Figure 8. In addition, the time required for that study to generate the mesh and seed part for the LCS model was about 12 h without running the FE simulation while the time required for creating the solid material model was about one second.

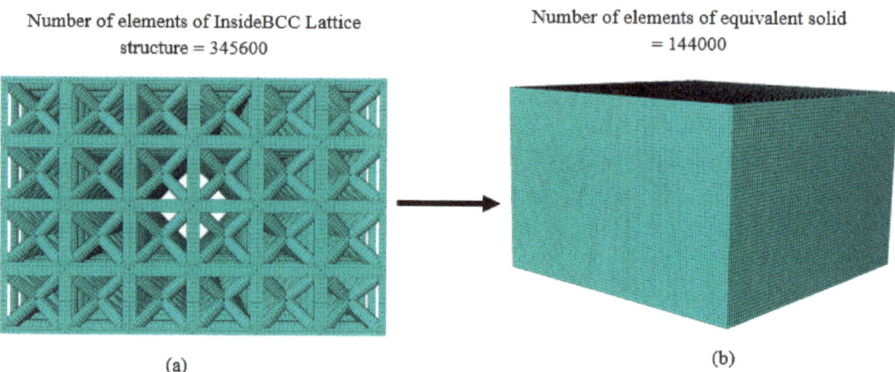

Figure 8. FEA mesh of full simulation for (**a**) the InsideBCC lattice cell structure (LCS) model and (**b**) the equivalent solid model.

The raw material properties used for the LCS (Figure 8a) simulation of the FE model are shown previously in Table 1, whereas the equivalent solid properties obtained from the previous section was used for the solid model shown in Figure 8b.

To mimic precise boundary conditions of the experimental test, the top and bottom faces of both the InsideBCC LCS in Figure 8a and the equivalent solid model in Figure 8b were tied to perfect rigid plates for all degrees of freedom. In that manner, a displacement load was applied on the top face of the model to move towards the bottom face. The stress–strain curves for both the fully lattice structure and equivalent solid model are illustrated in Figure 9 inclusive of the experimental results. In this figure, the experimental stress–strain plot was created based on the average modulus of elasticity of three LCS samples.

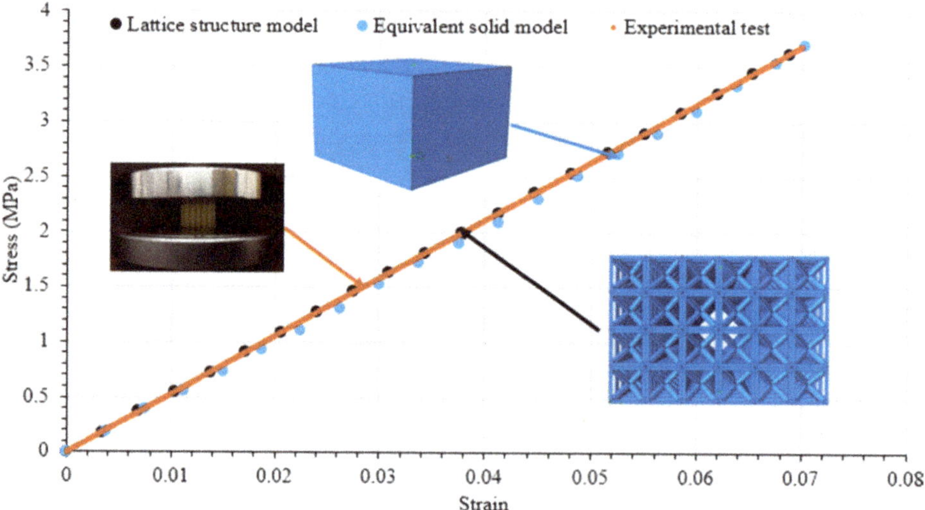

Figure 9. Comparison of the FEA simulation of the LCS model, experimental test, and equivalent solid model.

5. Results and Discussion

5.1. Load–Displacement Behavior

For InsideBCC configuration, load–displacement data from three samples (experimental) were plotted in one diagram to investigate the specimen to specimen variation under the compression load. The stress–stain behavior of three InsideBCC samples is illustrated in Figure 10. In this paper, the mechanical behavior of InsideBCC configuration under the compression load within the elastic limit was discussed. In other words, the elastic limit until the yield point was considered, which means the stages of lattice cell failure were not included in this study.

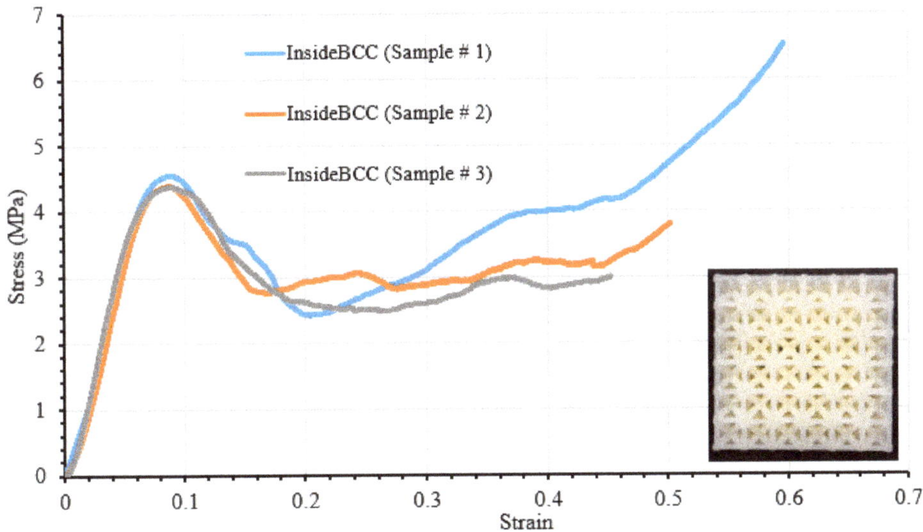

Figure 10. Stress–strain curves for three InsideBCC specimens.

5.2. Comparison between Experimental and FEA Model

To show the implementation of the proposed equivalent solid material model method, two separate finite element models were developed for specimens under compression simulation. One of them is the InsideBCC lattice structure model with the raw material properties and the other one was the completely solid material model with equivalent properties to the InsideBCC unit cell. The equivalent solid material properties obtained from unit cell models were E_e = 46.12 MPa, G_e = 21.77 MPa, and ν_e = 0.42. The outcomes of both FE models (the heterogeneous lattice structure model and equivalent solid material model) were validated with the experimental result of 3D printed LCS. The load–displacement data in Figure 10 were converted to stress–strain plots and the slope of the linear region was used to calculate modulus of elasticity of LCS. The experimental stress–strain plot in Figure 9 was based on the average modulus of elasticity from three LCS samples. The stress–strain plots acquired from the lattice cell structure of FEA (black solid circles), equivalent solid material (blue solid circles), and the experimental test of 3D printed LCS (brown line) were compared with each other as shown in Figure 9. The proposed equivalent solid material had the capability to capture the mechanical behavior of a large-scale heterogeneous lattice structure model. As it can be observed from Figure 9, there was good agreement within the linear elastic limit among the results of the overall lattice cell structure model, the equivalent solid material model, and the experimental work. Accordingly, the equivalent solid material model for a lattice-based structure gave accurate and acceptable capturing for the mechanical behavior of a full-scale heterogeneous lattice structure.

5.3. Comparison between BCC LCS and InsideBCC CLCS

Representative load–displacement behavior of both BCC and InsideBCC ABS samples were plotted together to compare the variation of compression behavior, which is illustrated in Figure 11. For the BCC sample the maximum failure or peak load was about 500 N while the InsideBCC sample was about 4000 N. Thus, it could be concluded that the InsideBCC lattice cell structures involving vertical and horizontal struts shows enormously higher stiffness and failure load than the BCC geometry. Additionally, from the area under the load–displacement curve, the strain energy absorption is lowest for the BCC feature, while the InsideBCC configuration had the largest one. Stiffness, failure load, and energy absorption for various parameters such as the strut diameter, cell size, and processing factors will be further investigated by developing a surrogate model using the intelligent method. The improved mechanical performance of InsideBCC LCS can be used for specific applications that required unique load–displacement characteristics.

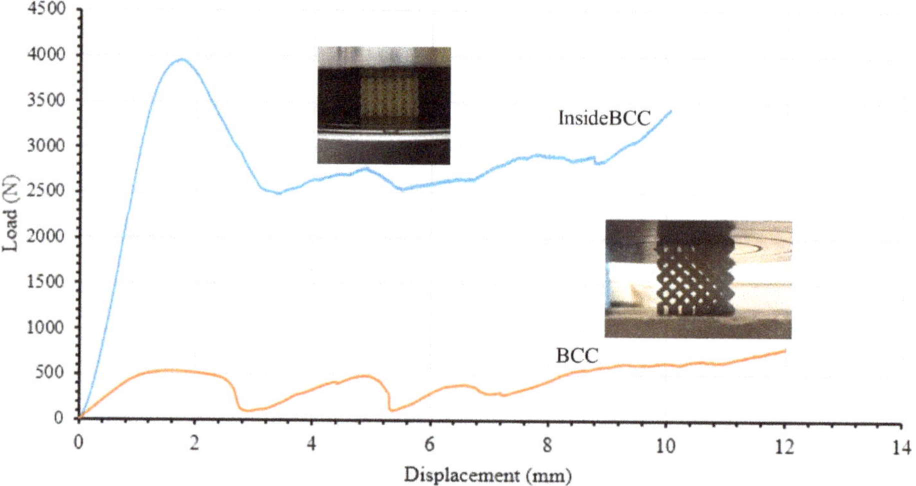

Figure 11. Comparison of load—displacement curves for BCC and InsideBCC configurations: BCC sample (brawn line) and InsideBCC sample (blue line).

6. Summary and Remarks

In this paper, the equivalent solid material model was developed such that the equivalent properties of heterogeneous unit cell configuration (InsideBCC) were used to mimic the behavior of the mechanical response of heterogeneous lattice structures involving vertical and horizontal struts for engineering design exploration. Consequently, the equivalent solid material model of InsideBCC configuration is very quick, accurate and practical when compared with a numerical model of the full-scale heterogeneous lattice structure. Therefore, developing an equivalent solid material not only demonstrates the computational time reduction from several days to few minutes but also provides an efficient analysis for FE simulation of InsideBCC LCS. Besides, one of the biggest challenges is to use FDM based 3D printing technology to create the InsideBCC lattice structure with vertical and horizontal struts. As a futuristic technique for this research, a surrogate model will be developed to determine the equivalent material properties for larger and more complicated combined LCS, with any arbitrary cell size, strut diameter, and type of material, which can be used for FEA simulations with a considerable reduction in the computational time.

Author Contributions: Conceptualization, T.A.A. and A.M.; methodology, T.A.A. and A.M.; formal analysis, T.A.A.; investigation, T.A.A. and A.M.; resources, A.M.; data curation, T.A.A.; writing—original draft preparation,

T.A.A.; writing—review and editing, A.M.; supervision, A.M.; project administration, A.M. All authors have read and agreed to the published version of the manuscript.

Funding: This research received no external funding.

Conflicts of Interest: The authors declare no conflict of interest.

References

1. Wilmoth, N. *Determining the Mechanical Properties of Lattice Block Structures*; NASA Technical Report Server; National Aeronautics and Space Administration: Cleveland, OH, USA, 2013; Available online: https://ntrs.nasa.gov/archive/nasa/casi.ntrs.nasa.gov/20130014836.pdf (accessed on 16 June 2020).

2. Bible, M.; Sefa, M.; Fedchak, J.A.; Scherschligt, J.; Natarajan, B.; Ahmed, Z.; Hartings, M.R. 3D-Printed Acrylonitrile Butadiene Styrene-Metal Organic Framework Composite Materials and Their Gas Storage Properties. *3D Print. Addit. Manuf.* **2018**, *5*, 63–72. [CrossRef]

3. Wong, M.; Owen, I.; Sutcliffe, C. Convective heat transfer and pressure losses across novel heat sinks fabricated by Selective Laser Melting. *Int. J. Heat Mass Transf.* **2009**, *52*, 281–288. [CrossRef]

4. Al-Saedi, D.; Masood, S.; Faizn, M. Mechanical properties and energy absorption capability of functionally graded F2BCC lattice fabricated by SLM. *Mater. Sci. Eng.* **2018**, *144*, 32–44. [CrossRef]

5. Mahmoud, D.; Elbestawi, M. Lattice Structures and Functionally Graded Applications in Additive Manufacturing of Orthopedic Implants. *Manuf. Mater. Process.* **2017**, *1*, 13. [CrossRef]

6. Ulah, L.; Brandt, M.; Feih, S. Failure and energy absorption characteristics of advanced 3D truss core structures. *Mater. Des.* **2016**, *92*, 937–948. [CrossRef]

7. Alsalla, H.; Hao, L.; Smith, C. Fracture toughness and tensile strength of 316 L stainless steel cellular lattice structures manufactured using the selective laser melting technique. *Mater. Sci. Eng.* **2016**, *669*, 1–6. [CrossRef]

8. Maskery, L.; Hussey, A.; Aremu, A.; Tuck, C.; Ashcroft, L.; Hague, R. An investigation into reinforced and functionally graded lattice structures. *J. Cell. Plast.* **2017**, *53*, 151–165. [CrossRef]

9. Al Rifaie, M.J. Resilience and Toughness Behavior of 3D-Printed Polymer Lattice Structures: Testing and Modeling. Master's Thesis, Wright State University, Dayton, OH, USA, 2017.

10. Fadeel, A.; Mian, A.; Al Rifaie, M.; Srinivasan, R. Effect of Vertical Strut Arrangements on Compression Characteristics of 3D Printed Polymer Lattice Structures: Experimental and Computational Study. *J. Mater. Eng. Perform.* **2018**, *1*, 1–8. [CrossRef]

11. Cetin, E.; Baykasoglu, C. Energy absorption of thin-walled tubes enhanced by lattice structures. *Int. J. Mech. Sci.* **2019**, *158*, 471–484. [CrossRef]

12. Turner, A.J.; Al Rifaie, M.; Mian, A.; Srinivasan, R. Low-Velocity Impact Behavior of Sandwich Structures with Additively Manufactured Polymer Lattice Cores. *ASM Int.* **2018**, *27*, 2505–2512. [CrossRef]

13. Lozanovski, B.; Leary, M.; Tran, P.; Qian, M.; Choong, P.; Brandt, M. Computational modelling of strut defects in SLM manufactured lattice structures. *Mater. Des.* **2019**, *171*, 107671. [CrossRef]

14. Smith, M.; Guan, Z.; Cantwell, W. Finite element modelling of the compressive response of lattice structures manufactured using the selective laser melting technique. *Int. J. Mech. Sci.* **2013**, *67*, 28–41. [CrossRef]

15. Koeppe, A.; Padilla, C.A.H.; Voshage, M.; Schleifenbaum, J.H.; Markert, B. Efficient numerical modeling of 3D-printed lattice-cell structures using neural networks. *Manuf. Lett.* **2018**, *15*, 147–150. [CrossRef]

16. Alwattar, T.A.; Mian, A. Development of an Elastic Material Model for BCC Lattice Cell Structures Using Finite Element Analysis and Neural Networks Approaches. *J. Compos. Sci.* **2019**, *3*, 33. [CrossRef]

17. *Study Part Module in ABAQUS CAE 6.17 Documentation*; DS Simulia: Providence, RI, USA, 2017.

18. Chakravorty, D. Craftcloud. 2019. Available online: http://all3dp.com/what-is-stl-file-format-extension-3d-printing (accessed on 16 June 2020).

19. Stratasys. uPrint SE Plus. 2017. Available online: https://support.stratasys.com/products/fdm-platforms/uprint (accessed on 16 June 2020).

20. Support Removal. Available online: https://www.oryxadditive.com/sca-1200ht (accessed on 16 June 2020).

21. Abdulhadi, H.S.; Mian, A. Effect of strut length and orientation on elastic mechanical response of modified body-centered cubic lattice structures. *J. Mater. Des. Appl.* **2019**, *233*, 2219–2233. [CrossRef]

22. Ravari, M.K.; Kadkhodaei, M.; Badrossamay, M.; Razaei, R. Numerical investigation on mechanical properties of cellular lattice structures fabricated fused deposition modeling. *Int. J. Mech. Sci.* **2014**, *88*, 154–161. [CrossRef]

23. Cuan-Urquizo, E.; Bhaskar, A. Flexural elasticity of woodpile lattice beams. *Eur. J. Mech. A Solids* **2018**, *67*, 187–199. [CrossRef]

24. Chen, W.; Lee, T.; Lee, P.; Lee, J.W.; Lee, S. Effects of internal stress concentrations in plantar soft-tissue—A preliminary three-dimensional finite element analysis. *Med. Eng. Phys.* **2010**, *32*, 324–331. [CrossRef]

25. Tadepalli, S.C.; Erdemir, A.; Cavanagh, P.R. Comparison of hexahedral and tetrahedral elements in finite element analysis of the foot and footwear. *J. Biomech.* **2011**, *44*, 2337–2343. [CrossRef]

26. Oliveira, B.; Sundnes, J. Comparison of Tetrahedral and Hexahedral Meshes for Finite Element Simulation of Cardiac Electro-Mechanics. In Proceedings of the VII European Congress on Computational Methods in Applied Sciences and Engineering, Crete Island, Greece, 5–10 June 2016; pp. 164–177.

Journal of
Composites Science

Article

A Machine Learning Model to Detect Flow Disturbances during Manufacturing of Composites by Liquid Moulding

Carlos González [1,2,*] and Joaquín Fernández-León [1,2]

1   IMDEA Materials, C/Eric Kandel 2, 28906 Getafe, Madrid, Spain; joaquin.fernandez@imdea.org
2   Departamento de Ciencia de Materiales, Universidad Politécnica de Madrid, E.T.S. de Ingenieros de Caminos, 28040 Madrid, Spain
*   Correspondence: carlosdaniel.gonzalez@imdea.org

Received: 20 May 2020; Accepted: 5 June 2020; Published: 10 June 2020

Abstract: In this work, a supervised machine learning (ML) model was developed to detect flow disturbances caused by the presence of a dissimilar material region in liquid moulding manufacturing of composites. The machine learning model was designed to predict the position, size and relative permeability of an embedded rectangular dissimilar material region through use of only the signals corresponding to an array of pressure sensors evenly distributed on the mould surface. The burden of experimental tests required to train in an efficient manner such predictive models is so high that favours its substitution with synthetically-generated simulation datasets. A regression model based on the use of convolutional neural networks (CNN) was developed and trained with data generated from mould-filling simulations carried out through use of OpenFoam as numerical solver. The evolution of the pressure sensors through the filling time was stored and used as grey-level images containing information regarding the pressure, the sensor location within the mould and filling time. The trained CNN model was able to recognise the presence of a dissimilar material region from the data used as inputs, meeting accuracy expectation in terms of detection. The purpose of this work was to establish a general framework for fully-synthetic-trained machine learning models to address the occurrence of manufacturing disturbances without placing emphasis on its performance, robustness and optimization. Accuracy and model robustness were also addressed in the paper. The effect of noise signals, pressure sensor network size, presence of different shape dissimilar regions, among others, were analysed in detail. The ability of ML models to examine and overcome complex physical and engineering problems such as defects produced during manufacturing of materials and parts is particularly innovative and highly aligned with Industry 4.0 concepts.

Keywords: machine learning; mould filling simulations; composite materials; liquid moulding

1. Introduction

Fibre reinforced polymer composites (PMCs) are nowadays widely used in applications that require lightweight materials such as those found in aerospace, automotive, energy, health-care and sports, among others. Structural PMCs are processed by the infiltration of a polymer matrix in the form of a thermoset or thermoplastic resin into a fabric preform, producing materials with optimal stiffness, strength, fatigue performance and resistance to environmental effects [1,2]. The high level of maturity reached as regards design and manufacture of advanced structural composites has enabled their extensive use in civil and military aircraft applications. For instance, the A350XWB contains as much as 53% of the structural weight of components, including carbon fuselage, wings or tail planes [3]. The automation of process steps, the increase of part integration level as well as the continuous demands towards zero-defect manufacturing, guided by Industry 4.0 concepts are the subsequent major steps that will undoubtedly allow future material optimization and cost reduction.

Liquid moulding of composites (LCM) starts with a dry fabric preform which is draped and placed into a mould for its impregnation with a liquid resin by prescribing a pressure gradient [4,5]. After the part is totally impregnated, the resin is cured, normally by the simultaneous action of heat sources until the part becomes solid and can be demoulded. Nowadays, LCM techniques can deliver to the industry high-quality and complex-shape composite articles, providing a solid out-of-autoclave alternative.

Resin transfer moulding (RTM) makes use of a closed solid mould in which the fabric is impregnated by a pressure gradient imposed between the inlet and outlet gates of the mould. Variations of such technologies include, among others, the replacement of a half part of the mold by a vacuum bag in vacuum infusion (VI), the use of auxiliary flow media to induce through-the-thickness flow of resin into the preform (Seemann composite resin infusion molding process, SCRIMP), or the use of flexible membranes as in light RTM (LRTM).

One of the major drawbacks in LCM arises from the inherent uncertainty as regards the flow patterns produced during resin impregnation which are strongly affected by different processing disturbances. For instance, variations of local permeability of the fabric preform triggered by uneven mould clamping, fabric shearing generated during draping operations or unexpected resin channels created at mould edges, also known as race-tracking, result in non-homogeneous resin flow far from theoretical predictions and consequently the formation of dry spots and porous areas. Counteracting against processing disturbances requires its early detection through use of the appropriate sensor networks as well as the implementation of the necessary corrective actions if the quality of the composite article is intended to be secured [6]. This later objective is precisely one of the differential concepts that emanates from Industry 4.0 smart factories which relies on the development of the appropriate cyber-physical systems able to perform automatically diagnosis and detect possible processing faults, as well as implement the necessary corrective actions.

Significant efforts have been made by the scientific and technical community in the last years to spread predictive models based on artificial intelligence (AI) and machine learning concepts (ML) to different sectors. Good examples can be found, for instance, in automated image recognition and computer vision algorithms used in autonomous car driving, fast text language translation and facial recognition, among others [7–9]. Such algorithms, when appropriately trained for each individual case, open revolutionary opportunities for other less explored fields such as continuum [10] and fluid mechanics or manufacturing. At this time, it is worth mentioning some interesting contributions close to the topic presented in this paper [11–13]. Wu et al. [11] developed a permeability surrogate model based on microstructural images through use of a convolutional neural network (CNN). A training set is generated first by these authors that contains synthetic images of porous microstructures while the corresponding effective permeability is computed, solving numerically the boundary value problem with Lattice Boltzmann methods. The CNN is then trained to learn and link the specific features of the microstructure (e.g., porosity) with the effective permeability value. Overall, surrogate models are viewed as very effective methods to overcome complex problems involving strong non-linearity, uncertainties, accelerating computation times in flow propagation through random media [14].

Although machine learning methods have been extensively used for surrogate modelling, the ability of these technologies to link input and output datasets without taking in consideration the underlying physics made them convenient tools for inverse modelling [12,13]. In this case, a forward model, the physical model, enables the generation of a synthetic output dataset based on the resolution of the governing equations (e.g., a fluids or mechanics problem) given the input dataset. Once the output dataset is generated, it is possible to relate the outputs with their corresponding inputs through use of regression machine learning tools. From an engineering viewpoint, such approximations are of major importance in practical problems such as tomographic reconstruction, computer imaging or sensors. Iglesias et al. [12] used concepts based on Bayesian inference to address the problem of uncertainties of fabric permeability in resin transfer moulding by using information coming from pressure sensors. Lähivaara et al. [13] developed a solution for the determination of the material porosity based on ultrasound tomography through use of CNN. A forward model is used to generate the theoretical response of the ultrasound sensors to a given material with known porosity parameters. The inverse problem is approximated with a regression based CNN which enables the determination of the effective porosity of a material from the direct inspection of the ultrasound transducers response.

The main purpose of this work was to provide a first exploration of machine learning methods to detect automatically flow disturbances caused by the presence of dissimilar permeability regions in liquid moulding of composites. The detection capabilities of the model fall on the analysis of pressure changes recorded by a distributed network of sensors. The model was developed to address the problem of rectangular dissimilar region in a squared flat RTM mould and has been kept intentionally simple with the purpose to explore its learning capabilities. The general description of the methodology is presented in Section 2 including the forward model based on the resolution of the flow propagation problem in porous media. The systematic generation of a fully-based simulation results set corresponding to the physical problem is presented in Section 3 while the general architecture of the neural network model, its training and deploying is described in Section 4. A general discussion on the applicability of the model presented is done in Section 5, while some remarks and conclusions are lastly drawn in Section 6.

2. Model Description

The general description of the model is sketched in Figure 1. An $L \times L$ square flat RTM mould containing a dissimilar material region of relative permeability $\beta = K/K_0$ is analyzed, where K and K_0 stand for the permeability of the fabric and the dissimilar region embedded, respectively. The position and size of this dissimilar region are defined by x_0, y_0 and $2b \times 2h$. The forward model uses the 5-tuple input $X_{input} = x_0, y_0, 2b, 2h, \beta$ to determine the virtual response of a set of pressure sensors Y_{output} distributed over the mould surface.

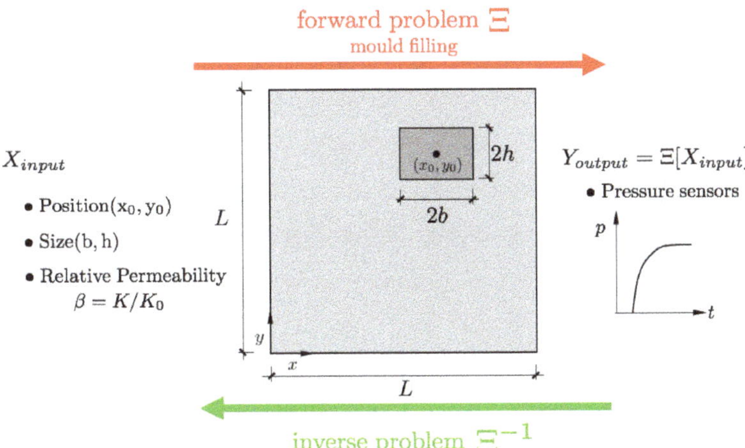

Figure 1. General description of the problem. Square RTM mould containing a rectangular dissimilar material region with relative permeability of $\beta = K/K_0$, size $2b \times 2h$ and centered at (x_0, y_0). K_0 and K stand for the permeability of the dissimilar region and the surrounding media, respectively.

The Latin Hypercube sampling was used to generate the different realizations of the inputs $x_0, y_0, 2b, 2h, \beta$ and the corresponding output dataset is generated, containing all the pressure sensor signals for different dissimilar material cases ($Y_{output} = \Xi[X_{input}]$). A supervised regression machine learning model based on convolutional neural network is used to approximate the solution of the inverse problem ($X_{input} = \Xi^{-1}[Y_{output}]$), thus enabling the position and size of the dissimilar material region to be estimated when the pressure sensor signal information is available. Such a kind of inverse approximation permits the on-line detection of flow disturbances and possible dry spot regions during the resin injection without accessing visually the interior of the mould and by using only the information of the pressure sensors readings.

3. Dataset Synthetic Generation

3.1. Mould Filling Model

The model used in this work is based on the resolution of the Darcy equation for the fluid flow through porous medium. The Darcy equation establishes a linear relation between the average fluid velocity through the fiber preform $\mathbf{v}(\mathbf{x}, t)$ and the pressure gradient $\nabla p(\mathbf{x}, t)$, with the proportionality factor being related to the fabric permeability tensor $\mathbf{K}(\mathbf{x})$ and the fluid viscosity μ as

$$\mathbf{v}(\mathbf{x}, t) = -\frac{\mathbf{K}(\mathbf{x})}{\mu} \nabla p(\mathbf{x}, t) \tag{1}$$

In this equation, $\mathbf{x}$ and t stand for the position of a given point in the fabric and the time, respectively. Assuming flow continuity, $\nabla \cdot \mathbf{v}(\mathbf{x}, t) = 0$, the governing equation for the pressure field can be obtained as

$$\nabla \cdot \left(-\frac{\mathbf{K}(\mathbf{x})}{\mu} \nabla p(\mathbf{x}, t) \right) = 0 \tag{2}$$

Initial ($t = 0$) and boundary conditions should be given to determine the evolution of the pressure and velocity fields during the time. Such a problem is normally defined as a moving boundary problem because the flow front position $\Gamma(\mathbf{x}, t)$ evolves during the time until the preform is completely filled. Normally, if the position of the flow front is known for given time t, the pressure and velocity fields are determined by standard finite element modelling. Once such information is acquired, updating the flow front position for time $t + \Delta t$ can be obtained. Several numerical techniques were developed in the past to solve such problems in liquid moulding manufacturing of composites. For instance, the finite element/control volume method uses regions associated with every node to update information about the filling factors through use of the flow rates obtained with the velocity fields. Such numerical methods are implemented in well-known simulation tools such as LIMS from Advani and co-authors [15,16] or PAM-RTM from ESI group. Other ways to simulate mould filling processes are based on the direct solving of the two-phase flow problem by using the Navier-Stokes equations for incompressible fluids. In this case, the continuity equation $\nabla \cdot \rho \mathbf{v}(\mathbf{x}, t) = 0$ is accompanied by the linear momentum equation reading as

$$\rho \frac{d\mathbf{v}}{dt} + \nabla \cdot (\rho \mathbf{v} \otimes \mathbf{v}) = -\nabla p + \mu \nabla^2 \mathbf{v} + \mathbf{S} \tag{3}$$

where $\mathbf{S}$ is known as the Darcy–Forchheimer sink term

$$\mathbf{S} = -\left(\mu \mathbf{D} + \frac{1}{2} \rho \, |\mathbf{v}| \, \mathbf{F} \right) \mathbf{v} \tag{4}$$

where $\mathbf{D}$ and $\mathbf{F}$ stand for material parameters. The second term in Equation (4) is related to inertial effects which are negligible for the case of liquid moulding of composites under low Reynolds number flow. In this situation, the inertial terms related to the velocity can be neglected, recovering the standard Darcy equation with the factor $\mathbf{D}$ being the inverse permeability of the fabric. In two-phase flow, the interface between the two fluids, namely resin and air in liquid moulding, is tracked by means of the volume of fluid (VOF) approach by using α as a phase variable. This variable ranges between $\alpha = 1$

and $\alpha = 0$ for the resin and air fluids, respectively. Within this approach, the α variable is continuously updated during simulation time by using the advection equation given by

$$\frac{d\alpha}{dt} + \nabla \cdot (\alpha\mathbf{v}) = 0 \tag{5}$$

OpenFoam® (Open source Field Operation And Manipulation) [17] is a free open source Computational Fluid Dynamics (CFD) software that can be used to solve the problems related with filling in liquid moulding of composites. OpenFoam includes `interFoam` a solver for two-phase incompressible, isothermal and immiscible fluids tracking interfaces with the VOF method. Details and performance of the algorithms used to track interfaces as well as pressure and velocity solvers for two-phase flow can be found in [18].

3.2. Mould Filling Simulations Containing Dissimilar Material Regions

The bi-dimensional problem of a square mould of dimensions $L \times L \times t$ containing a rectangular region with different permeability is modeled in this section as shown in Figure 2a. This geometrical definition constitutes the base of the mould filling forward problem to be solved by computational fluid mechanics. For simplicity, macroscopic unidirectional flow is induced by the application of a constant pressure condition p_0 at $x = 0$ while $p = 0$ is set in the opposite edge at $x = L$. Slip-free conditions are applied in the remaining faces of the mould.

A rectangular region with centre $(x_0 = \alpha_1 L, y_0 = \alpha_2 L)$ and size $(b = \alpha_3 L, h = \alpha_4 L)$ with dissimilar permeability $K = \beta K_0$ is inserted in the model, where $K_0 = 10^{-12}$ m^2 stands for the permeability of the surrounding material. For simplicity, both permeabilities, K_0 and K, were assumed to be isotropic and representative of angle-ply 2D woven preforms although simulations can be carried out by assuming anisotropic behavior without any loss of generality.

Accordingly, the set of non-dimensional numbers $\alpha_1, \alpha_2, \alpha_3, \alpha_4$ and β corresponds to the position, size and relative permeability of the dissimilar region defined as a 5-tuple $(\alpha_1, \alpha_2, \alpha_3, \alpha_4, \beta)$ object. A uniform brick cell discretization of the domain was used with in-plane dimensions $L/100$ and with a single cell in the mould through-the-thickness direction. Thus, the models contain 10,000 brick cells which were judged fine enough to capture accurately the flow front position evolution during time. The two fluids selected for the `interFoam` solver corresponded to water and air, for simplicity. Their density and kinematic viscosity were set to 1000 Kg/m^3 and $\nu = 10^{-6}$ m^2/s for the water, and 1 Kg/m^3 and $\nu = 10^{-8}$ m^2/s for the air, respectively. Despite the simplicity of the model presented in this work, more complicated problems including different mould shapes, inlet and outlet configurations among others can be addressed and used for synthetic training of the artificial intelligence method.

To fully explore the problem space, the involved variables were presented in non-dimensional form as $\hat{x} = x/L$, $\hat{p} = p/p_0$ and $\hat{t} = t/t_0$ where $t_0 = \mu L^2/2Kp_0$ and p_0 the injection pressure, respectively. This latter value of t_0 corresponds exactly to the mould filling time for a perfectly homogeneous distribution of the permeability ($\beta = 1$). A network of 3×3 pressure probes equally distributed in the surface $L \times L$ of the mould is used to record the time evolution $\hat{p}_{i=1,9}(\hat{t})$ of the fluid pressure (Figure 2b).

Figure 3 shows the position of the front flow for different times obtained for a case with $\alpha_1 = \alpha_2 = 0.25$ and $\alpha_3 = \alpha_4 = 0.125$. The relative permeability in this case was set to $\beta = 0.1$. In the early stages, the flow progresses uniformly until it reaches the position of the first dissimilar material region. As the permeability inside the dissimilar region is lower than the surrounding media ($\beta = 0.1$), the flow delays with respect to it. Lastly, the flow progresses until the outlet gate but first in the upper part of the mould far from the dissimilar material region. Such a non-uniform flow presented in Figure 3 is reflected in the probe pressure evolution as shown in Figure 4a.

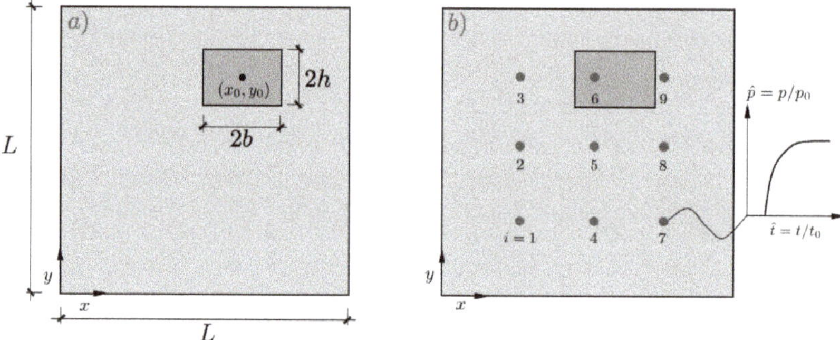

Figure 2. (**a**) Definition of the model containing a dissimilar rectangular region with center coordinates (x_0, y_0), size $2b \times 2h$ and permeability $K = \beta K_0$, (**b**) 3×3 network of pressure sensors equally distributed on the mould surface.

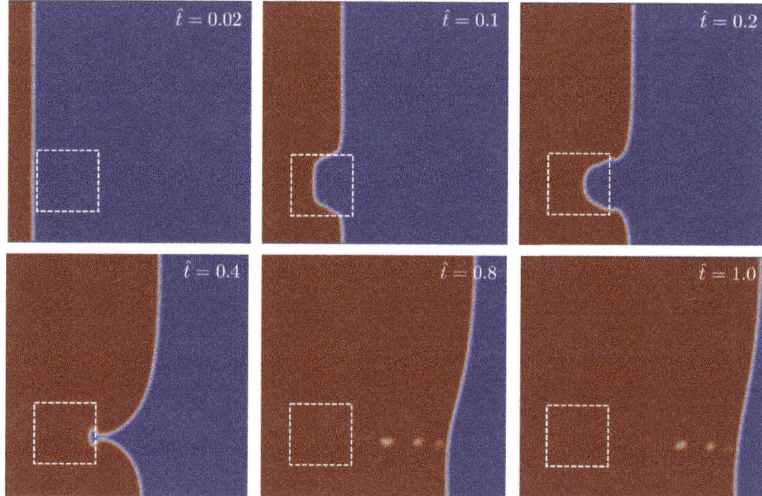

Figure 3. Snapshots of the flow progress through the mould for different non-dimensional times $\hat{t} = 0.02, 0.1, 0.2, 0.4, 0.8$ and 1.0. The mould contains a square region of size $0.25L$ centered at $x_0 = 0.25L$ and $y_0 = 0.25L$ with relative permeability of $\beta = 0.1$. Flow progress from left to right. Red and blue colours correspond to phase values of $\alpha = 0$ and $\alpha = 1$, respectively. For the sake of clarity, white dashed lines corresponding to the dissimilar material region were superimposed.

The evolution of the nine pressure sensors evenly distributed over the surface of the mould is presented in Figure 4, again expressed as non-dimensional numbers $\hat{p} = p/p_0$ and $\hat{t} = t/t_0$. Results for three different cases were presented in this figure and correspond to a square region of size $0.25L$ located at three positions (a) $\alpha_1 = 0.25$, $\alpha_2 = 0.25$, (b) $\alpha_1 = 0.5$, $\alpha_2 = 0.5$ and (c) $\alpha_1 = 0.75$, $\alpha_2 = 0.75$. The shape of the pressure sensor evolution curve was similar in all of them. A sudden increase is produced when the fluid reaches the sensor position and progressively stabilizes to a steady-state value consistent with the pressure gradient induced between the inlet and outlet gates. The dashed lines in Figure 4 were obtained by assuming no dissimilar material region by setting $\beta = 1$.

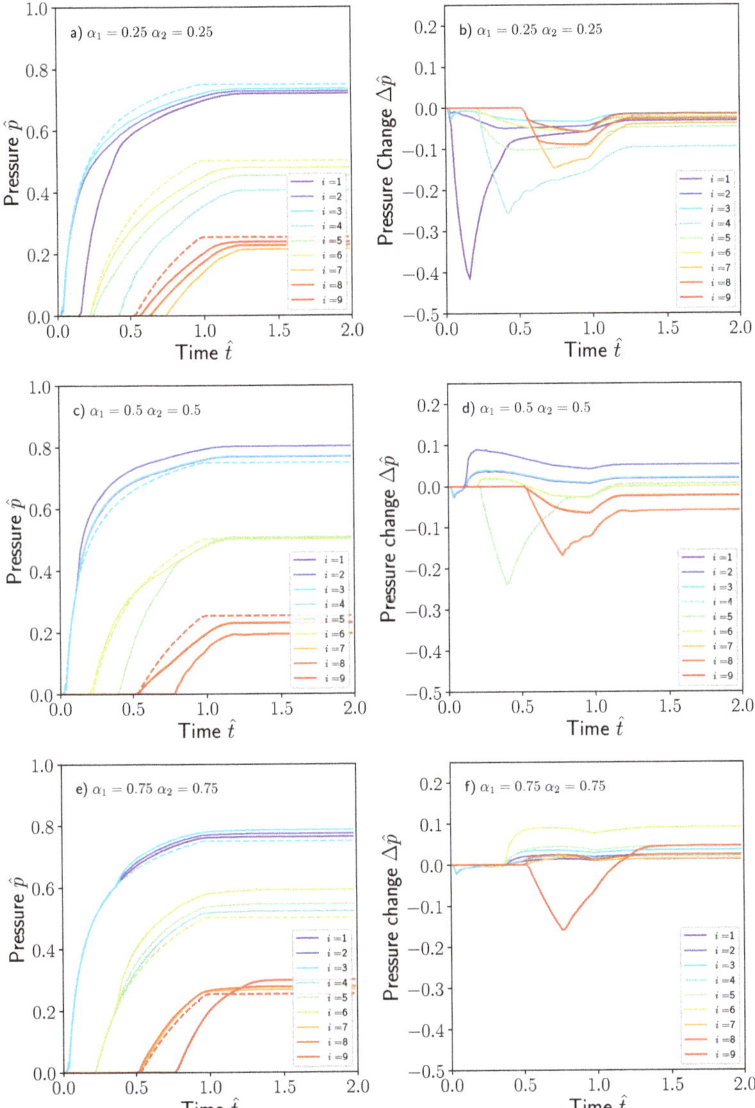

Figure 4. Non-dimensional pressure evolution $\hat{p}(\hat{t})$ for the nine sensors distributed in the mould containing a $0.25L$ square dissimilar material region at (**a**) $\alpha_1 = 0.25$, $\alpha_2 = 0.25$, (**c**) $\alpha_1 = 0.5$, $\alpha_2 = 0.5$ and (**e**) $\alpha_1 = 0.75$, $\alpha_2 = 0.75$. Dashed lines corresponds with the pressure sensor evolution in the absence of dissimilar material region. The pressure changes $\Delta\hat{p}(\hat{t})$ in (**b**,**d**,**f**) are as the difference, or perturbation, between the local pressure value with and without dissimilar material region. For visualization purposes, the non-dimensional time was arbitrarily extended to $\hat{t} = 2$.

The pressure perturbation $\Delta\hat{p}$ is also presented in the same figure, defined as the difference between the local pressure $\hat{p} = p/p_0$ obtained in the problem containing the dissimilar region and the problem without dissimilar region. The pressure evolution and/or the perturbations caused by the presence of the dissimilar region could help to ascertain the position, size and intensity of the defect.

However, given that its complexity with respect to the random position and size make the problem intractable in terms of mathematical complexity, data science or statistical techniques could be more appropriate for such kind of inverse problems.

The position of the sensor with respect to the location of the dissimilar material region influences the evolution of the pressure. If the sensor is placed upstream from the dissimilar material region, an increase in the pressure is produced as compared with the case without the defect. This effect can be observed in sensors 1, 2 and 3 in Figure 4c,e. However, if the sensor is placed downstream from position of the dissimilar material region, the pressure is delayed to some extent as shown in sensors 1, 2 and 3 in Figure 4a. Both effects are reflected in the pressure perturbation $\Delta\hat{p}(\hat{t})$ in Figure 4b,d,f, exhibiting a positive change if the sensor is placed upstream and negative in the opposite case. The information of the pressure $\hat{p}$, or the pressure perturbation $\Delta\hat{p}$, recorded by the sensors was stored as a footprint image and is presented in Figure 5a,b for the three cases presented previously by using a pressure sensor network of 3×3.

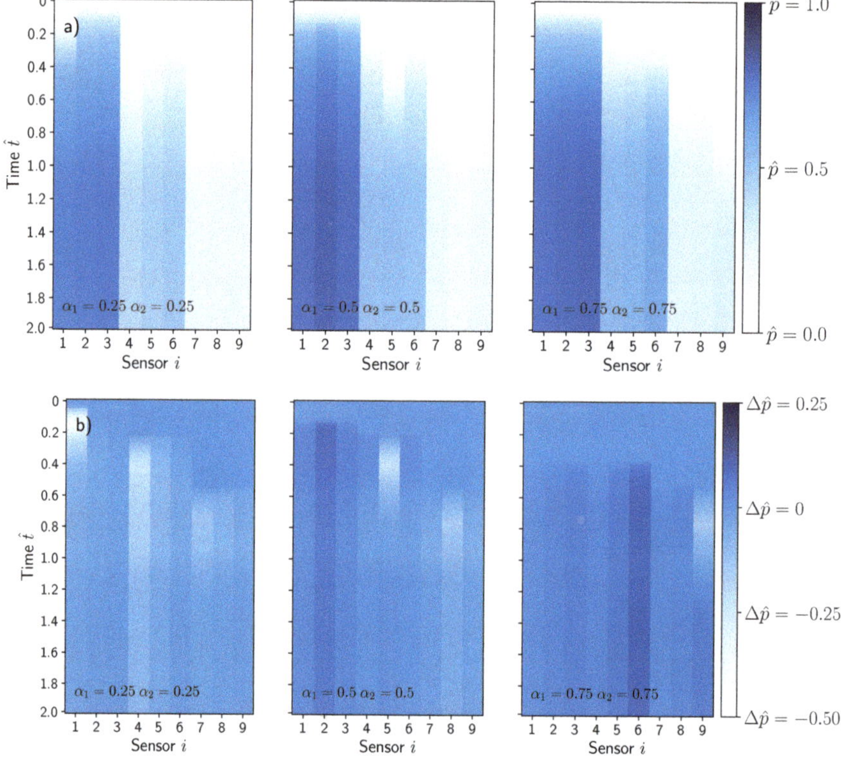

Figure 5. (a) Sensor footprints corresponding to the plot cases presented in Figure 4. The vertical and horizontal axis of the images corresponded with the non-dimensional time $\hat{t}$ and sensor position $i = 1, 9$. The color intensity reflects the sensor non-dimensional pressure $\hat{p}$, (b) Pressure perturbation $\Delta\hat{p}(\hat{t})$ evolution footprint. For visualization purposes, the non-dimensional time was arbitrarily extended to $\hat{t} = 2$.

3.3. Dataset Generation

Once the forward problem is defined, the complete dataset should be generated. A uniform covering of the whole variable space of the model, namely $\alpha_1, \alpha_2, \alpha_3, \alpha_4$ and β, results in intractable problem size. For simplicity, the variables were assumed to follow uniform random probability distributions as $U_{\alpha_1,\alpha_2}(0.150, 0.850)$, $U_{\alpha_3,\alpha_4}(0.075, 0.150)$, $U_\beta(10^{-2}, 10^{-1})$. U_{α_1,α_2} and U_{α_3,α_4} distributions ensure that the dissimilar material region is entire contained in the mould surface. Assuming, for instance, ten different random realizations of the aforementioned variables, the number of combinations is 10^5 which seems unfeasible from a practical viewpoint. Thus, instead of trying to cover uniformly the whole variable space, more efficient techniques for variable sampling should be used. In this work, a set of 3000 simulation cases was generated by means of the Latin Hypercube sampling technique by using PyDOE software package [19]. This number of simulations were enough to maintain the accuracy of the model. Mould filling simulations were run sequentially by OpenFoam for each of the $\alpha_1, \alpha_2, \alpha_3, \alpha_4$ and β combinations provided with the Latin Hypercube method. The automation of the computational process was carried out by using `PyFoam`, a Python library that manipulates and control OpenFoam running cases. Instances of the problem corresponding to each of the $\alpha_1, \alpha_2, \alpha_3, \alpha_4$ and β combinations were generated by modifying OpenFoam dictionaries `topoSetDict` and `controlDict`. Once an individual simulation finished, normally in a few minutes in a regular laptop, the pressure probes dataset is stored. This process is followed by a new job submission until the whole dataset is created. Lastly, the dataset generated including the pressure sensor signals are stored as a bi-dimensional array that contains time and sensor position together with their corresponding 5-tuple model variables $\alpha_1, \alpha_2, \alpha_3, \alpha_4$ and β. Datasets were serialized into a Python `pickles` to provide easy access in subsequent training tasks.

4. Building, Training and Deploying Machine Learning Models

ML scripts were built by using the Python Keras® API (application programming interface) and their main features are presented in this section. Keras is a high-level neural network API, written in Python and capable of running on top of TensorFlow®, CNTK®, or Theano® [20]. This work has selected, among those available in the published literature, a kernel architecture made with a CNN. The main purpose is to describe a novel methodology to build and deploy a machine learning algorithm able to track flow disturbances produced during manufacturing of composites by liquid moulding rather than establish the most effective and robust model architecture that can be used.

4.1. CNN Machine Learning Networks

CNN machine learning networks are preferred for classifying images in computer vision problems depending on specific image features [21,22]. However, this work seeks to predict a set of continuous scalar variable by regression rather than use algorithms as classifiers. The most effective CNN architecture is somewhat difficult to establish from a priori statements. The number of layers, their interrelations, kernels and filters are often result from trial-and-error numerical experiments driven by an accuracy criterion. A typical CNN architecture is formed by a sequential set of convolutions and pooling operations carried out over the image dataset. In this paper, the dataset is composed of the pressure probes footprint images obtained with the mould filling simulations. The results of the convolution process are transmitted to the inputs by using a fully connected neural network (FCNN).

The CNN architecture used in this work is sketched in Figure 6, and more precisely detailed in Table 1. More information is given below:

- Convolution2D (`Conv2D`). This corresponds to an image operation based on the application of a given set of filters to enhance specific features of the image. If the input image is **A** (see Figure 5 with footprint of 9×100) the 2D convolutions of this individual image may be obtained, namely **B**, by applying the kernel function **F**, as

$$\mathbf{B}(i, j) = \mathbf{F} * \mathbf{A} = \sum_k \sum_l F_{k,l} A_{i-k,j-l} \tag{6}$$

where **F** stands for the filter applied and $*$ the convolution operator. The operation can be parametrized by using different filter sizes, strides, paddings, activation functions or kernel regularization. Filters of size $(n_k, n_k, n_{channel})$ are intended to highlight specific features in time and space produced by the presence of the dissimilar material region in the mould. Input image dimensions are given by (n_w, n_h, n_{canal}), where $n_{channel} = 1$ for greyscale images, and output image dimensions are calculated by $(n_w - n_k + 1, n_h - n_k + 1, 1)$. For instance, the first `Conv2D` layer in Figure 6 uses a $(4, 4, 1)$ kernel with an input grayscale image object of $(9, 100, 1)$. This filter operation produces an output image of $(9, 50, 32)$ for this first convolution. Filters normally make use of a certain step, named `stride`, that move the convolution filter along the x and y axis of the input image. Padding is the parameter which maintains the size of the output images resulting from the convolution **B**. Keras uses the `padding='same'` technique to avoid image edge trimming. Lastly, when convolution is completed, a ReLu (Rectified Linear Unit $f(x) = max(0, x)$) is used as cut-off function thus avoiding negative outputs that can be generated.

- MaxPooling (`MaxPooling2D`). This is applied to reduce the dimensions of the convolution filtered images with the purpose of obtaining more efficient and robust characteristics. The model uses a 2×2 pooling filter, taking the maximum value of the pixels in the neighborhood as the result for a given point. For instance, Figure 6 shows MaxPooling operations used to reduce the image size from 9×50 to 8×25 in the first convolution.
- Batch normalization (`BatchNormalization`). This seeks to alleviate the movements produced in the distributions of internal nodes of the network with the intention of accelerating its training. Those movements are avoided via a normalization step, constraining means and variances of the layer inputs. Furthermore, it reduces the need for dropout, local response normalization and other regularization techniques [23].
- Flatten (`Flatten`). This operation splits up the characteristics, transforming them and preparing for obtaining a vector-type object [24]. It is used as training input of the subsequent fully-connected neural network (FCNN).
- Dense (`Dense`). The fully-connected layer is implemented by one or several dense functions. Each layer obtains n inputs from the precedent layer or, in case of the first dense layer, n inputs from the `Flatten` layer. Then, these inputs are balanced by the neural network weights and bias, and transformed into a set of outputs for the following layer according to their specific activation functions. The final dense layer contains a five-component vector-type to perform the regression on the values of $\alpha_1, \alpha_2, \alpha_3, \alpha_4$ and β. Particularly, the neural network used in this work contains three fully connected layers, containing 128, 256 and 512 neurons respectively.

Table 1. Convolutional Neural Network structure and parameters used.

Layers	Specifications
Conv2D	32 4×4 filters with 2×1 stride and padding same activation ReLu + L_2 Regularization 5×10^{-4}
Batch Normalization	-
MaxPooling2D	2×2 filter with 2×1 stride
Conv2D	64 4×4 filters with 2×1 stride and padding same activation ReLu + L_2 Regularization 5×10^{-4}
Batch Normalization	-
MaxPooling2D	2×2 filter with 2×1 stride
Flatten	-
Fully Connected	128 (activation ReLu+ Dropout 0.4)
Fully Connected	256 (activation ReLu+ Dropout 0.25)
Fully Connected	512 (activation Softmax+ Dropout 0.5)
Fully Connected	5 ($\alpha_1, \alpha_2, \alpha_3, \alpha_4, \beta$)

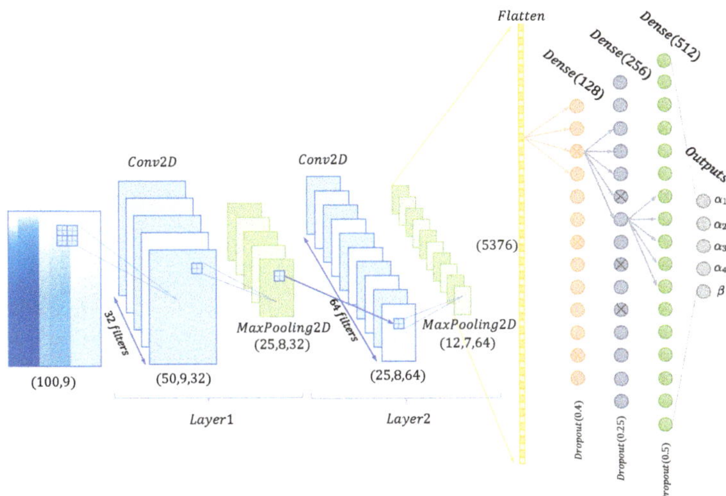

Figure 6. Sketch of the convolutional neural network (CNN) used in this work.

4.2. Training Machine Learning Models

Numerical training experiments started with the pressure probe footprints generated by using mould filling simulations. A set of 3000 images of pressure footprints was used in the CNN model with the purpose of predicting the five variables of position, α_1, α_2, size α_3, α_4 and the relative permeability β. Both datasets, pressure probe footprints, as well as the the five variables were already normalized in the $[0, 1]$ interval so no extra treatments were required.

The first step was to split randomly the dataset generated into two pieces usually known as training and test sets. Training and test sets contained all the data in a ratio of 80/20. The test set was qualified as a never-before-seen dataset with the intention to evaluate the model under new data not used during training.

The subsequent step deals with training of the CNN model by using Keras. The network described in the previous section was coded in Keras by a sequential linking of two convolutional layers and three dense layers (see Figure 6 for more details). Each layer applies some tensor operations with the input data, and these operations make use of weight and bias factors. The weight and bias factors are the intrinsic attributes of the different layers used and are considered the parameters where the learning capacity of the network resides. A total of 860517 parameters was used in the CNN model, Table 1. The determination of the network parameters is carried out by minimization of a norm defined as the sum of the squared differences between the truth values of the variables ($X_i = \alpha_1, \alpha_2, \alpha_3, \alpha_4, \beta$) and the predicted ones by using the CNN network. This Mean Squared Error (MSE) is used in this work as loss function to minimize ($MSE = 1/N \sum_{i=1}^{N} (X_i^{pred} - X_i^{true})^2$ with N the size of the dataset). The iterative method for minimizing the loss function in combination with a gradient descent called `Adadelta` were used to this end. The exact rules governing a specific use of gradient descent are defined by the `Adadelta` [25] Keras optimizer. Training was carried out after not more than 5000 epochs by using 64 as the batch size and lasted around 16 hours with a 10 cores Intel Xeon W-2155CPU-3.30 GHz machine. The evolution of training and test losses against the number of epoch training cycles is presented in Figure 7. The best model configuration obtained produces a minimum MSE of $\approx 10^{-2}$ after training which was judged accurately enough for modelling purposes.

It is worth highlighting the similarity of MSE loss curves obtained for both training and set datasets which is an indicator of reasonable model performance for unseen data. Highly dissimilar behaviour of these two curves usually indicates overfitting, which is a common problem in machine learning. If the complexity of the network and number of network parameters is too high, not in correspondence with the datasets size, overfitting is produced. In this case, the accuracy obtained after training can be excellent but the error corresponding to the test dataset could be still unacceptable and indicates a deficient model generalization for new unseen data. Several strategies were implemented in this work to alleviate possible overfitting problems according to recommendations found in the literature, namely data-augmentation, L_2 regularization and dropout rate in fully-connected layers.

An augmented dataset is generated from the pressure sensor signal by adding a white noise to each image of the training set. The white noise generated follows a normal distribution $N(0, 0.001)$ with zero mean and 0.001 standard deviation. The augmented dataset contains then a total of 14,400 images, with 2400 being from the original set computed with OpenFoam and the remaining 12,000 the augmented one. L_2 regularization techniques add a constraining term to the MSE loss function which is proportional, with a regularization factor of $\lambda = 5 \times 10^{-4}$, to the total sum of the squared values of the parameters of the network ($MSE = 1/N \sum_{i=1}^{N} [(X_i^{pred} - X_i^{true})^2 + \lambda \epsilon_i^2]$). Thus, the presence of data outliers is penalized preventing possible overfitting. Lastly, dropout rates were applied in the fully-connected neuron layers, entailing random dropping out, setting to zero, a number of output features of the layer during training producing a less regular structure. The loss curves that correspond to the case without any of the strategies used to alleviate overfitting are also presented in Figure 7. Although the training loss in this case was excellent, and close to 10^{-4}, the differences with the test loss were unacceptable. Thus, the model in this case was unable to generalize with the same precision level with new unseen data.

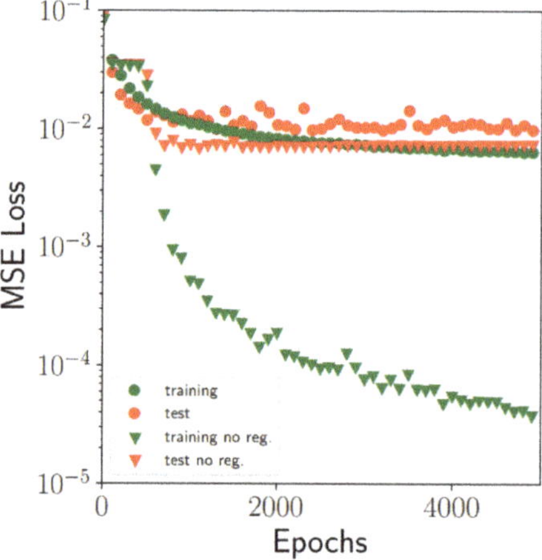

Figure 7. Training and test MSE (mean square error) losses evolution against the number of epochs training cycles. The data include losses obtained with and without the application of techniques to alleviate overfitting (data augmentation, L_2 regularization and dropout).

The comparisons between the ground truth values of the variables, $\alpha_1, \alpha_2, \alpha_3, \alpha_4$ and β, and the predicted ones through use of the CNN are gathered in Figure 8. The figure includes both training and test datasets. As a first approximation, the correlation between predicted and ground truth values was fairly good. This was especially true for the position of the dissimilar material region α_1, α_2, Figure 8a,b. The network in this situation was able to learn in a highly efficient manner from the given footprint by using only the features associated with the rise up of the pressure signals. However, the accuracy attained for the remaining variables was, in general, more modest although the overall trends were perfectly captured, Figure 8c,d,f. A plausible explanation for this accuracy reduction could be attributed to the similarity of the pressure fields generated by the presence of the dissimilar material region. Two regions defined with similar values of size and/or relative permeability produce very close fluid pressure fields almost indistinguishable, producing nearly no single-valued pressure footprints. This reduction of accuracy was more evident in the case of the relative permeability parameter β which is essentially controlled by the pressure gradients. Figure 8e shows the previous statement. Pressure field differences for two small values of β may differ only slightly when the macroscopic flow reached the outlet gates, thus producing again almost equal pressure footprints. Nonetheless, the accuracy was judged to be reasonable for the automatic detection of the position and severity of the dissimilar material region.

The histograms of the individual absolute errors computed as $X^{pred} - X^{true}$ corresponding to the five variables are also presented in Figure 8f. As mentioned previously, the prediction of the two position variables (α_1, α_2) was excellent and the error in this case exhibits a Dirac-like type function with 90% of the data lying within an absolute error band of less than 3%. It should be pointed out that the model variables were expressed in non-dimensional and normalized form and thus, the absolute errors were expressed in percentage. The error distribution for the remaining variables was, of course, more flatten and the plausible reasons were discussed previously. The fraction of the total data corresponding to predictions with absolute error lying in the $\pm 10\%$ and $\pm 20\%$ error band are presented in Table 2 for the sake of completion.

Table 2. Fraction of the total data with absolute error falling within the $\pm 10\%$ and $\pm 20\%$ bands.

Error	α_1	α_2	α_3	α_4	β
$\pm 10\%$	0.5%	3.0%	49%	35%	42%
$\pm 20\%$	0.2%	0.0%	16%	3%	15%

Figure 9a is presented to illustrate the overall performance of the model. The plots contain some dissimilar material regions selected randomly together with the corresponding predictions by using the test dataset and the 3×3 sensor network size. As discussed previously, the accuracy of the predictions was fairly good, thus showing the ability of the proposed model to capture the presence of dry regions during liquid moulding.

The accuracy of the model was also addressed for some additional cases including different pressure network sizes of 2×2, 3×3 (baseline) and 4×4 corresponding to 4, 9 and 16 equally spaced pressure sensors. It should be noted that as OpenFoam simulations were run a single time, saving the pressure probe evolution at the locations corresponding to each specified network, there was no need for further recalculations. The three models were trained by using the same procedure previously explained and the corresponding MSE losses obtained for the 2×2, 3×3 and 4×4 networks were 0.016, 0.011 and 0.012, respectively. The MSE losses obtained for the 3×3 and 4×4 networks were very similar between them. Such results seem to indicate that the dissimilar material region size used in this study, following the uniform distribution $U_{\alpha_3,\alpha_4}(0.075, 0.150)$, is perfectly captured even with the 3×3 network. Increasing the number of sensors to 4×4 will not result in a better accuracy of the model for such dissimilar material region size. Accordingly, the sensor network size should be previously determined if a minimum dissimilar material size is sought. The predictions for the ground truth cases presented in Figure 9a by using the 2×2, 3×3 and 4×4 network are now summarized in Figure 9b for the sake of completion.

Lastly, the flow progress predictions for the case presented in Figure 3 are now shown in Figure 10. This case corresponded to a $2b = 2h = 0.25L$ square region centred in $x_0 = 0.25L$, $y_0 = 0.25L$ and relative permeability of $\beta = 0.1$. The pressure footprint presented in Figure 4a was used as input to predict the position, size and relative permeability yielding the 5-tuple $(0.258, 0.241, 0.118, 0.117, 0.083)$ from the ground truth values of $(0.250, 0.250, 0.125, 0.125, 0.1)$. OpenFoam simulations were run subsequently and the corresponding flow patterns gathered in Figure 10. The agreement between the ground truth flow patterns shown in Figure 3 and the predicted ones was excellent, $4.9 \cdot 10^{-2}$ MSE, considering that the only information used comes from a discrete network of pressure sensors.

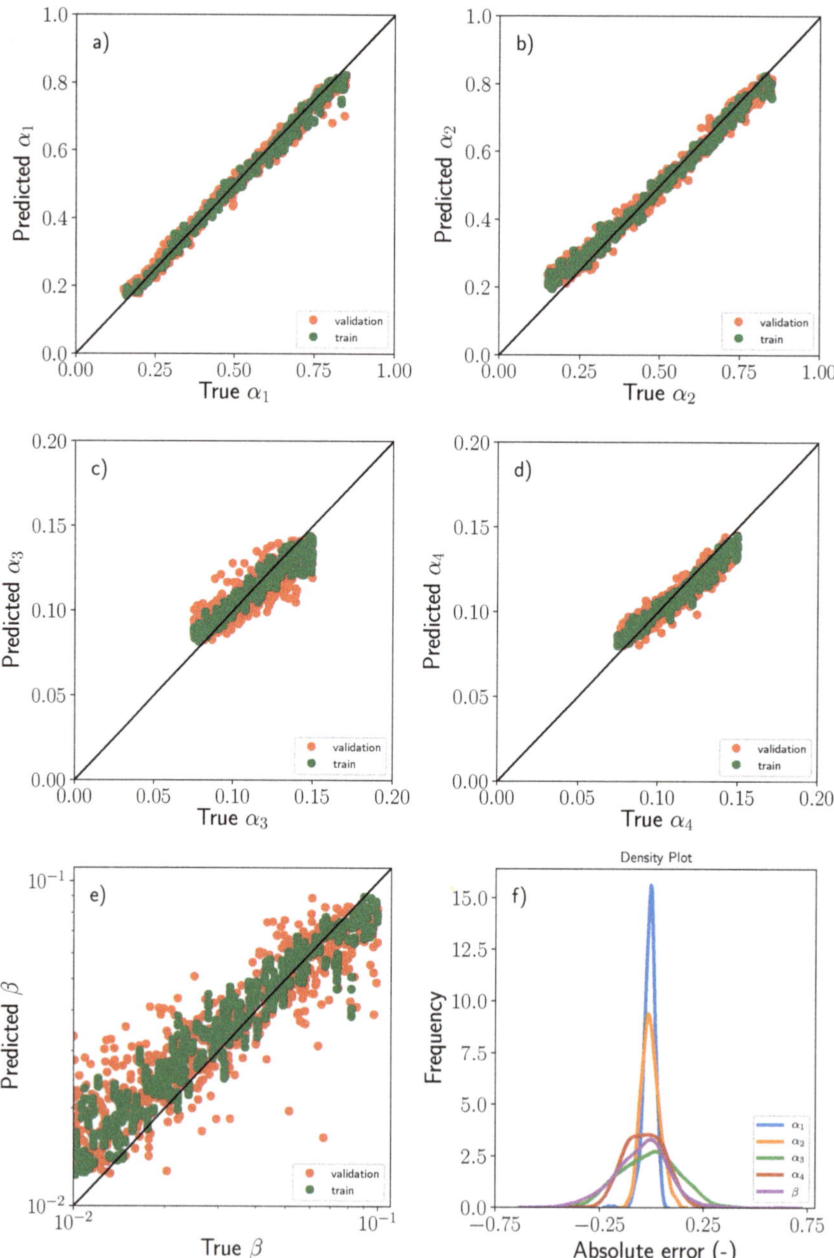

Figure 8. Comparisons between the ground truth values of the variables, (**a**) α_1, (**b**) α_2, (**c**) α_3, (**d**) α_4 and (**e**) β including training (green) and test (red) data sets, (**f**) Histograms of the absolute error corresponding to each of the variables used in the regression.

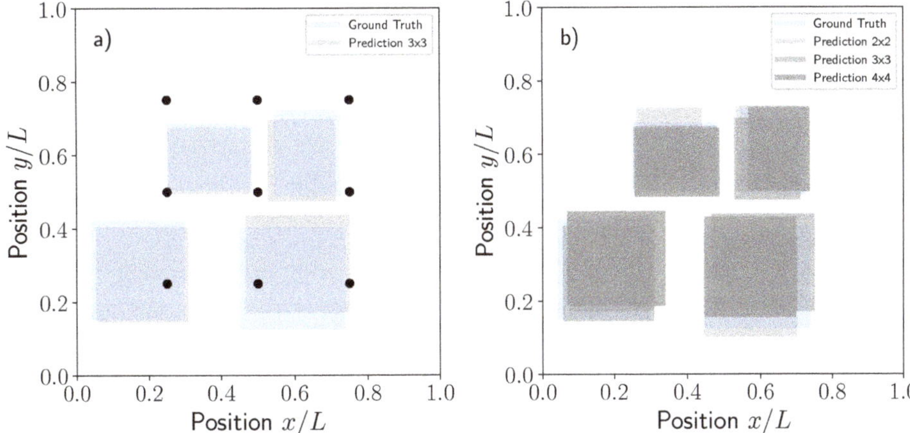

Figure 9. Comparisons between the ground truth and predictions for five randomly selected dissimilar material regions in terms of position and (**a**) size for the test data set using the 3×3 sensor network size, (**b**) Same results as previously shown but with different sensor network of 2×2, 3×3 and 4×4.

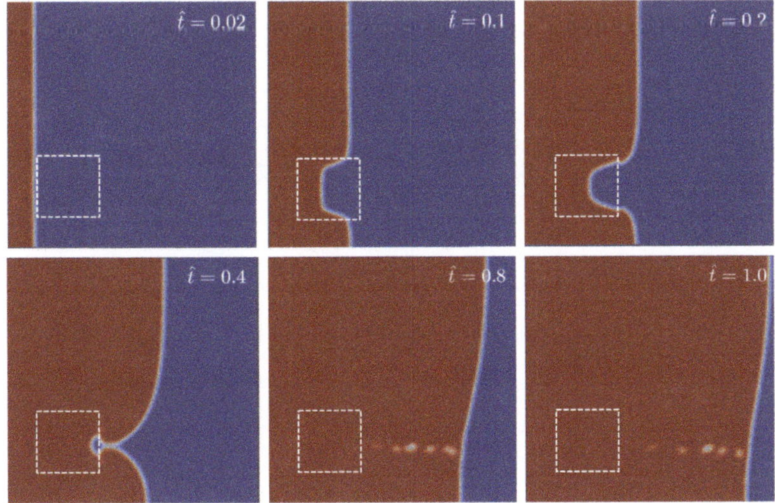

Figure 10. Snapshots of the flow progress through the mould for different non-dimensional times $\hat{t} = 0.02$, 0.2, 0.4, 0.8 and 1.0 corresponding to the predicted values for the case shown in Figure 3. Flow progress from left to right. Red and blue colours correspond to phase values of $\alpha = 0$ and $\alpha = 1$, respectively. For the sake of clarity, white dashed lines corresponding to the dissimilar material region were superimposed.

4.3. Deploying Machine Learning Models

The CNN model and the weights resulting after training were saved for subsequent predictions with new unseen data that could come, for instance, from the manufacturing laboratory. To this end, the model architecture and the corresponding weights were stored by using json format (JavaScript Object Notation) and hdf5 Hierarchical Data Format (HDF) respectively. Once the network was trained, it could be deployed subsequently and evaluated at any time by loading the model, weights and the new unseen data without the need of retraining.

5. Discussion

Model Robustness

The previous section was devoted to describing a new machine learning model to detect automatically the presence of a dissimilar material region during manufacturing of composites by liquid moulding. Examination of the model architecture, training procedure, accuracy and the ability to generalize a response under new unseen data was analyzed in detail. However, a deeper assessment of its robustness should be conducted to address other relevant effects and uncertainties that could potentially arise during the manufacturing process. Among others, these involve the presence of increasing pressure signal noise, as well as possible sensor malfunctions or the size of the sensor array used. Additionally, the response of the trained model through use of different type of unseen data was evaluated. For instance, the response to other dissimilar material region shapes instead of a rectangular one, the rectangle size out of the distribution used for training, as well as the presence of simultaneous regions during filling, will be analyzed.

As described previously, a white noise signal following a Gaussian distribution $N(0, 0.001)$ was originally superposed on the pressure footprints to improve the overfitting response of the model and its capability to generalize for new unseen data inputs. Once the model was trained, the prediction ability with white noise corrupted data was checked. Figure 11a shows the pressure evolution corresponding to the nine sensors installed in the model with white noise of $N(0, 0.01)$ for the case presented in Figure 4a. Such corrupted signals were used to determine the lack of accuracy due to the presence of noise and the results presented in Figure 11b in terms of MSE. It was evident that for increasing noise standard deviation, some of the pressure signals may overlap hampering the adequate prediction of the model. It should be pointed out again, that the main sources of model deviations from ground truth values were related to the α_3, α_4 and β variables, and more specifically to β related with the permeability. Unsurprisingly, the position of the dissimilar material region was still well predicted by the model even if the noise signal level was increased.

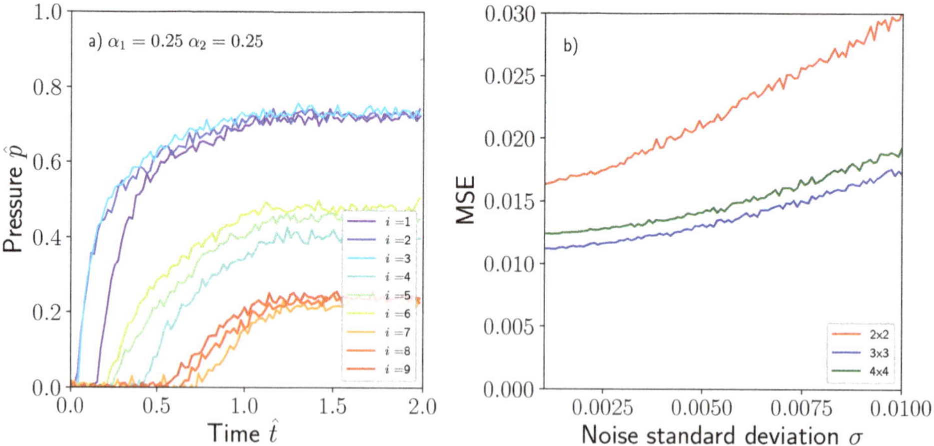

Figure 11. (**a**) Corrupted pressure signal with white noise following the Gaussian distribution $N(0, 0.01)$, (**b**) MSE as a function of white noise standard deviation σ.

Another possible source of lack of prediction capacity of the model is sensor malfunction. During manufacturing, it is not unusual that sensor signals can be lost due to a malfunction problem or inadequate wiring. In this case, if a sensor signal is lost from the early beginning of the test, the sensor footprint image will be undoubtedly altered creating a zero-pressure reading in the columns presented in Figure 5a. Under such circumstances, the trained model was not able to recognize the new patterns created by the signal lost and the accuracy of the model was destroyed. Although this can be considered an important shortcoming of the model, it can be easily circumvented if the network is previously trained with data including a signal lost which is time inexpensive for the limited number of sensors used. For instance, pressure footprint datasets used for the 3×3 network can be trained removing one of the sensor readings leading to a $(8, 100)$ size images rather than the baseline $(9, 100)$ ones.

Figure 12 shows the robustness of the trained model against new unseen data with a structure different than those data used for training. For this exercise, the model was evaluated by using pressure signals obtained with circular dissimilar material regions, see Figure 12a and double size rectangle regions following the $U_{\alpha_3,\alpha_4}(0.150, 0.300)$ distributions, see Figure 12b. Although the geometry used in these two cases produced, of course, different pressure footprints than those used for training of the baseline problem, the results obtained from the model were still reasonable in location and area indicating an adequate detection of specific patterns even if non-standard images are used.

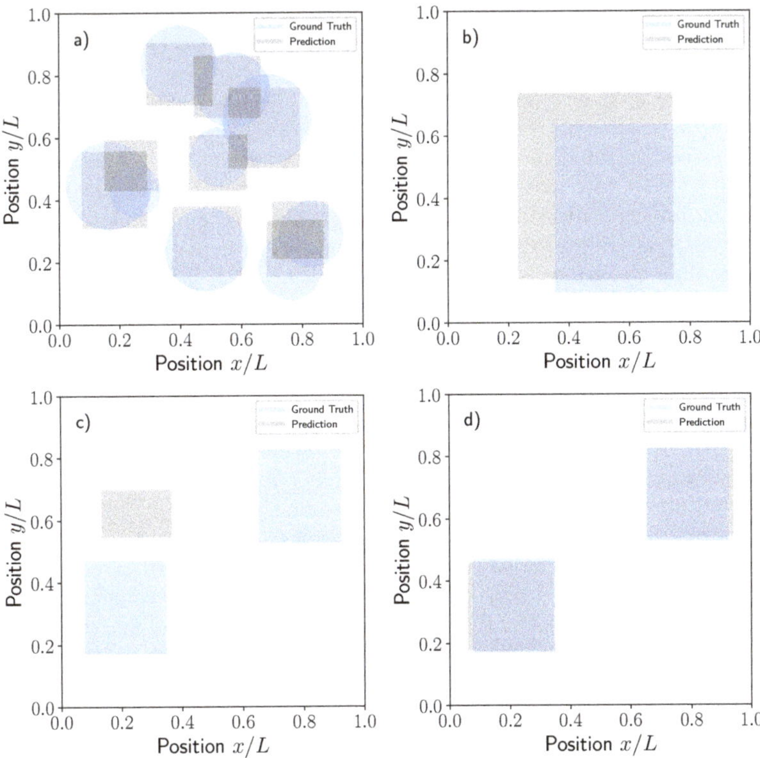

Figure 12. Comparisons between the ground truth and predictions for different dissimilar material realizations: (**a**) Circular, (**b**) Double size rectangular, (**c**) Two equal rectangles, (**d**) Two equal rectangles with CNN model trained with a new double rectangle data set.

The case of two rectangular regions deserves additional and specific comments. In this case, two equal rectangle dissimilar material regions were generated randomly but assuming centered x-y symmetry so the problem variables used for the regression remain the same, Figure 12c. Independent pressure footprints corresponding to two rectangle problems were generated with OpenFoam and the baseline model with a 3×3 sensor network size was used to predict the position of an unique equivalent dissimilar material region. Unfortunately, in this case, the response was misleading and the model was unable to detect the specific features of this outlier case. The prediction of the model was obviously a single rectangular region that neither matches the averaged position nor the area of the two input rectangles, Figure 12c. As mentioned previously, such problems were easily circumvented by additional training of the CNN model with the new data set corresponding to 3000 random realizations of the two rectangular dissimilar material regions. The CNN was trained again and the new predictions presented in Figure 12d for the same individual case presented in c. Now, the model was able to detect perfectly the two rectangles with a precision level meeting the expectations (see error histograms in Figure 8f). This simple exercise demonstrated the prediction capability of the CNN architecture trained with the appropriate dataset.

6. Conclusions and Final Remarks

A supervised regression machine learning model based on a convolutional neural network is presented in this work to predict the position, size and relative permeability of a dissimilar material region inside a square mould subjected to a macroscopic one dimensional flow in liquid moulding. The presence of the dissimilar material region produces distortions of the pressure field and flow patterns which can be detected by a discrete network of sensors equally distributed over the mould surface. The use of a fully-based experimental approach to build machine learning models in materials manufacturing would be challenging due to cost factors. To avoid this issue, dataset augmentation based on simulations becomes crucial to success. This was the approach used in this paper which makes use of extensive modelling of mould filling through use of OpenFoam as the fluid mechanics solver.

The forward model described in the paper is able to generate the pressure probes evolution during the filling time and these data were properly stored as a pressure footprints. A total of 3000 random realizations for the five variables describing the position, size and relative permeability of a rectangular dissimilar region were generated by computer. The inverse problem to predict the five variables from the regression of the individual pressure footprints was based on the use of a convolutional neural network which was able to learn from specific features of the artificial images generated. The CNN had two sequential convolutional layers and three fully-connected neuron layers to relate the pressure footprints with the corresponding position, size and relative permeability of the dissimilar material regions. The model architecture and training were implemented in Keras, a high-level neural network API. The determination of the network parameters was carried out by minimizing the mean square error loss over an increasing number of epochs. Training was carried out by using some numerical strategies to alleviate overfitting, producing a model able to generalize the response under new unseen data not used during training. The model yielded excellent accuracy for the center position of the dissimilar material region while the one corresponding to its size and relative permeability was good but more modest as shown in the error histograms distribution in Figure 8f. Taking into account the difficulties and scatter associated with the permeability measurement, the accuracy attained by the model was judged to be coherent with the experimental limitations. Lastly, the model robustness against new data with structure not used during training is analyzed in detail. In addition, the effects of sensor malfunctions, noise signals, presence of simultaneous dissimilar regions, different shapes among others were studied. It should be remarked that the purpose of this work was to establish a general framework for fully-synthetic-trained machine learning models to address the occurrence of manufacturing disturbances without placing emphasis on its performance, robustness and optimization. In summary, the following conclusions can be drawn:

- Machine learning and artificial intelligence strategies open revolutionary opportunities in material science and, more specifically, in materials and parts manufacturing. The ability of these technologies to deal and overcome complex physical and engineering problems that relate different datasets, as sensors and dissimilar material regions in this work, is highly aligned with Industry 4.0 concepts. In the future, this will enable the development of efficient cyber-physical systems to detect defects automatically during manufacturing while guaranteeing the implementation of the appropriate corrective actions.

- Some of the major concerns and drawbacks of the application of machine learning methods in manufacturing of structural composites are related with enormous costs and development times associated with the experimental generation of the large datasets required for training. A way of easing the burden on experimental tasks is to involve the cooperative help of simulation results to create augmented datasets. Of course, it could be argued that the accuracy of processing simulation tools to predict the involved variables is dubious, specially if manufacturing disturbances should be taken into account. This paper has presented a fully simulation-based machine learning model, although the purpose in future implementations will be to combine both experimental and model datasets.

- Many modelling techniques involve machine learning algorithms and include different architectures and methods. It should be noted that similar or even better results could be obtained by using other approaches not examined here. The aim of the paper was not to provide the optimum model configuration but to analyze the ability of machine learning to detect automatically flow disturbances occurring during liquid moulding by testing a simple problem. A future objective will entail examining of integration into a complete system able to detect manufacturing disturbances while implementing automatically the required corrective actions to maintain a constant quality of the manufactured part.

Author Contributions: Conceptualization, C.G. and J.F.-L.; methodology, C.G. and J.F.-L.; software, C.G. and J.F.-L.; validation, C.G. and J.F.-L.; formal analysis, C.G. and J.F.-L.; investigation, C.G. and J.F.-L.; resources, C.G.; data curation, C.G. and J.F.-L.; writing–original draft preparation, J.F-L; writing–review and editing, C.G.; visualization, C.G. and J.F.-L.; supervision, C.G.; project administration, C.G.; funding acquisition, C.G. All authors have read and agreed to the published version of the manuscript.

Funding: This research was funded by the Regional Government of Madrid through the research and innovation Innovation Hubs 2018 TEMACOM project.

Conflicts of Interest: The authors declare no conflict of interest.

References

1. González, C.; Vilatela, J.; Molina-Aldareguía, J.; Lopes, C.; LLorca, J. Structural composites for multifunctional applications: Current challenges and future trends. *Prog. Mater. Sci.* **2017**, *89*, 194–251. [CrossRef]

2. Llorca, J.; González, C.; Molina-Aldareguía, J.; Segurado, J.; Seltzer, R.; Sket, F.; Rodríguez, M.; Sádaba, S.; Muñoz, R.; Canal, L. Multiscale modeling of composite materials: A roadmap towards virtual testing. *Adv. Mater.* **2011**, *23*. [CrossRef] [PubMed]

3. Roth, Y.C.; Weinholdt, M.; Winkelmann, L. Liquid Composite Moulding - Enabler for the automated production of CFRP aircraft components. In Proceedings of the 16th European Conference on Composite Materials, ECCM 2014, Seville, Spain, 22–26 June 2014; pp. 22–26.

4. Advani, S.; Hsiao, K.T. (Eds.) 1—Introduction to composites and manufacturing processes. In *Manufacturing Techniques for Polymer Matrix Composites (PMCs)*; Woodhead Publishing Series in Composites Science and Engineering; Woodhead Publishing: Sawston, UK, 2012; pp. 1–12. [CrossRef]

5. Rudd, C.D.; Long, A.C.; Kendall, K.N.; Mangin, C.G.E. *Liquid Moulding Technologies*; Woodhead Publishing: Sawston, UK, 1997. [CrossRef]

6. Siddig, N.A.; Binetruy, C.; Syerko, E.; Simacek, P.; Advani, S. A new methodology for race-tracking detection and criticality in resin transfer molding process using pressure sensors. *J. Compos. Mater.* **2018**, *52*, 4087–4103. [CrossRef]

7. Zhou, B.; Lapedriza, A.; Xiao, J.; Torralba, A.; Oliva, A. Learning deep features for scene recognition using places database. *Adv. Neural Inf. Process. Syst.* **2014**, *1*, 487–495.

8. Xiao, J.; Hays, J.; Ehinger, K.A.; Oliva, A.; Torralba, A. SUN database: Large-scale scene recognition from abbey to zoo. In Proceedings of the IEEE Computer Society Conference on Computer Vision and Pattern Recognition, San Francisco, CA, USA, 13–18 June 2010; pp. 3485–3492. [CrossRef]

9. Mezzasalma, S.A. Chapter Three The General Theory of Brownian Relativity. *Interface Sci. Technol.* **2008**, *15*, 137–171. [CrossRef]

10. Bock, F.E.; Aydin, R.C.; Cyron, C.J.; Huber, N.; Kalidindi, S.R.; Klusemann, B. A review of the application of machine learning and data mining approaches in continuum materials mechanics. *Front. Mater.* **2019**, *6*. [CrossRef]

11. Wu, J.; Yin, X.; Xiao, H. Seeing permeability from images: Fast prediction with convolutional neural networks. *Sci. Bull.* **2018**, *63*, 1215–1222. [CrossRef]

12. Iglesias, M.; Park, M.; Tretyakov, M.V. Bayesian inversion in resin transfer molding. *Inverse Probl.* **2018**, *34*, 1–49. [CrossRef]

13. Lähivaara, T.; Kärkkäinen, L.; Huttunen, J.M.J.; Hesthaven, J.S. Deep convolutional neural networks for estimating porous material parameters with ultrasound tomography. *J. Acoust. Soc. Am.* **2018**, *143*, 1148–1158. [CrossRef] [PubMed]

14. Zhu, Y.; Zabaras, N. Bayesian deep convolutional encoder–decoder networks for surrogate modeling and uncertainty quantification. *J. Comput. Phys.* **2018**, *366*, 415–447. [CrossRef]

15. Bruschke, M.V.; Advani, S.G. A finite element/control volume approach to mold filling in anisotropic porous media. *Polym. Compos.* **1990**, *11*, 398–405. [CrossRef]

16. Bruschke, M.V.; Advani, S.G. A numerical approach to model non-isothermal viscous flow through fibrous media with free surfaces. *Int. J. Numer. Methods Fluids* **1994**, *19*, 575–603. [CrossRef]

17. Weller, H.G.; Tabor, G.; Jasak, H.; Fureby, C. A tensorial approach to computational continuum mechanics using object-oriented techniques. *Comput. Phys.* **1998**, *12*, 620–631. [CrossRef]

18. Deshpande, S.S.; Anumolu, L.; Trujillo, M.F. Evaluating the performance of the two-phase flow solver interFoam. *Comput. Sci. Discov.* **2012**, *5*, 014016. [CrossRef]

19. PyDOE. The Experimental Design Package for Python. Available online: https://pythonhosted.org/pyDOE/ (accessed on 9 June 2020).

20. Chollet, F.; KERAS. Available online: https://github.com/fchollet/keras (accessed on 9 June 2020).

21. Cun, Y.L.; Boser, B.; Denker, J.S.; Howard, R.E.; Habbard, W.; Jackel, L.D.; Henderson, D. Handwritten Digit Recognition with a Back-propagation Network. In *Advances in Neural Information Processing Systems 2*; Morgan Kaufmann Publishers Inc.: San Francisco, CA, USA, 1990; pp. 396–404.

22. LeCun, Y.; Bengio, Y.; Hinton, G. Deep learning. *Nature* **2015**, *521*, 436–444. [CrossRef]

23. Ioffe, S.; Szegedy, C. Batch normalization: Accelerating deep network training by reducing internal covariate shift. *arXiv* **2015**, arXiv:1502.03167.

24. Jin, J.; Dundar, A.; Culurciello, E. Flattened convolutional neural networks for feedforward acceleration. *arXiv* **2014**, arXiv:1412.5474.

25. Zeiler, M.D. ADADELTA: An Adaptive Learning Rate Method. *arXiv* **2012**, arXiv:abs/1212.5701.

Journal of
Composites Science

Article

A Study of the Interlaminar Fracture Toughness of Unidirectional Flax/Epoxy Composites

Yousef Saadati [1,*], Jean-Francois Chatelain [1], Gilbert Lebrun [2], Yves Beauchamp [3], Philippe Bocher [1] and Nicolas Vanderesse [1]

[1] Mechanical Engineering Department, École de technologie supérieure, 1100 Notre-Dame West, Montreal, QC H3C 1K3, Canada; jean-francois.chatelain@etsmtl.ca (J.-F.C.); philippe.bocher@etsmtl.ca (P.B.); nicolas.vanderesse@ens.etsmtl.ca (N.V.)

[2] Mechanical Engineering Department, Université du Québec à Trois-Rivières (UQTR), 3351 Boul. des Forges, Trois-Rivières, QC G9A 5H7, Canada; gilbert.lebrun@uqtr.ca

[3] McGill University, 845 Sherbrooke Street West, Montreal, QC H3A 0G4, Canada; yves.beauchamp@mcgill.ca

* Correspondence: yousef.saadati.1@ens.etsmtl.ca

Received: 14 May 2020; Accepted: 2 June 2020; Published: 5 June 2020

Abstract: Having environmental and economic advantages, flax fibers have been recognized as a potential replacement for glass fibers as reinforcement in epoxy composites for various applications. Its widening applications require employing failure criteria and analysis methods for engineering design, analysis, and optimization of this material. Among different failure modes, delamination is known as one of the earliest ones in laminated composites and needs to be studied in detail. However, the delamination characteristics of unidirectional (UD) flax/epoxy composites in pure Mode I has rarely been addressed, while Mode II and Mixed-mode I/II have never been addressed before. This work studies and evaluates the interlaminar fracture toughness and delamination behavior of UD flax/epoxy composite under Mode I, Mode II, and Mixed-mode I/II loading. The composites were tested following corresponding ASTM standards and fulfilled all the requirements. The interlaminar fracture toughness of the composite were determined and validated based on the specific characteristics of natural fibers. Considering the variation in the composite structure configuration and its effects, the results of interlaminar fracture toughness fit in the range of those reported for similar composites in the literature and provide a basis for the material properties of this composite.

Keywords: flax-epoxy composite; interlaminar fracture energy; fracture toughness; delamination; Mode I; Mode II and Mixed-mode I–II interlaminar fracture; critical energy release rate

---

1. Introduction

In the recent past decades since the 1990s, the increasing public environmental awareness and commencement of new legislation are demanding more sustainable and ecologically efficient products [1–7]. Synthetic glass and carbon fiber-reinforced polymer composites (FRPCs) have high performance and are extensively applied in different industrial sectors, but they are not environment-friendly [7–9]; thus, environmentally sustainable replacements are required. Numerous research efforts have been dedicated to finding a sustainable substitute for synthetic FRPCs, and have confirmed that plant-based natural fiber-reinforced polymer composites (NFRPCs) are the most promising alternatives (to petroleum-based FRPCs) for many applications [2,3,5,10–17]. Having many advantages over man-made FRPCs such as improved sustainability and competitive properties at a lower cost, NFRPCs have attracted much attention worldwide and have penetrated the composites world with a rapidly extending industrial applications from packaging to structural construction components [1,6,13,16,18–21]. Replacing the engineered FRPCs with fully biodegradable composites is not yet an economically and mechanically desirable solution in many applications; nevertheless, the synthetic polymers reinforced with natural fibers to produce economic and partially biodegradable composites can cover the requirements of a wide variety of applications and have been a viable alternative for petroleum-based FRPCs [5,7,22]. For example, flax, hemp, and ramie fibers, placed within the category of plant-based bast fibers, are the most widely used reinforcement in NFRPCs, offering comparable mechanical properties to glass fiber-reinforced polymer composites (GFRPCs) [10,19,23–25]. Among them, flax fibers have excellent performance, and many manufacturers are interested in using them in NFRPCs for various applications [8,12,17,26–36]. Also, epoxy resin, a thermoset polymer matrix, offers high mechanical performance and durability and is the most commonly used matrix to produce flax fiber-reinforced polymer composites (FFRPCs) [29]. Furthermore, the application of a unidirectional (UD) or optimally woven reinforcement structure is a primary factor in maximizing the structural performance of FFRPCs [37,38]; therefore, UD flax fiber-reinforced epoxy composites (FFRECs) are of outstanding importance and are used ubiquitously in the industry.

There are, however, some downsides to extending their application field, for instance, workpiece defects in the machining process, required for complex components, and in-service degradation when exposed to a humid (harsh) environment [6,28]. Accordingly, FFRECs have been the subject of several pieces of research, trying to explore their properties and enhance their performance and durability to resolve these drawbacks, aiming at facilitating their applications [2,29,32,39–42].

One of the commonly observed and major damage phenomena in laminated composites is the separation of two adjacent laminae (plies), which is known as interlaminar failure or delamination damage. This failure mode is the primary challenge for the laminated FRPCs that generally affects the in-service functionality of the component by reducing its overall stiffness and load-bearing capacity and subsequently leading to the failure of the entire component [43–45]. It might form through the production process and exist as a defect in the composite, occur under different loadings and environmental conditions, or incur during machining operations, particularly drilling [46–48]. In fracture mechanics, the interlaminar failure, according to the relative displacement of crack surfaces, has been classified into three fracture modes: opening or tensile mode (Mode I), in-plane shear or sliding mode (Mode II), and out-of-plane shear or tearing mode (Mode III) [43,49]. Figure 1 [49] illustrates the crack propagation caused by different modes of failure and the corresponding loadings acting on the cracks. Delamination can initiate and propagate due to any of these fracture modes or any combinations of them (mixed modes).

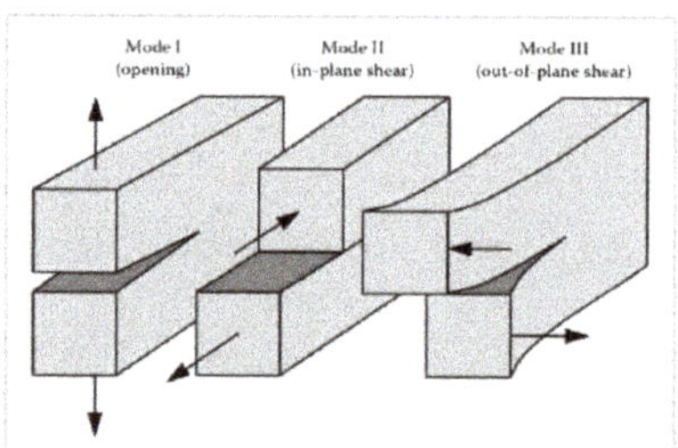

Figure 1. Fracture modes and applied loadings [49].

The resistance of the material to crack propagation, or to delamination damages in composite laminates, is known as interlaminar fracture toughness (IFT) [49]. The Linear Elastic Fracture Mechanics (LEFM) approach, initially developed for isotropic materials, has been successfully adopted and applied as a practice to characterize the IFT of the laminated FRPCs [50,51]. For fracture analysis, LEFM assumes an elastic material behavior for the whole body, except for a contained damage zone around the crack tip (which is a provision for the validity of the results). The IFT is a fundamental material property and can be determined based on either the Stress Intensity Factor (K_C) or the Critical Strain Energy Release

Rate (G_C) of the material; the latter is commonly named Interlaminar Fracture Energy (IFE). They are, respectively, a measure of the stress intensity level and energy required for an increment of the crack growth [49]. For the isotropic, homogeneous, and relatively brittle metallic materials, IFT depends on the energy dissipation at the crack tip and can be characterized by K_C. However, for FRPCs, the stress field in the vicinity of the crack tip is more complicated, and IFT is affected by the result of energy losses of different failure mechanisms such as matrix cracking, fiber fracture, fiber pullout, and fiber-matrix debonding [50,52]. Therefore, G_C is commonly used for the determination of IFT and the study of delamination in FRPC laminates for different failure modes [44,52–55]. By comparing the energy release rate (G) for a loading case to the G_C of the material, the capability of the body to resist crack extension can be assessed.

Ever-growing applications of laminated NFRPCs and their penetration into structural applications, as well as the vital role of delamination in these composites, necessitate the investigation of the interlaminar fracture behavior and determination of IFE of these materials. IFT, as a primary property of a material, is required for engineering design purposes as well as numerical analysis methods such as the Finite Element Method (FEM). The interlaminar fracture of the synthetic FRPCs under different modes of failure has been extensively investigated, and its state of the art has been comprehensively reviewed by many authors [43,44,50,52,56]. However, in the literature, limited work has been documented addressing the interlaminar fracture behavior of the laminated NFRPCs, particularly FFRECs that show better toughness [57]. Wong et al. [58] studied the influence of polymer modification on IFT of a randomly oriented-flax/poly(L-lactic acid) composite and noted that by adding hyperbranched polymers (HBPs) to the matrix, G_{IC} of the composite significantly increased: from 38.9 J/m^2 for the original composite to 115 J/m^2 for the composite with the highest volume of HBP. Investigating the properties of a hybrid flax/glass UD-fabric (twisted flax yarns) reinforced phenolic resin composite, Zhang et al. [59] measured G_{IC} = 550 J/m^2 for flax composite that increased by hybridization to 560 J/m^2 for [(flax/ glass)$_n$]$_s$ composite. Their study shows that adding off-center plies of different fibers to the laminate will change the IFT, although the delamination is happening between the same type of composite plies. Li et al. [60] investigated the effect of fiber surface treatment on the fracture properties of sisal-fabric/vinyl-ester composites with a fiber volume fraction (V_f) of around 32% ($V_f \approx 0.32$) and observed significant improvements in fracture energies of Mode I (G_{IC}) and small increases in that of Mode II (G_{IIC}) in comparison to the untreated system. While G_{IC} of the permanganate-treated fiber experienced the most significant improvement. The authors followed ASTM D5528 and measured the G_{IC} values of 320.5/1158.7 [888.0/1682.2] J/m^2, and G_{IIC} values of 2141.0 [2384.7] J/m^2 for the crack initiation/propagation of untreated and [permanganate-treated] sisal composites. In an experimental study, Chen et al. [45] investigated the potential of carbon nanotube buckypaper interleaf for improving the IFT of flax-fabric/phenolic resin composites with $V_f \approx 0.73$. They followed ASTM D5528 and tested some partially and fully interleaved laminates at the mid-plane as well as laminates without interleaf for IFT. They observed a 26% [22%] and 29% [51%] increase in initiation G_{IC}, and [propagation G_{IC}], respectively, for partially and fully interleaved composite in comparison to those of the original one ($\approx$4.4 and [$\approx$5.6] kJ/m^2) and related them to the extensive favorable fibrillations observed in the interleaved laminates. Trying to enhance the IFT of flax-fabric/epoxy composite of V_f = 0.60, Li et al. [61] interleaved them at the mid-plane with chopped flax yarns of different lengths and content and followed ASTM D5528 to evaluate their G_{IC}. The authors observed a 4% to 31% improvement in the propagation G_{IC}, with the maximum increase for the moderate length and content of the chopped yarns, compared to that of the original composite, G_{IC} =1.40 kJ/m^2. It should be noted that they have used Modified Compliance Calibration method but have stated that used Modified Beam Theory (MBT). Ravandi et al. [62–64] conducted experimental investigations, according to ASTM D5528, and numerical analysis, using FEM, to study and predict the effects of through-the-thickness stitching and reinforcement architecture on the G_{IC} of the UD ($V_f \approx 0.31$) and woven fabric ($V_f \approx 0.40$) FFRECs. The authors tested double cantilever beams (DCBs), according to ASTM D5528, to determine the G_{IC} of the FFRECs as well as UD-glass/epoxy

(V_f = 0.60) composites, and observed that average G_{IC} of the UD-glass/epoxy was less than half of that of unstitched UD-FFRECs ($G_{IC} \approx 1.3$ kJ/m^2). Also, their findings showed that the average G_{IC} value of the UD-FFREC was three times lower than that of woven-FFRECs ($G_{IC} \approx 3.2$ kJ/m^2), and generally, stitching with flax yarns induced continuous improvement in IFT of the laminates, increasing with stitch areal fraction to a maximum of 21% for the highest investigated level (of aerial fraction). Besides, they reported that their FEM results agreed with the experimental ones giving the G_{IC} values of 0.771 and 1.25 kJ/m^2, respectively, for the crack initiation and propagation of UD-FFRECs as well as 0.671 kJ/m^2 for the G_{IC} value of glass/epoxy composite. Bensadoun et al. [57] tested some FFRECs reinforced with seven different fiber configurations; i.e., one plain weave, two twill weaves (low/high twist), a twisted-yarn-fabric cross-ply and an untwisted-fiber cross-ply architecture (cross-ply laminates were tested in both 0°/90° directions), to investigate the effect of the reinforcement architectures on the G_{IC} and G_{IIC} of the composites with V_f = 0.40. In terms of delamination fracture, their composites with untwisted-fiber cross-ply reinforcement denoted as UD [90,0], has a 0°/0° ply interface at the mid-plane of the laminate and is the closest one to the UD-FFRECs assessed in the current study. They reported initiation G_{IC} values between 457 J/m$_2$ for plain woven and 777 J/m^2 for twisted-yarn-fabric cross-ply composite when tested in [90,0] direction, as well as G_{IC} = 496 J/m^2 for UD [90,0] tests. While the propagation values of G_{IC} vary from 663 J/m^2 for UD [90,0] to 1597 J/m^2 for the low-twist twill-woven composites. Besides, they obtained G_{IIC} values between 728 J/m^2 for UD [90,0] and 1872 J/m^2 for plain weave laminates. The authors named the cross-plied twisted-yarn-fabric (with 90% of yarns in longitudinal direction and 10% along the transverse direction) and the UD-untwisted-fiber architecture respectively "quasi-UD" and "UD" architecture, whereas, their interlaminar fracture behavior could be different with those of UD laminates, where all plies lay in one direction. Therefore, the results may not coincide exactly with the results of UD laminates. In two different works, Almansour et al. reported their findings of the effect of water absorption on the G_{IC} [47] and G_{IIC} [48] of the woven flax (V_f = 0.31) and woven flax/basalt (basalt plies in the skin of the laminates and overall V_f = 0.33) reinforced vinyl-ester composites in [±45°] arrangement. Their DCB test results showed that upon immersion in water, the initiation and propagation G_{IC} of the flax/vinyl-ester, as well as the initiation G_{IC} of flax/basalt hybrid composites, decreased by 27%, 10%, and 23%, respectively. In contrast, the propagation G_{IC} of flax/basalt hybrid composites increased by 15%. They have documented the average initiation/propagation G_{IC} values of 3870/12093 J/m^2, and 4431/9738 J/m^2, respectively, for flax and flax-basalt hybrid composites in dry condition. The concerning issue in this study is that the crack did not propagate in the mid-plane of the laminate and deviated from that by crossing the layers; it thus cannot be considered a pure interlaminar failure. Along with the high G_{IC} value of vinyl-ester (410 J/m^2 compared with 69–150 J/m^2 for epoxy [57]), this might be another reason for higher measured values of IFT. Also, in spite of the increase in initiation G_{IC}, it seems that for the flax/basalt hybrid composites, the shorter final deviated-crack growth combined with the ever-increasing R-curve resulted in a smaller propagation G_{IC}. Whereas, comparing the propagation G_{IC} associated with the last crack length exhibited by the hybrid composites with that of flax composite, an increase in propagation G_{IC} can be observed. In terms of G_{IIC} values, testing end-notched-flexure (ENF) specimens, the authors observed that hybridization by basalt fibers (adding basalt composite plies to the skin of the flax laminates) improved the initiation G_{IIC} value of flax composite by 58%; also, moisture absorption improved the initiation G_{IIC} of flax composite by 29% and that of hybrid composite by 20%, due to improved matrix ductility. Their published average initiation G_{IIC} values for the dry flax and flax/basalt hybrid composites are 253 J/m^2, and 400 J/m^2, respectively. Nevertheless, considering that the basalt plies are located at the surface of the laminate, while the IFT is being evaluated between flax plies at the mid-plane, the basalt layers have the same function that the stiffeners (bonded as tabs to the skin of DCB and ENF test specimens to avoid large deformations and specimens arm failure) used by some researchers measuring the IFT of flax composites have [57,62–65]. Accordingly, the findings of these studies show that the stiffeners affect the values of both Mode I and Mode II IFT, which contradicts the assumption of these researchers. As a result, stiffening the arms of the DCB and ENF specimens will affect the IFT values and is not a proper solution, instead, increasing the thickness or changing the initial crack length to improve the specimen stiffness seems desirable, as recommended by the corresponding ASTM standards [53,54]. Trying to improve the IFT of UD-flax/gliadin composites, Vo Hong et al. [65] investigated the influences of some processing conditions, fiber surface treatment, and matrix plasticization on IFT by fabricating and testing flax/gliadin composite laminates with $V_f = 0.40$. Testing the original composite DCB specimens according to ISO 15024 standard, the authors reported $G_{IC} = 50$ to 100 J/m^2 and $G_{IC} = 450$ to 550 J/m^2, respectively, for crack initiation and propagation measurements. Their findings show that the optimum value of IFT, $G_{IC} \approx 1000$ J/m^2, was obtained with a combination of all of the investigated processing and materials parameters, i.e., fiber treatment, adding glycerol to the matrix, and medium cooling rate. Rajendran et al. [66] investigated the behavior of twill-weave flax/epoxy composites in Mode I and Mode II delamination (no V_f reported). The authors tested DCB and ENF specimens and, following the experimental calibration method (ECM), calculated a $G_{IC} = 485$ J/m^2 and $G_{IIC} = 962$ J/m^2 for the composite. In another report [67], they published their findings of the Mode I, Mode II, and Mixed-mode I/II interlaminar fracture toughness of the same composite with $V_f \approx 0.44$. In this work, they followed ASTM D5528 and D7905 standards and tested standardized DCB and ENF specimens to determine the IFT in Mode I and Mode II, respectively, while, they tested single-leg-bending (SLB) specimens, which is not standardized, to measure the Mixed-mode I/II fracture toughness ($G_{(I/II)C}$). Their obtained values in this recent report are $G_{IC} = 363$ J/m^2, $G_{IIC} = 962$ J/m^2 and $G_{(I/II)C} = 649$ J/m^2 (for a $G_{II}/(G_I + G_{II}) = 0.43$), which shows a discrepancy in G_{IC} value for the same material from two different reports. In a recent study, Saidane et al. [68] used the acoustic emission (AE) method to investigate the Mode I fracture toughness of flax, glass, and hybrid flax/glass woven-fiber/epoxy composites, flax laminates had a $V_f = 0.40$. They conducted DCB tests according to ASTM D5528 standard but used AE to detect the onset of delamination and the corresponding critical load to calculate G_{IC}. The initiation G_{IC} values of 1079 J/m^2, 945 J/m^2 and 923 J/m^2 were determined respectively for flax, hybrid flax/glass and glass composites. Nevertheless, the method used in this work to detect the delamination initiation and the corresponding critical load used in calculation of G_{IC} influences the obtained values, thus, this fact should be considered when using the results for comparison purposes. The interesting point in this study is that the propagation G_{IC} value of flax composites (R-curve) continuously increased with delamination length which contradicts the findings of previously cited studies, including those of Zhang et al. [59] for the similar materials which show a stabilized plateau value after a certain crack length. Also, their findings for hybrid flax/glass composites regarding the behavior of R-curves and the decrease in G_{IC} is in contrast with those of Zhang et al. [59] for the similar materials.

As summarized above, the literature review shows that the previous works addressing the IFT of flax composites have mostly studied flax-fabric reinforced composites, while few researchers investigated UD-FFRECs and only in Mode I [62]. Also, to the knowledge of the authors, no study has, in particular, investigated their delamination behavior in Mode II and Mixed-mode I/II. Therefore, the damage tolerance and behavior knowledge of UD-FFRECs, as the basic building block of the FFRECs, particularly in Mode II and Mixed-mode I/II are missing. This is of paramount importance in using flax/epoxy composites in recently demanded structural applications as well as employing widely-used and reliability-proven FEM numerical analysis method to study this environmentally and economically-advantageous and ubiquitously used material. Consequently, an in-depth understanding of their delamination behavior in different modes is essential both for improving their material properties and for generating demanded knowledge and mechanical properties in the engineering design field and for future studies employing analytic and numerical resolutions. Because of the variations in the fabrication processes, origins and properties of raw materials, and the mechanical properties evaluation methods of these composites, the use of well-defined and precisely controlled uniform composites as well as standard test methods is a pre-requisite to achieve reliable results.

The aim here is to investigate the interlaminar fracture behavior of the UD-FFRECs in three different modes of failure, namely, Mode I, Mode II, and Mixed-mode I/II and determine the corresponding interlaminar fracture toughness. For this purpose, UD-FFRECs are fabricated using identical raw materials and processing techniques with a precisely controlled fiber volume fraction. The commonly used ASTM standard test procedures are followed to evaluate the composites under different modes of loading to determine their IFT and produce valid and reliable data by respecting the defined validity criteria.

2. Materials and Methods

2.1. Material System and Test Specimen Preparation

Unidirectional tapes of flax fibers, FLAXTAPE™ 200 (from LINEO, Valliquerville, France, https://eco-technilin.com), with an areal density of 200 gr/m^2, were cut in 300 mm × 300 mm dimensions and stacked up to form the UD reinforcement preform, Figure 2a. The thermosetting Marine 820 Epoxy System, mixed with 18 Wt.% Marine 824 hardener (from ADTECH® Plastic Systems, Madison Heights, MI, USA, www.axson-technologies.com), was used to impregnate the fibers employing the resin transfer molding (RTM) process. The number of layers in [0]$_n$ layup and the thickness of laminates were accurately computed according to the ASTM D3171 procedure and controlled to result in a unique V_f of 41% for all composite plates and meet the requirements of the employed ASTM test procedures regarding the thickness of test specimens. In order to induce a pre-crack as the initiation site for delamination, a 13 μm-thick polytetrafluoroethylene (PTFE) film, was accommodated on the mid-plane of the laminate during layup of the preform; the schematic view of the plate including the insert film is shown in Figure 2b.

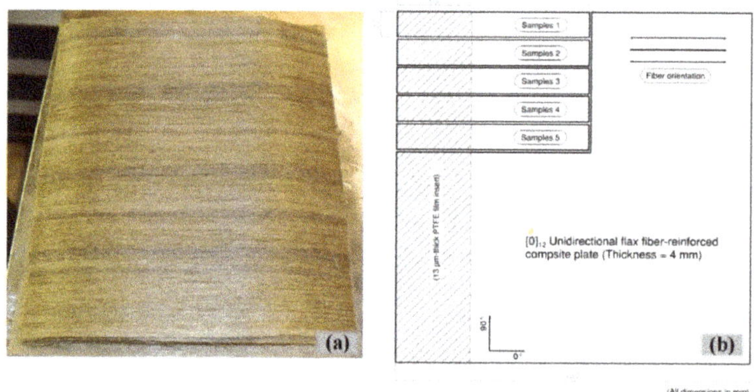

Figure 2. (a) Unidirectional flax fiber preform and (b) configuration of flax/epoxy pre-cracked laminate.

After fabrication and post-curing of the composite laminate, the PTFE film was located via X-ray imaging, and the laminate was trimmed to reach the dimension of the insert that is recommended by the ASTM standards, as shown in Figure 1b. Next, the rectangular UD-laminated composite specimens with uniform thickness and containing a PTFE non-adhesive insert at one end on the mid-plane were cut to their desired dimensions by a 10-inch/90-tooth, DIABLO's cutting saw blade with high-density carbide tooltips, as illustrated in Figure 2b.

2.2. Mechanical Testing

Three different types of fracture tests were conducted to evaluate the interlaminar fracture toughness of the composite in pure Mode I and Mode II as well as in Mixed-mode I/II loading conditions. Numerous test methods and specimen configurations have been developed over the years to evaluate the IFT of the laminated FRPCs; however, only some of them have been standardized [43,69].

In this study, the most commonly used standardized test methods by ASTM will be employed. All of the tests were carried out at ambient conditions, and the specimens were exposed to the test conditions at least 48 hours before the experiments. The details of the corresponding specimens, test procedures, and data reduction methods are described as follows.

2.2.1. Mode I Interlaminar Fracture Toughness

Following the ASTM D5528 standard test method, DCB specimens were prepared and tested to determine the Mode I interlaminar fracture toughness of the composite. The thickness, h, and initial delamination length, a_0, of the composite laminate were designed to fulfill the requirements of the standard. After cutting the test coupons to their desired dimensions, two piano hinges were bonded to the delaminated arms of the DCB specimen to introduce the tensile opening load and induce a Mode I delamination fracture; the configuration and nominal dimensions of the DCB specimen are illustrated in Figure 3a. To facilitate the detection of the onset and propagation of the crack, the side edge of the specimens was coated with white correction fluid, and the approximate location of the pre-crack tip and required final crack length were marked, as shown in Figure 3b. The DCB tests were conducted on an MTS Landmark 370 hydraulic machine equipped with a 1 kN-capacity load cell, and MTS 647/10 hydraulic grips. The specimens were loaded and unloaded, respectively, at a constant crosshead speed of 3 and 25 mm/min in two loading/unloading cycles. The initial loading was stopped after a crack growth of 3 to 5 mm, and the specimens were unloaded; this was a procedure to create a natural Mode I pre-crack in the DCB specimens. They were then reloaded until the delamination crack (a) propagated past 50 mm from the tip of pre-crack, and then they were unloaded.

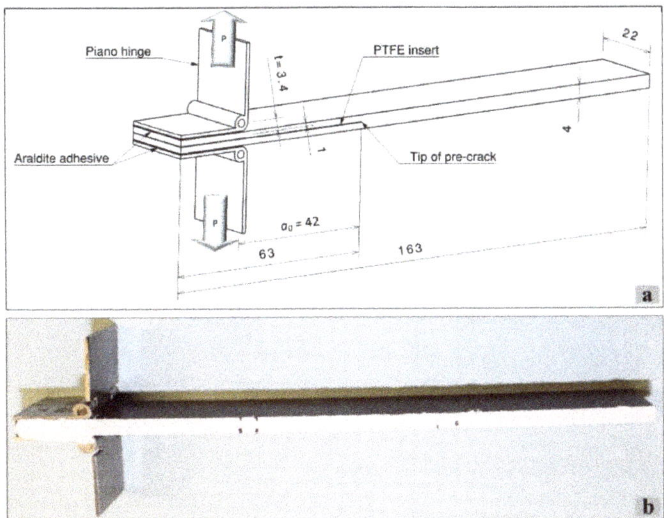

Figure 3. Double Cantilever Beam (DCB) specimen: (**a**) configuration and dimensions, (**b**) final test specimen.

A Digital Image Correlation (DIC) system was employed to observe and track the delamination front; the test setup is shown in Figure 4. DIC is an optical non-contact deformation measurement technique [70,71] with adequate performance, low cost, and handy experimental setup that has been used in many research fields for strain analysis studies, including crack initiation [72,73]. The principle of DIC and theoretical basis for the calculation of stains using consecutive images has been reported by many researchers and is available in the literature [74–76]. The recorded images were synchronized with the load and displacement signals of the MTS machine, the crack opening displacement (COD) was estimated as the crosshead displacement, and the data was recorded for the entire loading/unloading cycles. Then, the registered information was processed to associate the load and displacement values at the intervals of delamination growth defined by the standard. These values were used to calculate G_{IC} for different delamination lengths and to form the delamination resistance curves (R-curves). Eight DCB specimens were tested to obtain at least five valid tests to determine G_{IC} of the composite. The piano hinges were extended with a bracket to avoid the interference that occurred between the arms of the specimen with regular piano hinges and the jaws of the grip. In order to more precisely locate the end of the insert film, the first failed sample was completely opened, and the end of the film was marked to measure a_0 and used for marking the other samples.

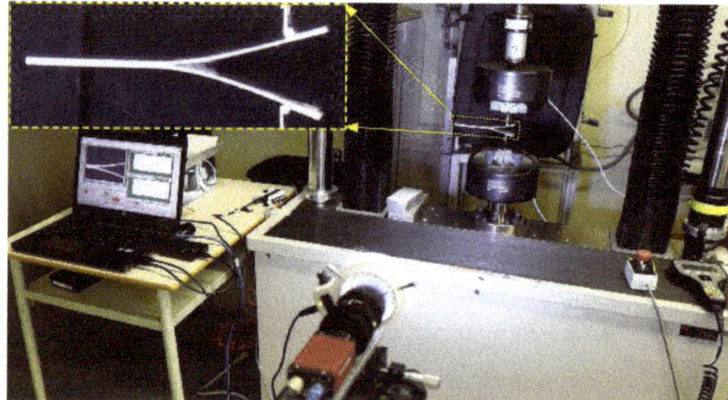

Figure 4. Mode I delamination test setup with a DIC system to follow crack propagation.

There are various data reduction methods for calculating G_{IC} values, three of which are presented in the ASTM D5528 standard; however, the MBT method, yielding to the most conservative values, has been recommended. In this study, MBT was implemented to calculate G_{IC} values based on the processed data and using Equation (1):

$$G_I = \frac{3P\delta}{2b(a+|\Delta|)} \tag{1}$$

where P is the applied load, δ is the COD on the load line, b is the width of the specimen, a represents the delamination crack length, and Δ is the crack length correction factor. Δ can be determined experimentally as the x-intercept of linear least squares plot of $C^{1/3}$ versus delamination length, where C is the compliance of the specimen ($C = \delta/P$). G_{IC} is the IFT of the composite, associated with critical load (P_C) determined as follows.

The crack initiation value of G_{IC} (G_{IC}^{ini}) was determined using three definitions presented in the standard, namely, at the deviation from linearity point in the load-displacement curve (NL), when delamination is visually observed (VIS) and at 5%-increased-compliance point or the maximum load, as illustrated in Section 3.1. The NL G_{IC}^{ini} is the lowest and recommended for damage tolerance analysis, and the first propagation value of G_{IC} is known as Mode I pre-crack G_{IC} [53]. Concerning deflections of the specimen arms due to relatively low flexural modulus of the FFREC, the values of h and a_0 were calculated according to the recommendations outlined in the ASTM standard and based on the elastic properties of the composite, leading to a shorter $a_0 = 42$ mm compared to $a_0 = 50$ mm, recommended for synthetic FRPCs. Nevertheless, for some delamination lengths, the ratio of the load-line displacement to delamination length was higher than 0.4 ($\frac{\delta}{a_0} > 0.4$). Consequently, according to the standard, a large-displacement correction factor (F) was calculated using Equation (2) and incorporated in the data reduction method by multiplying F by G_{IC} values calculated using Equation (1):

$$F = 1 - \frac{3}{10}\left(\frac{\delta}{a}\right)^2 - \frac{3}{2}\left(\frac{\delta t}{a^2}\right) \tag{2}$$

where t was measured as defined in the standard and is shown in Figure 3.

2.2.2. Mode II Interlaminar Fracture Toughness

End-notched flexure (ENF) tests were conducted following the procedure outlined in ASTM D7905 to determine the Mode II interlaminar fracture toughness, G_{IIC}, of the UD-FFRECs using the compliance calibration (CC) method, which is presented as the only acceptable data reduction method for this test [54]. The specimens were cut from the same laminates used for DCB specimens mentioned above (Figure 1) and with the same tool to the dimensions that fall within the range of allowable lengths and widths specified in the standard. Both sides of the specimens were painted with a correction liquid to help the detection of crack initiation, and the end of pre-crack, as well as the required loading points, was marked. Figure 5a illustrates the configuration and dimensions, and Figure 5b shows a picture of the ENF specimen with markings. The ENF specimens were mounted in a three-point-bending (TPB) fixture with a 100-mm-span and loaded on the same machine fitted with the 1kN-loadcell that was used for DCB specimens, as shown in Figure 6. The coupons were loaded and unloaded at 0.5 mm/min crosshead speed, and the data acquisition was performed for the entire fracture tests and the loading phase of the CC tests.

According to the standard, the non-pre-cracked (NPC) and pre-cracked (PC) CC tests, at $a_i = 20$ and 40 mm, and fracture tests, at $u_0 = 30$ mm, were carried out on the same specimen to obtain the compliance plots as well as the maximum loads (P_{Max}), to be used for determination of G_{IIC}. The initial CC forces (P_i) were approximated using the flexural modulus (E_f) of the same material [77] and G_{IIC} of a similar material [66], then, the result of ENF tests with accepted G_{IIC} was used to update the P_C and P_i. The calculated crack length after NPC tests (a_{calc}) was calculated using unloading data of the NPC fracture test, as mentioned in the test procedure.

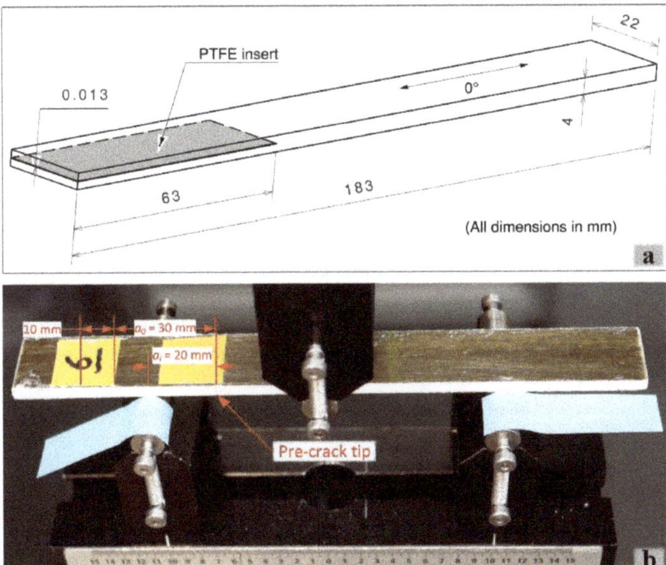

Figure 5. ENF specimen for Mode II delamination: (**a**) configuration and dimensions, (**b**) specimen and loading fixture.

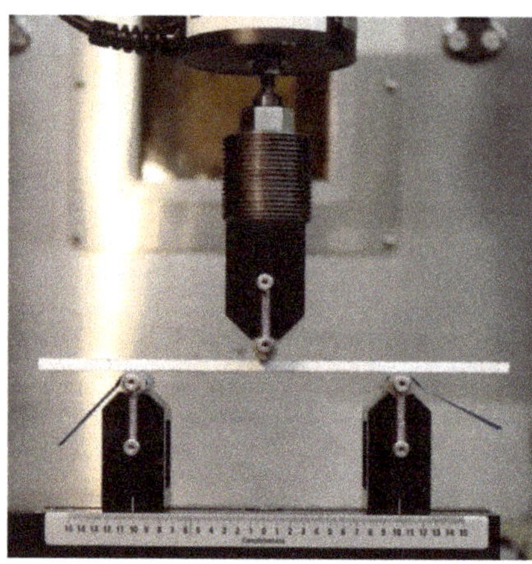

Figure 6. Mode II delamination test setup.

At each crack length, for NPC and PC tests, the compliance (C) was specified as the slope of the load-point displacement versus load curve by linear least-squares regression analysis (LLSRA). Then, C values were plotted versus cubed crack length (a^3), and the CC coefficients, A and m in Equation (3), were determined using LLSRA:

$$C = A + ma^3 \tag{3}$$

G_{IIC} values were calculated using Equation (4) and validated based on the criteria of the ASTM standard:

$$G_{IIC} = \frac{3mP_{Max}^2 a_0^2}{2B} \tag{4}$$

where m is the CC coefficient, a_0 and P_{Max} are respectively the crack length used and the maximum force measured in the fracture test, and B is the specimen width. Six ENF specimens were tested to obtain adequate results for the calculation of G_{IIC}. The value of G_{IIC} is considered as IFT, only if it satisfies the criteria defined in the standard.

2.2.3. Mixed-Mode I/II Interlaminar Fracture Toughness

The Mixed-Mode Bending (MMB) tests were carried out according to ASTM D6671 standard test method [55] to obtain the Mixed-mode I/II fracture toughness ($G_{(I/II)C}$) of the composite. Following the guidelines of the standard, piano hinges, with drilled holes matching to the hinge clamp of the MMB test fixture, were adhesively bonded to the pre-cracked specimens cut from the same plate as for DCB and ENF specimens, to form an initial crack length of $a_0 = 28$ mm. The length of the initial crack was calculated based on the criteria defined in the standard and using the thickness, mechanical properties, and an estimated value of fracture energy, $G_{(I/II)C}^{est}$, to result in an ultimate allowable deflection. Also, the side edges of the specimens were painted white and marked. Figure 7a presents the configuration, dimensions, and load application points of the MMB specimen. Figure 7b shows the MMB apparatus with a mounted specimen and the finalized MMB specimen. Mixed-mode delamination under a mode mixture ratio of $G_{II}/(G_I + G_{II}) = 0.55$ was applied by setting the length of the lever as $c = 47.4$ mm (Figure 7b). The tests were performed at a constant crosshead speed of 0.5 mm/min on an MTS-322.31 machine equipped with a 5 kN loadcell, and the force versus displacement data was recorded. A moveable microscope was synchronized with the load-displacement data and used to observe the crack tip to detect the onset and follow the propagation of the delamination; the test setup is shown in Figure 8. The loading of the specimen was continued until the delamination extended past 25 mm. The required elastic and shear properties were used from our previous work [77], and the out-of-plane shear modulus G_{13} was assumed to be equal to G_{12}. Also, for the calculations of the estimated force and deflection, the estimated Mixed-mode IFT, $G^{est}_{(I/II)C}$, was considered as the average values of the G_{IC} and G_{IIC} of the composite measured in this present study. Six MMB tests were carried out to achieve a minimum of five valid tests by controlling the failure mode and maximum allowable deflection of the specimens according to the criteria of the standard.

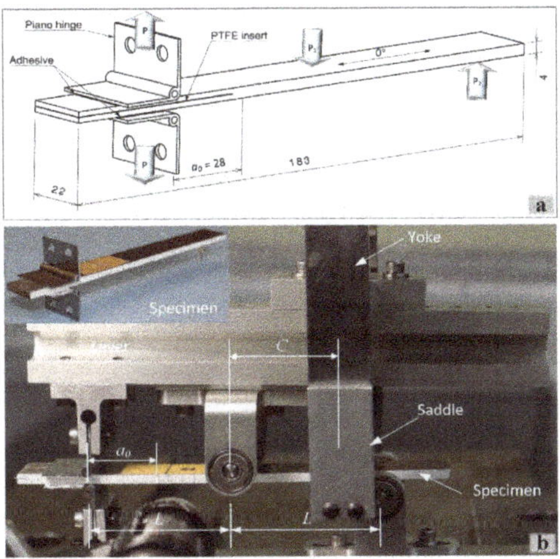

Figure 7. Mixed-Mode Bending (MMB) test specimen for Mixed-mode I/II delamination: (**a**) configuration and dimensions, (**b**) test fixture and final test specimen.

$G_{(I/II)C}$ values were calculated using the Equations (5)–(7), as mentioned in the ASTM D6671:

$$G_{IC} = \frac{12P_C^2(3c - L)^2}{16b^2h^3L^2E_{1f}}(a_0 + \chi h)^2 \tag{5}$$

$$G_{IIC} = \frac{9P_C^2(c + L)^2}{16b^2h^3L^2E_{1f}}(a_0 + 0.42\chi h)^2 \tag{6}$$

$$G_{(I/II)C} = G_{IC} + G_{IIC} \tag{7}$$

where G_{IC} and G_{IIC} are the Mode I and Mode II components of the Mixed-mode IFT, respectively, P_C is the critical load, in this study, it is the load associated with the VIS point, c is the lever length, b and h are respectively the specimen width and half-thickness, a_0 is the delamination length, L is the half-span length, and χ is the crack correction parameter defined in the standard.

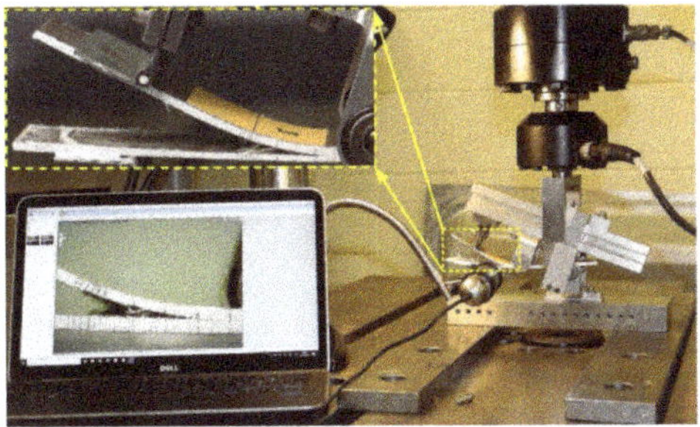

Figure 8. Mixed-mode I/II delamination test setup.

3. Results and Discussion

3.1. Mode I Interlaminar Fracture Toughness

The load-COD curve presented in Figure 9a shows the response of a typical DCB specimen to the crack opening load during the second loading cycle (after inducing the natural pre-crack in the first cycle). This curve shows that the UD-FFREC DCB specimen has a perfect linear behavior up to 95% of the P_{max}, followed by a short nonlinear curve up to the P_{max} point. Upon reaching the peak point, which differs from the point at which the delamination initiates, the load drops slowly and continuously with crack propagation, indicating that stable crack propagation is happening. Figure 9b confirms unique behavior for all of the DCB specimens and shows good repeatability of the tests for a NFRPC. This behavior of the DCBs perfectly agrees with the observations of Ravanid et al. [62–64] for woven and UD-FFREC DCBs, which were stiffened by GFRP/CFRP tabs on both sides, as well as with those of Bensadoun et al. [57] for cross-ply (UD [0 90]$_{4s}$) FFRECs laminate stiffened likewise and those of Saidaine et al. [68] for woven flax/epoxy composite. Investigating woven/vinyl-ester DCB specimens, Almansour et al. [47] also observed an initial linear response followed by a nonlinear curve up to the peak load. However, this is in marked contrast to the observations of Rajendran et al. [66,67] for the flax fabric/epoxy DCBs., who reported a high nonlinear response up to the maximum load. Ravandi et al. [62] and Rajendran et al. [67] tested epoxy composite DCBs made of relatively similar flax fabrics supplied by the same supplier with the specifications as described in the Introduction part. However, they reported an entirely different load-displacement response and IFT values; the average G_{IC} value reported by the former ($\approx$2000 J/m^2) is much higher than that of the latter (363 J/m^2).

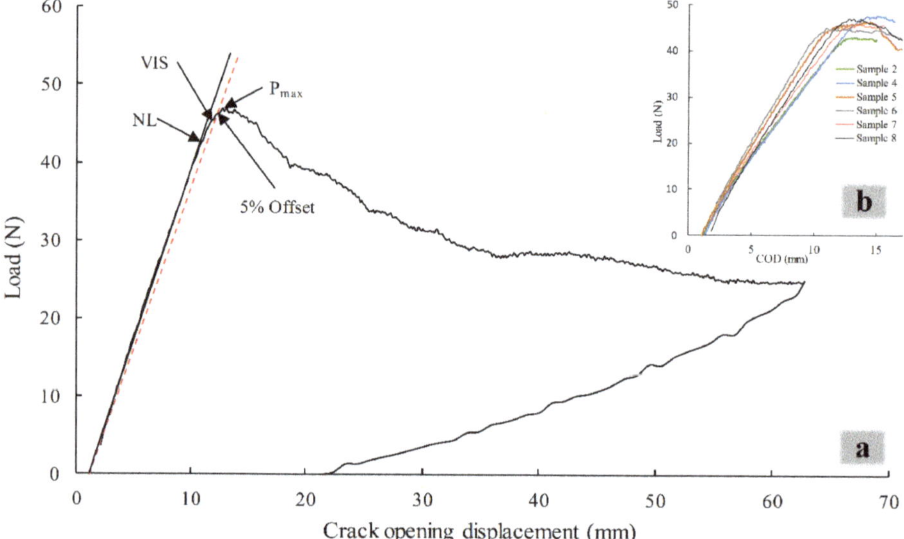

Figure 9. Load vs. crack opening displacement (COD) of the DCB specimen: (**a**) Typical loading/unloading cycle and initial G_{IC} value measurement points, and (**b**) overall initial behavior.

The calculated G_{IC} values for a typical DCB specimen are drawn as a function of delamination length (Figure 10) to generate the delamination resistance curve, known as R-curve, and to determine the initiation and propagation values of G_{IC}. As can be seen in Figure 10, the G_{IC} value determined at NL is the minimum followed by those at VIS and 5% Offset points, which happened before the P_{max} point for all DCB tests on the load-COD curve (Figure 9a). Figure 10 shows that G_{IC} increases by delamination growth to a maximum value and then drops to stabilize at a plateau value. Several authors observed similar R-curves, and this is a commonly occurred phenomenon for UD-FRPC DCB specimens when the delamination grows parallel to the fibers between two UD plies that is attributed to the development of fiber bridging across the crack [45,53,57,62,64,78]. This resistance-type fracture behavior of UD DCB composites is an incidence of properly implantation of the delamination insert and validates the tests. The scale of the fiber bridging is much more extensive for UD-NFRPCs compared to their synthetic fiber counterparts [59,62–64,68]. The authors related this phenomenon to the high strength of the longer synthetic fibers which tend to remain unbroken and less mobile, whereas, UD natural fibers are composed of shorter technical fibers with irregular geometry and alignment that helps them bridge two opening plies. They also believe that the higher G_{IC} value determined for UD-FFECs compared to those for UD glass/epoxy composites is associated with the larger scale of fiber bridging.

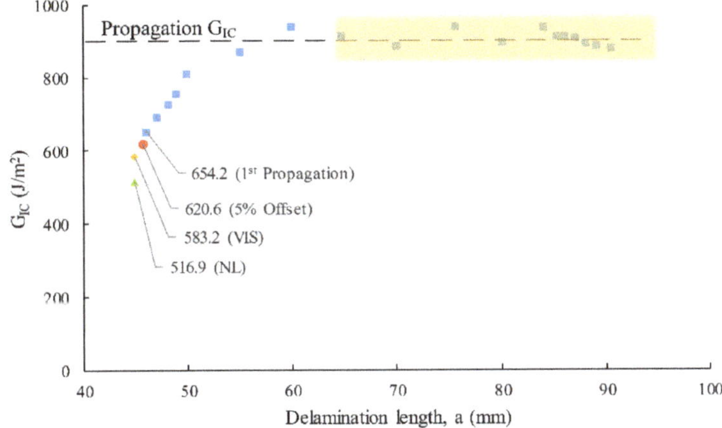

Figure 10. Crack Resistance Curve (R-Curve) of a typical UD-FFREC DCB, the plateau region is highlighted.

The G_{IC} values, determined based on different definitions of onset of delamination, are presented in Table 1. However, combining some damage detection techniques, such as Radiography, strain gauge, and AE, with DCB tests, reveal that the delamination starts in the interior of the specimen width very close to NL point on the load-COD curve [53,68,79], and as can be seen in Figure 10, this point is far from propagation values, thus is considered as initiation G_{IC} value, as recommended by ASTM D5528. The plateau value of G_{IC} in the R-curve, the average value within the highlighted zone in Figure 10, is considered as the propagation G_{IC} value and summarized in Table 1.

Table 1. Mode I fracture toughness values for UD FFREC.

Mode I G_{IC}	G_{IC} (NL Point)	G_{IC} (VIS Point)	G_{IC} (5%/Max)	G_{IC} (Propagation)
Average (J/m^2)	574	641	644	903
Standard deviation	38	31	26	55
Coefficient of variation (%)	6.6	4.8	4.0	6.1

The relevant DCB tests conducted for FFRPCs and their corresponding G_{IC} values available in the literature are summarized in Table 2. As can be seen in Table 2, the most similar tests have been performed by Ravandi et al. [64] for an analogous UD-FFECs ([0]$_{16}$) laminate containing a similar UD reinforcement with a lower surface density (110 g/m^2) compared to 200 g/m^2 used in the under-study [0]$_{12}$ laminate. However, the obtained initiation G_{IC} value at VIS point (641 J/m^2), as well as the propagation G_{IC} value (903 J/m^2) in the present study, are about 20% lower than those reported by these authors. Considering that they stiffened their DCB specimens with GFRP, this difference may be explained by referring to the work of Almansour et al. [47], where replacing the out-most layers of flax fabric composites by the 4-fold stiffer basalt fiber plies, equivalent to stiffening flax composite DCBs, augmented the G_{IC} values by around 19%. Therefore, the obtained results in the current study are in excellent agreement with their results. The experiments of Bensadoun et al. [57] for the cross-ply DCB ([90,0] specimens), where the delamination occurs at 0°//0° interface, also have similar failure conditions to the present DCB tests; however, it has been shown that G_{IC} value measured for delamination at 0°//0° interface within a UD laminate is different with that within a multidirectional laminate and replacing some off-center 0° plies with 90° layers reduce the G_{IC} value [80,81]. Therefore, in view of this fact, the lower G_{IC} values reported by these authors, presented in Table 2, are also in good agreement with those of the present work. In general, from the studies summarized in Table 2, it can be deduced that the composites reinforced with women fibers have higher G_{IC} values compared to those with UD reinforcements. As explained before, the huge difference observed for the results of Almansour et al. [47] can be due to the very high toughness of matrix (that is the dominant factor in this failure mode) used in their study in comparison to that used in the current study (410 J/m^2 of vinyl-ester against 69–150 J/m^2 of epoxy [57]) as well as the woven reinforcement structure used in their composite that exhibits higher G_{IC} compared to UD laminate as shown in Table 2 and [57,82]. However, for propagation values, clearly, the deviation of crack from the mid-plane of the laminates is a significant reason. Despite the similarity of the material, the value reported by Rajendran et al. [67] seems relatively low. In addition to other differences listed in Table 2, and as discussed earlier, their load-COD curve is entirely in contrast with the findings of this study and all other authors. Moreover, they reported a 34% higher value for G_{IC} of the same material in another work [66].

Overall, considering the differences in reinforcements architecture/source, matrix type, processing, testing, and data reduction method, the obtained results in the present study are generally consistent with the literature data given in Table 2. Therefore, the determined and validated values for the initiation and propagation G_{IC} for UD-FFRECs can be considered as the material properties of these composites and be confidently applied in future research and engineering fields.

Table 2. Mode I interlaminar fracture toughness tests and values for FFRPCs.

	Current Study	Ravandi et al. [62]	Ravandi et al. [64]	Bensadoun et al. [57]	Bensadoun et al. [57]	Almansour et al. [47]	Rajendran et al. [66,67]	Li et al. [61]	Chen et al. [45]	Saidane et al. [68]	Zhang et al. [59]	Vo Hong et al. [65]
Reinforcement	UD-flax (200 g/m^2) $[0]_{12}$	4 × 4 W-flax fabric (500 g/m^2)	UD-flax (110 g/m^2) $[0]_{16}$	0°/0° flax-plies $[90,0]_{2s}$	W-flax fabrics	(±45°) W-flax fabric	2 × 2 W-flax fabric (200 g/m^2)	UD-T-flax fabric (200 g/m^2)	UD-T-flax fabric	2 × 2 W-flax fabric	UD-T-flax fabric	UD-flax
Matrix	Epoxy	Epoxy	Epoxy	Epoxy	Epoxy	Vinyl ester	Epoxy	Epoxy	Phenolic resin	Epoxy	Phenolic resin	Gliadin powder
Stiffener	-	CFRP	GFRP	GFRP	GFRP	-	-	-	-	-	-	GFRP
Composite fabrication	RTM ‡	VARI †	VARI †	RTM ‡	RTM ‡	VI ±	Hand layup	CM +	VARI †	CM +	CM +	Hand layup
V_f (%)	41	31	40	40	40	31	44	60	73	40	67	40
Test /Data reduction method	ASTM D5528/MBT	ASTM D5528/MBT	ASTM D5528/MBT	ASTM D5528/MBT	ASTM D5528/MBT	ASTM D5528/MBT	CC	ASTM D5528/CC *	ASTM D5528/CC	ASTM D5528/MBT	ASTM D5528/-	ISO 15024/MBT
Damage initiation point	NL	-	VIS	NL	NL	VIS	SG detected/P_{max}	-	VIS	AE detected	-	NL
G_{IC} (J/m^2) Initiation	574	≈2000	771	496	457–754	3579	363/485	-	≈440	1079	280	50–60
G_{IC} (J/m^2) Propagation	903	≈3200	1250	663	1151–1597	11789	-	1400	≈580	≈2400	550	450–550

† Vacuum-assisted resin infusion, ‡ Resin transfer molding, ± Vacuum infusion, + Compression molding, * In the paper mentioned MTB (by mistake), AE: Acoustic Emission, W: Woven, UD: Unidirectional, T: Twisted, SG: Strain gauge, CC: Compliance Calibration, MBT: Modified Beam Theory.

3.2. Mode II Interlaminar Fracture Toughness

Figure 11 presents the load-displacement curve of a typical ENF fracture test. As can be seen, the composite shows an overall linear response with a clear and sudden load drop, from which, respectively, the compliance of the specimen and the peak load can be determined. This is a normal and known behavior exhibited by ENF where, under an in-plane shear loading, the surfaces of the mid-plane plies slide over each other and lead to unstable crack growth and sudden load drop [54]. This failure allows only the measurement of crack initiation IFT. Almansour et al. observed an analogous behavior for woven flax/vinyl-ester [48] and non-woven flax mat/vinyl-ester [83], whereas Rajendran et al. [66,67] reported a complete nonlinear load-displacement curve for woven flax/epoxy composites. The experience of the latter, likewise to their DCB tests, is in stark contrast with the observations of previously cited authors (Almansour et al.) and many other authors [84–86].

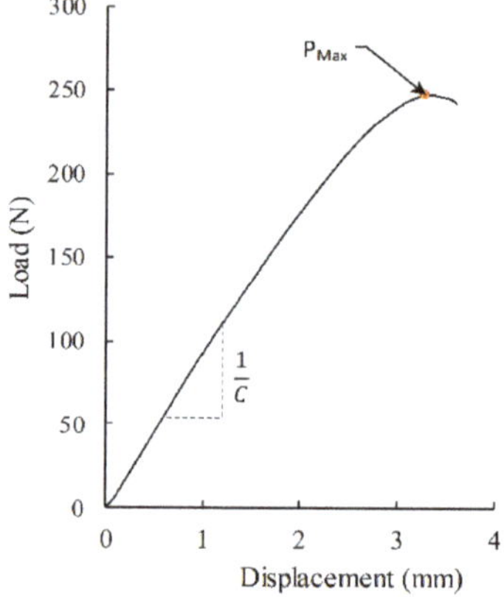

Figure 11. Load-displacement curve of ENF test and illustration of compliance (*C*) and P_{Max} determination.

The Mode II interlaminar fracture energy, G_{IIC}, was calculated using the data reduction method explained in the previous section, and the averaged results of five valid NPC and PC tests are presented in Table 3. As can be noticed, considering the variations, the averaged results are in the same range and overlapping, however, NPC results have a larger variation.

Table 3. Mode II fracture toughness values for UD FFREC.

	G_{IIC} **(NPC)**	G_{IIC} **(PC)**	G_{IIC} **(Average)**
Mean (J/m^2)	401	378	390
Standard deviation	18	7	18
Coefficient of variation (%)	4.5	1.9	4.6

There are very limited similar ENF test data available in the literature for comparison and validation purposes. In fact, there is no available ENF test being conducted and treated exactly in the same way for UD FFRECs; nevertheless, the closest ones are summarized in Table 4.

Table 4. Mode II interlaminar fracture toughness tests and values for FFRPCs.

	Current Study	Bensadoun et al. [57]		Almansour et al. [48]	Almansour et al. [83]	Rajendran et al. [67]
Reinforcement	UD-flax (200 g/m^2)	0°//0° UD-flax plies in [90,0]$_{2s}$	W-flax fabrics	(±45°) W-flax fabric	Non-woven flax mat	W-flax fabric (200 g/m^2)
Matrix	Epoxy	Epoxy	Epoxy	vinyl ester	vinyl ester	Epoxy
Stiffener	-	GFRP	GFRP	–/basalt	–/basalt	-
Composite fabrication	RTM	RTM	RTM	VARTM	Hand layup + CM	Hand layup
Fiber content	$V_f = 0.41$	$V_f = 0.40$	$V_f = 0.40$	100 Wt.%	$V_f = 0.23$	$V_f = 0.44$
Test /Data reduction method	ASTM D7905/ CC/SBT	SBT	SBT	SBT	SBT	CC
G_{IIC} (J/m^2)	378/612	728	1315-1872	266/430	1940/2173	962

UD: Unidirectional, W: Woven, RTM: Resin transfer molding, VARTM: Vacuum-assisted resin transfer molding, CM: Compression molding, SBT: Classical Simple Beam Theory, CC: Compliance Calibration.

In Table 4, the obtained G_{IIC} value for the under-study UD-FFREC laminate is placed within the range of those reported for similar tests. However, it is clearly lower than that of the closest material, i.e., [90,0]$_{2s}$ laminate with 0°//0° UD-flax/epoxy plies in the midplane [57]. This difference can not only be associated with the variation existing within the properties of natural fibers, it can also be due to the difference in the reinforcement configuration of the laminates, the GFRP tabs used for stiffening [90,0]$_{2s}$ laminate and the different data reduction method employed for calculation of G_{IIC}. The results of Almansour et al. [48,83] show that replacing the out-most plies of the flax composite with high-stiffness basalt plies, which is equivalent to tabbing ENF specimens, augmented the G_{IIC} value of the flax composites up to 62%. Therefore, tabbing can be one of the reasons for obtaining higher value for the compared composite. In addition, for DCB tests, it is proven that alterations in the off-center plies (orientation and material) of the UD laminates with the same delamination interface layers affect G_{IIC} values [68,80,81]. There is no research result for ENF tests; however, this might be true for Mode II delamination, as well. Furthermore, the ASTM D7905 standard is followed in the current study; however, Bensadoun et al. used the classical simple beam theory method (SBT). While using different data reduction methods yields variations in the obtained G_{IIC} values, for instance, applying the classical SBT used in their study to the test data of the current study, a higher value (612 J/m^2) is obtained, Table 4. The findings of some authors [57,87] revealed that woven reinforcements result in higher G_{IIC} values compared to UD ones; thus, the difference between the G_{IIC} values of UD laminates and those of woven-fabric reinforced composites given in Table 4 is reasonable. Some authors believe that normally G_{IIC} values are higher than G_{IC} [57,88], whereas some observed the opposite [47,48,67], and some believe that both can be true for laminated composites [89]. As listed in Tables 3 and 4, depending on the applied data reduction method, both cases can happen for the UD-FFREC. For the most similar instance ([90,0]$_{2s}$ laminate) discussed above, using ASTM D7905 standard for G_{IC}, and SBT for G_{IIC} results in the same trend.

3.3. Mode II Interlaminar Fracture Toughness

The mixed-mode loading behavior of the composite is plotted in Figure 12. For clarity, a single curve specimen is presented in Figure 12a and shows that in general, MMB specimens of FFREC exhibit an initial linear behavior followed by a nonlinear curve up to the peak load, after which, a plateau type curve with gradual load drop can be seen till the end of the test. Having plotted the curves of all 5 MMB specimens in Figure 12b, it can be seen that they exhibit a similar and consistent response with acceptable repeatability for the natural fiber composites.

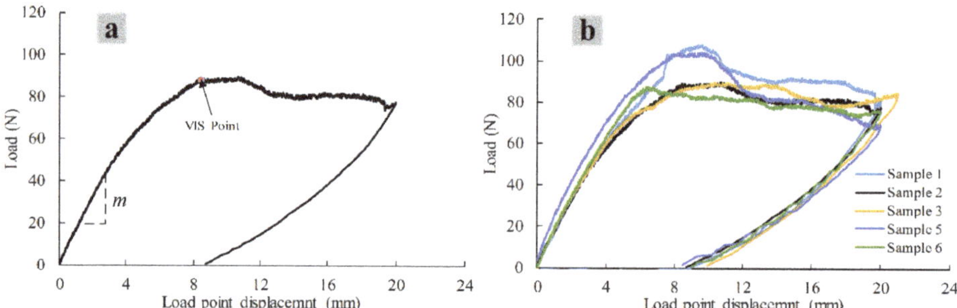

Figure 12. Load-displacement curves of MMB tests: (**a**) determination of m and P_C, and (**b**) all curves of the tests.

The slope of the load-displacement curve (m) and the critical load (P_C) need to be determined for the calculation of mixed-mode I/II fracture energy of the composite. m was directly calculated from the linear part on the curve, as shown in Figure 12a. However, there are some definitions for the P_C in the standard; NL, VIS, and 5%/max points, as define for the calculations of G_{IC}, and the most suitable one should be used. Due to the large nonlinearity part in the load-displacement curves, NL and the calculated 5% offset points reach too early for most of the tests on this curve, so that do they not seem to be the failure point and using the corresponding load would result in an underestimated fracture energy. On the other hand, comparing the VIS and the maximum force points, it was observed that the former happened earlier and right before peak load. This was also observed by Rajendran et al. [67], who used strain gauge to detect the onset of failure in similar tests. Accordingly, the VIS point was considered as the onset of delamination, and $G_{(I/II)C}$ was calculated based on the associated load at this point. For all the tests, the maximum deflection was smaller than the allowable deflection that was determined based on the criterion of the standard, and they were validated.

The obtained average mixed-mode I/II fracture energy value was $G_{(I/II)C}$ = 414 J/m^2 with a 15% coefficient of variation. The variation of results seems relatively high; however, considering the nature of the natural fiber composites, the complexity of this test, and the subjective visual solution used for detecting the initiation of failure, it is reasonable. To the knowledge of the authors, MMB tests reported by Rajendran et al. [67] are the only literature data available for flax/epoxy composites. These authors tested three single-leg-bending specimens ($G_{IIC}/G_{(I/II)C}$ = 0.43) of woven-flax/epoxy composite with V_f = 0.44, and using compliance calibration method, obtained $G_{(I/II)C}$ = 649.1 J/m^2. Comparison is limited to the result of this work that is obviously higher than the value obtained in the present study; nevertheless, it should be noted that the reinforcement, manufacturing method, mode mixture (mixed-mode ratio), $G_{II}/G_{(I/II)C}$, and test method in their study was different with the current one. Furthermore, in the previous section, it was seen that woven reinforcement configurations produced higher IFT, and for the same material, in another work, the authors reported a much higher G_{IIC}; therefore, this difference is reasonable. Considering that there is no other pertinent data for comparison, and compared to the obtained Mode I and Mode II fracture toughness, which were determined and validated earlier, this value of the mixed-mode fracture energy is the only value obtained for UD-FFRECs and reasonably acceptable. Therefore, it can be considered as the representative of the material property of this composite and used in engineering design and numerical simulation studies.

The B-K delamination criterion was used and fitted to the obtained data to determine the B-K fitting parameter, η, and to predict the interlaminar toughness for various mixed-mode ratios [90]:

$$G_{(I/II)C} = G_{IC} + (G_{IIC} - G_{IC})\left(\frac{G_{II}}{G_{(I/II)C}}\right)^{\eta} \tag{8}$$

where $G_{II}/G_{I(I/II)C}$ is the mixed-mode ratio, and η is the B-K fitting parameter. Using Equation (8), $\eta = 0.35$ is computed for the prediction of the $G_{I(I/II)C}$ with other mixed-mode ratios.

4. Conclusions

The present study investigated the interlaminar fracture energy of unidirectional flax epoxy composites in Mode I, Mode II, and Mixed-mode I/II delamination failure. The composite laminates were manufactured via RTM method and were tailored to have a constant $V_f = 0.44$, and a 13 μm-thick PTFE was placed in the midplane to induce pre-crack, it was then located by X-ray in the laminate. All the DCB, ENF, and MMB tests were conducted and validated according to corresponding ASTM standards to evaluate the fracture energies in Mode I, Mode I, and Mixed-mode I/II, respectively. For Mode I, $G_{IC} = 574$ and $G_{IC} = 903$ J/m^2 were obtained respectively for crack initiation and propagation values. The obtained value based on the standard method for Mode II delamination is $G_{IIC} = 378$ J/m^2, which in contrast to the results of previous authors, calculated via the SBT data reduction method, is lower than G_{IC}. However, when following the SBT method, $G_{IIC} = 612$ J/m^2 is obtained, which is consistent with their results. MMB tests resulted in $G_{(I/II)C} = 414.4$ J/m^2 for Mixed-mode I/II fracture energy. Despite the limited, and in some cases lack of data available in the literature for comparison, the results are compared to the existing similar data and confirmed. In view of the fact that there is limited or no interlaminar fracture energy for these composites, the obtained values are valuable material properties to be further used in the design engineering field and numerical simulation methods.

Author Contributions: Conceptualization, Y.S., J.-F.C., G.L. and Y.B.; methodology, Y.S., J.-F.C., G.L.,Y.B., P.B. and N.V.; experiments, Y.S and N.V. resources, G.L., J. F.C., Y.B., P.B.; data curation, Y.S, N.V,; writing—original draft preparation, Y.S.; writing—review and editing, J.-F.C., G.L., Y.B., P.B., N.V.; supervision, J.-F.C., G.L., Y.B.; project administration, Y.S.; funding acquisition, J.-F.C., G.L., Y.B. All authors have read and agreed to the published version of the manuscript.

Funding: This research was funded by the Natural Sciences and Engineering Research Council of Canada (NSERC), grant number RGPIN-2017-04305.

Acknowledgments: We sincerely thank Ahmed Maslouhi and Charly Batigne (Université de Sherbrooke), who provided access to their testing equipment as well as Bradley View from SAINT-GOBAIN PERFORMANCE CORPORATION, for supplying the PTFE films, that greatly assisted us in this research.

Conflicts of Interest: The authors declare no conflict of interest. The funders had no role in the design of the study; in the collection, analyses, or interpretation of data; in the writing of the manuscript, or in the decision to publish the results.

References

1. Sanjay, M.R.; Siengchin, S.; Parameswaranpillai, J.; Jawaid, M.; Pruncu, C.I.; Khan, A. A comprehensive review of techniques for natural fibers as reinforcement in composites: Preparation, processing and characterization. *Carbohydr. Polym.* **2019**, *207*, 108–121. [CrossRef]
2. Pil, L.; Bensadoun, F.; Pariset, J.; Verpoest, I. Why are designers fascinated by flax and hemp fibre composites? *Compos. Part A Appl. Sci. Manuf.* **2016**, *83*, 193–205. [CrossRef]
3. Gurunathan, T.; Mohanty, S.; Nayak, S.K. A review of the recent developments in biocomposites based on natural fibres and their application perspectives. *Compos. Part A Appl. Sci. Manuf.* **2015**, *77*, 1–25. [CrossRef]
4. Mohanty, A.; Misra, M.; Hinrichsen, G. Biofibres, biodegradable polymers and biocomposites: An overview. *Macromol. Mater. Eng.* **2000**, *276*, 1–24. [CrossRef]
5. Ramesh, M.; Palanikumar, K.; Reddy, K. Plant fibre based bio-composites: Sustainable and renewable green materials. *Renew. Sustain. Energy Rev.* **2017**, *79*, 558–584. [CrossRef]
6. Lotfi, A.; Li, H.; Dao, D.V.; Prusty, G. Natural fiber–reinforced composites: A review on material, manufacturing, and machinability. *J. Thermoplast. Compos. Mater.* **2019**, 1–47. [CrossRef]
7. Netravali, A.N.; Chabba, S. Composites get greener. *Mater. Today* **2003**, *6*, 22–29. [CrossRef]
8. Rajak, D.K.; Pagar, D.D.; Menezes, P.L.; Linul, E. Fiber-Reinforced Polymer composites: Manufacturing, properties, and applications. *Polymers* **2019**, *11*, 1667. [CrossRef] [PubMed]
9. Shekar, H.S.S.; Ramachandra, M. Green Composites: A Review. *Mater. Today Proc.* **2018**, *5*, 2518–2526. [CrossRef]

10. Pickering, K.L.; Efendy, M.G.A.; Le, T.M. A review of recent developments in natural fibre composites and their mechanical performance. *Compos. Part A Appl. Sci. Manuf.* **2016**, *83*, 98–112. [CrossRef]

11. Chandrasekar, M.; Ishak, M.; Sapuan, S.; Leman, Z.; Jawaid, M. A review on the characterisation of natural fibres and their composites after alkali treatment and water absorption. *Plast. Rubber Compos.* **2017**, *46*, 119–136. [CrossRef]

12. Chandrasekar, M.; Shahroze, R.M.; Ishak, M.R.; Saba, N.; Jawaid, M.; Senthilkumar, K.; Kumar, T.S.M.; Siengchin, S. Flax and sugar palm reinforced epoxy composites: Effect of hybridization on physical, mechanical, morphological and dynamic mechanical properties. *Mater. Res. Express* **2019**, *6*, 105331. [CrossRef]

13. Dicker, M.P.M.; Duckworth, P.F.; Baker, A.B.; Francois, G.; Hazzard, M.K.; Weaver, P.M. Green composites: A review of material attributes and complementary applications. *Compos. Part A Appl. Sci. Manuf.* **2014**, *56*, 280–289. [CrossRef]

14. Bodros, E.; Pillin, I.; Montrelay, N.; Baley, C. Could biopolymers reinforced by randomly scattered flax fibre be used in structural applications? *Compos. Sci. Technol.* **2007**, *67*, 462–470. [CrossRef]

15. Joshi, S.V.; Drzal, L.; Mohanty, A.; Arora, S. Are natural fiber composites environmentally superior to glass fiber reinforced composites? *Compos. Part A Appl. Sci. Manuf.* **2004**, *35*, 371–376. [CrossRef]

16. Cristaldi, G.; Latteri, A.; Recca, G.; Cicala, G. Composites based on natural fibre fabrics. *Woven Fabr. Eng.* **2010**, *17*, 317–342.

17. Yan, L.; Wang, B.; Kasal, B. Can plant-based natural flax replace basalt and e-glass for fiber-reinforced polymer tubular energy absorbers? A comparative study on quasi-static axial crushing. *Front. Mater.* **2017**, *4*, 42. [CrossRef]

18. Yashas Gowda, T.G.; Sanjay, M.R.; Subrahmanya Bhat, K.; Madhu, P.; Senthamaraikannan, P.; Yogesha, B. Polymer matrix-natural fiber composites: An overview. *Cogent Eng.* **2018**, *5*, 1446667. [CrossRef]

19. Fuqua, M.; Huo, S.; Ulven, C. Natural Fiber Reinforced Composites. *Polym. Rev.* **2012**, *52*, 259–320. [CrossRef]

20. Shah, D.U. Developing plant fibre composites for structural applications by optimising composite parameters: A critical review. *J. Mater. Sci.* **2013**, *48*, 6083–6107. [CrossRef]

21. Habibi, M.; Laperrière, L.; Lebrun, G.; Toubal, L. Combining short flax fiber mats and unidirectional flax yarns for composite applications: Effect of short flax fibers on biaxial mechanical properties and damage behaviour. *Compos. Part B Eng.* **2017**, *123*, 165–178. [CrossRef]

22. John, M.J.; Thomas, S. Biofibres and biocomposites. *Carbohydr. Polym.* **2008**, *71*, 343–364. [CrossRef]

23. Ahmad, F.; Choi, H.S.; Park, M.K. A review: Natural fiber composites selection in view of mechanical, light weight, and economic properties. *Macromol. Mater. Eng.* **2015**, *300*, 10–24. [CrossRef]

24. Ramamoorthy, S.K.; Skrifvars, M.; Persson, A. A review of natural fibers used in biocomposites: Plant, animal and regenerated cellulose fibers. *Polym. Rev.* **2015**, *55*, 107–162. [CrossRef]

25. Ramesh, M. Flax (Linum usitatissimum L.) fibre reinforced polymer composite materials: A review on preparation, properties and prospects. *Prog. Mater. Sci.* **2019**, *102*, 109–166. [CrossRef]

26. Shah, D.U.; Schubel, P.J.; Clifford, M.J. Can flax replace E-glass in structural composites? A small wind turbine blade case study. *Compos. Part B Eng.* **2013**, *52*, 172–181. [CrossRef]

27. Moudood, A.; Rahman, A.; Öchsner, A.; Islam, M.; Francucci, G. Flax fiber and its composites: An overview of water and moisture absorption impact on their performance. *J. Reinf. Plast. Compos.* **2018**, *38*, 323–339. [CrossRef]

28. Moudood, A.; Rahman, A.; Khanlou, H.M.; Hall, W.; Öchsner, A.; Francucci, G. Environmental effects on the durability and the mechanical performance of flax fiber/bio-epoxy composites. *Compos. Part B Eng.* **2019**, *171*, 284–293. [CrossRef]

29. Yan, L.; Chouw, N.; Jayaraman, K. Flax fibre and its composites–A review. *Compos. Part B Eng.* **2014**, *56*, 296–317. [CrossRef]

30. Mohammad Khanlou, H.; Hall, W.; Woodfield, P.; Summerscales, J.; Francucci, G. The mechanical properties of flax fibre reinforced poly(lactic acid) bio-composites exposed to wet, freezing and humid environments. *J. Compos. Mater.* **2018**, *52*, 835–850. [CrossRef]

31. Khanlou, H.M.; Woodfield, P.; Summerscales, J.; Francucci, G.; King, B.; Talebian, S.; Foroughi, J.; Hall, W. Estimation of mechanical property degradation of poly(lactic acid) and flax fibre reinforced poly(lactic acid) bio-composites during thermal processing. *Measurement* **2018**, *116*, 367–372. [CrossRef]

32. Khanlou, H.M.; Woodfield, P.; Summerscales, J.; Hall, W. Consolidation process boundaries of the degradation of mechanical properties in compression moulding of natural-fibre bio-polymer composites. *Polym. Degrad. Stab.* **2017**, *138*, 115–125. [CrossRef]

33. Moudood, A.; Hall, W.; Öchsner, A.; Li, H.; Rahman, A.; Francucci, G. Effect of moisture in flax fibres on the quality of their composites. *J. Nat. Fibers* **2019**, *16*, 209–224. [CrossRef]

34. Goudenhooft, C.; Bourmaud, A.; Baley, C. Flax (Linum usitatissimum L.) fibers for composite reinforcement: Exploring the link between plant growth, cell walls development, and fiber properties. *Front. Plant Sci.* **2019**, *10*, 411. [CrossRef] [PubMed]

35. Habibi, M.; Lebrun, G.; Laperrière, L. Experimental characterization of short flax fiber mat composites: Tensile and flexural properties and damage analysis using acoustic emission. *J. Mater. Sci.* **2017**, *52*, 6567–6580. [CrossRef]

36. Habibi, M.; Laperrière, L.; Lebrun, G.; Chabot, B. Experimental investigation of the effect of short flax fibers on the permeability behavior of a new unidirectional flax/paper composite. *Fibers* **2016**, *4*, 22. [CrossRef]

37. Goutianos, S.; Peijs, T. The Optimisation of Flax Fibre Yarns for the Development of High-Performance Natural Fibre Composites. *Adv. Compos. Lett.* **2003**, *12*, 096369350301200602. [CrossRef]

38. Goutianos, S.; Peijs, T.; Nystrom, B.; Skrifvars, M. Development of flax fibre based textile reinforcements for composite applications. *Appl. Compos. Mater.* **2006**, *13*, 199–215. [CrossRef]

39. Zhu, J.; Zhu, H.; Njuguna, J.; Abhyankar, H. Recent development of flax fibres and their reinforced composites based on different polymeric matrices. *Materials* **2013**, *6*, 5171–5198. [CrossRef] [PubMed]

40. Yan, L.; Chouw, N.; Yuan, X. Improving the mechanical properties of natural fibre fabric reinforced epoxy composites by alkali treatment. *J. Reinf. Plast. Compos.* **2012**, *31*, 425–437. [CrossRef]

41. Lefeuvre, A.; Bourmaud, A.; Baley, C. Optimization of the mechanical performance of UD flax/epoxy composites by selection of fibres along the stem. *Compos. Part A Appl. Sci. Manuf.* **2015**, *77*, 204–208. [CrossRef]

42. El Sawi, I.; Bougherara, H.; Zitoune, R.; Fawaz, Z. Influence of the Manufacturing Process on the Mechanical Properties of Flax/Epoxy Composites. *J. Biobased Mater. Bioenergy* **2014**, *8*, 69–76. [CrossRef]

43. Tay, T. Characterization and analysis of delamination fracture in composites: An overview of developments from 1990 to 2001. *Appl. Mech. Rev. ASME* **2003**, *56*, 1–32. [CrossRef]

44. Nasuha, N.; Azmi, A.; Lih, T. A review on mode-I interlaminar fracture toughness of fibre reinforced composites. *J. Phys. Conf. Ser.* **2017**, *908*, 012024. [CrossRef]

45. Chen, C.; Li, Y.; Yu, T. Interlaminar toughening in flax fiber-reinforced composites interleaved with carbon nanotube buckypaper. *J. Reinf. Plast. Compos.* **2014**, *33*, 1859–1868. [CrossRef]

46. Nassar, M.; Nassar, M.M.A.; Arunachalam, R.; Alzebdeh, K. Machinability of natural fiber reinforced composites: A review. *Int. J. Adv. Manuf. Technol.* **2017**, *88*, 2985–3004. [CrossRef]

47. Almansour, F.; Dhakal, H.; Zhang, Z.Y. Effect of water absorption on Mode I interlaminar fracture toughness of flax/basalt reinforced vinyl ester hybrid composites. *Compos. Struct.* **2017**, *168*, 813–825. [CrossRef]

48. Almansour, F.; Dhakal, H.; Zhang, Z.Y. Investigation into Mode II interlaminar fracture toughness characteristics of flax/basalt reinforced vinyl ester hybrid composites. *Compos. Sci. Technol.* **2018**, *154*, 117–127. [CrossRef]

49. Anderson, T.L. *Fracture Mechanics: Fundamentals and Applications*, 4th ed.; CRC Press: Boca Raton, FL, USA, 2017.

50. Zhu, Y. Characterization of Interlaminar Fracture Toughness of a Carbon/epoxy Composite Material. Master's Thesis, Pennsylvania State University, University Park, PA, USA, 2009.

51. Walker, C.A.; Jamasri. Mixed-mode stress intensity factors in finite, edge-cracked orthotropic plates. *J. Strain Anal. Eng. Des.* **1995**, *30*, 83–90. [CrossRef]

52. Prasad, M.S.; Venkatesha, C.; Jayaraju, T. Experimental methods of determining fracture toughness of fiber reinforced polymer composites under various loading conditions. *J. Miner. Mater. Charact. Eng.* **2011**, *10*, 1263. [CrossRef]

53. ASTM D5528. *Standard Test Method for Mode I Interlaminar Fracture Toughness of Unidirectional Fiber-Reinforced Polymer Matrix Composites*; ASTM International: West Conshohocken, PA, USA, 2013. [CrossRef]

54. ASTM D7905. *Standard Test Method for Determination of the Mode II Interlaminar Fracture Toughness of Unidirectional Fiber-Reinforced Polymer Matrix Composites*; ASTM International: West Conshohocken, PA, USA, 2019. [CrossRef]

55. ASTM D6671. *Standard Test Method for Mixed Mode I-Mode II Interlaminar Fracture Toughness of Unidirectional Fiber Reinforced Polymer Matrix Composites*; ASTM International: West Conshohocken, PA, USA, 2019. [CrossRef]

56. Ooi, C.; Tan, C.; Azmi, A. Experimental study towards inter-laminar fracture toughness of different fibre reinforced polymer composites. *J. Phys.: Conf. Ser.* **2019**, *1150*, 012029. [CrossRef]

57. Bensadoun, F.; Verpoest, I.; Van Vuure, A.W. Interlaminar fracture toughness of flax-epoxy composites. *J. Reinf. Plast. Compos.* **2017**, *36*, 121–136. [CrossRef]

58. Wong, S.; Shanks, R.A.; Hodzic, A. Mechanical behavior and fracture toughness of poly(L-lactic acid)-Natural fiber composites modified with hyperbranched polymers. *Macromol. Mater. Eng.* **2004**, *289*, 447–456. [CrossRef]

59. Zhang, Y.; Li, Y.; Ma, H.; Yu, T. Tensile and interfacial properties of unidirectional flax/glass fiber reinforced hybrid composites. *Compos. Sci. Technol.* **2013**, *88*, 172–177. [CrossRef]

60. Li, Y.; Mai, Y.-W.; Ye, L. Effects of fibre surface treatment on fracture-mechanical properties of sisal-fibre composites. *Compos. Interfaces* **2005**, *12*, 141–163. [CrossRef]

61. Li, Y.; Wang, D.; Ma, H. Improving interlaminar fracture toughness of flax fiber/epoxy composites with chopped flax yarn interleaving. *Sci. China Technol. Sci.* **2015**, *58*, 1745–1752. [CrossRef]

62. Ravandi, M.; Teo, W.; Tran, L.; Yong, M.; Tay, T. The effects of through-the-thickness stitching on the Mode I interlaminar fracture toughness of flax/epoxy composite laminates. *Mater. Des.* **2016**, *109*, 659–669. [CrossRef]

63. Ravandi, M.; Teo, W.; Tran, L.; Yong, M.; Tay, T. Mode I Interlaminar Fracture Toughness of Natural Fiber Stitched Flax/Epoxy Composite Laminates–Experimental and Numerical Analysis. In Proceedings of the American Society for Composites: Thirty-First Technical Conference, Williamsburg, VA, USA, 19–22 September 2016.

64. Ravandi, M.; Teo, W.S.; Yong, M.S.; Tay, T.E. Prediction of Mode I interlaminar fracture toughness of stitched flax fiber composites. *J. Mater. Sci.* **2018**, *53*, 4173–4188. [CrossRef]

65. Vo Hong, N.; Beckers, K.; Goderis, B.; Van Puyvelde, P.; Verpoest, I.; Willem Van Vuure, A. Fracture toughness of unidirectional flax fiber composites with rigid gliadin matrix. *J. Reinf. Plast. Compos.* **2018**, *37*, 1163–1174. [CrossRef]

66. Rajendran, T.S.; Johar, M.; Hassan, S.; Wong, K.J. Mode I and Mode II Delamination of Flax/Epoxy Composite Laminate. In *MATEC Web of Conferences (AAME 2018)*; EDP Sciences: Les Ulis, France, 2018; p. 01002. [CrossRef]

67. Rajendran, T.S.; Johar, M.; Low, K.O.; Abu Hassan, S.; Wong, K.J. Interlaminar fracture toughness of a plain weave flax/epoxy composite. *Plast. Rubber Compos.* **2019**, *48*, 74–81. [CrossRef]

68. Saidane, E.H.; Scida, D.; Pac, M.-J.; Ayad, R. Mode-I interlaminar fracture toughness of flax, glass and hybrid flax-glass fibre woven composites: Failure mechanism evaluation using acoustic emission analysis. *Polym. Test.* **2019**, *75*, 246–253. [CrossRef]

69. Davies, P.; Blackman, B.R.K.; Brunner, A.J. Standard test methods for delamination resistance of composite materials: Current status. *Appl. Compos. Mater.* **1998**, *5*, 345–364. [CrossRef]

70. Fathi, A.; Keller, J.-H.; Altstaedt, V. Full-field shear analyses of sandwich core materials using Digital Image Correlation (DIC). *Compos. Part B Eng.* **2015**, *70*, 156–166. [CrossRef]

71. Wang, H. Marker identification technique for deformation measurement. *Adv. Mech. Eng.* **2013**, *5*, 246318. [CrossRef]

72. Xu, D.; Cerbu, C.; Wang, H.; Rosca, I.C. Analysis of the hybrid composite materials reinforced with natural fibers considering digital image correlation (DIC) measurements. *Mech. Mater.* **2019**, *135*, 46–56. [CrossRef]

73. Tekieli, M.; De Santis, S.; de Felice, G.; Kwiecień, A.; Roscini, F. Application of Digital Image Correlation to composite reinforcements testing. *Compos. Struct.* **2017**, *160*, 670–688. [CrossRef]

74. Anzelotti, G.; Nicoletto, G.; Riva, E. Mesomechanic strain analysis of twill-weave composite lamina under unidirectional in-plane tension. *Compos. Part A Appl. Sci. Manuf.* **2008**, *39*, 1294–1301. [CrossRef]

75. Takano, N.; Zako, M.; Fujitsu, R.; Nishiyabu, K. Study on large deformation characteristics of knitted fabric reinforced thermoplastic composites at forming temperature by digital image-based strain measurement technique. *Compos. Sci. Technol.* **2004**, *64*, 2153–2163. [CrossRef]

76. Wang, H.; Hou, Z. Application of Genetic Algorithms in a surface deformation measurement technique. In Proceedings of the 2010 Sixth International Conference on Natural Computation, Yantai, China, 10–12 August 2010; Volume 2010, pp. 2291–2295. [CrossRef]

77. Saadati, Y.; Lebrun, G.; Chatelain, J.-F.; Beauchamp, Y. Experimental investigation of failure mechanisms and evaluation of physical/mechanical properties of unidirectional flax–epoxy composites. *J. Compos. Mater.* **2020**, 0021998320902243. [CrossRef]

78. Airoldi, A.; Dávila, C.G. Identification of material parameters for modelling delamination in the presence of fibre bridging. *Compos. Struct.* **2012**, *94*, 3240–3249. [CrossRef]

79. De Kalbermatten, T.; Jäggi, R.; Flüeler, P.; Kausch, H.H.; Davies, P. Microfocus radiography studies during mode I interlaminar fracture tests on composites. *J. Mater. Sci. Lett.* **1992**, *11*, 543–546. [CrossRef]

80. Robinson, P. The effects of starter film thickness, residual stresses and layup on GIc of a 0°/0° interface. *Adv. Compos. Lett.* **1996**, *5*, 159–163. [CrossRef]

81. Shokrieh, M.M.; Heidari-Rarani, M. Effect of stacking sequence on R-curve behavior of glass/epoxy DCB laminates with 0°//0° crack interface. *Mater. Sci. Eng. A* **2011**, *529*, 265–269. [CrossRef]

82. Pinto, M.; Chalivendra, V.B.; Kim, Y.K.; Lewis, A.F. Improving the strength and service life of jute/epoxy laminar composites for structural applications. *Compos. Struct.* **2016**, *156*, 333–337. [CrossRef]

83. Almansour, F.A.; Dhakal, H.N.; Zhang, Z.Y.; Ghasemnejad, H. Effect of hybridization on the mode II fracture toughness properties of flax/vinyl ester composites. *Polym. Compos.* **2015**, *38*, 1732–1740. [CrossRef]

84. Al-Khudairi, O.; Hadavinia, H.; Waggott, A.; Lewis, E.; Little, C. Characterising mode I/mode II fatigue delamination growth in unidirectional fibre reinforced polymer laminates. *Mater. Des.* **2015**, *66*, 93–102. [CrossRef]

85. Johar, M.; Israr, H.A.; Low, K.O.; Wong, K.J. Numerical simulation methodology for mode II delamination of quasi-isotropic quasi-homogeneous composite laminates. *J. Compos. Mater.* **2017**, *51*, 3955–3968. [CrossRef]

86. Wang, W.-X.; Nakata, M.; Takao, Y.; Matsubara, T. Experimental investigation on test methods for mode II interlaminar fracture testing of carbon fiber reinforced composites. *Compos. Part A Appl. Sci. Manuf.* **2009**, *40*, 1447–1455. [CrossRef]

87. Compston, P.; Jar, P.Y.B. Comparison of interlaminar fracture toughness in unidirectional and woven roving marine composites. *Appl. Compos. Mater.* **1998**, *5*, 189–206. [CrossRef]

88. Lens, L.N.; Bittencourt, E.; d'Avila, V.M.R. Constitutive models for cohesive zones in mixed-mode fracture of plain concrete. *Eng. Fract. Mech.* **2009**, *76*, 2281–2297. [CrossRef]

89. Dillard, D.A.; Singh, H.K.; Pohlit, D.J.; Starbuck, J.M. Observations of Decreased Fracture Toughness for Mixed Mode Fracture Testing of Adhesively Bonded Joints. *J. Adhes. Sci. Technol.* **2009**, *23*, 1515–1530. [CrossRef]

90. Benzeggagh, M.L.; Kenane, M. Measurement of mixed-mode delamination fracture toughness of unidirectional glass/epoxy composites with mixed-mode bending apparatus. *Compos. Sci. Technol.* **1996**, *56*, 439–449. [CrossRef]

Journal of
Composites Science

Article

Finite Element Modeling of the Fiber-Matrix Interface in Polymer Composites

Daljeet K. Singh [1,2], Amol Vaidya [3], Vinoy Thomas [2], Merlin Theodore [4], Surbhi Kore [5] and Uday Vaidya [5,6,7,]*

[1] NSF International, Engineering Labs, Plastics, Ann Arbor, MI 48105, USA; daljeet.kr.singh@gmail.com
[2] Department of Materials Science & Engnineering, University of Alabama at Birmingham, Birmingham, AL 35294, USA; vthomas@uab.edu
[3] Owens Corning, integrated Applications, Testing & Modeling (i-ATM), Granville, OH 43023, USA; amol.vaidya@owenscorning.com
[4] Energy and Transportation Science Division, Carbon Fiber Technology Facility, Oak Ridge National Laboratory, Oak Ridge, TN 37830, USA; theodorem@ornl.gov
[5] Mechanical, Aerospace & Biomedical Engineering, The University of Tennessee, Knoxville, TN 37996, USA; skore@vols.utk.edu
[6] Energy and Transportation Science Division, Manufacturing Demonstration Facility, Oak Ridge National Laboratory, Oak Ridge, TN 37830, USA
[7] Institute for Advanced Composites and Manufacturing Innovation, Knoxville, TN 37932, USA
*   Correspondence: uvaidya@utk.edu

Received: 13 April 2020; Accepted: 15 May 2020; Published: 20 May 2020

Abstract: Polymer composites are used in numerous industries due to their high specific strength and high specific stiffness. Composites have markedly different properties than both the reinforcement and the matrix. Of the several factors that govern the final properties of the composite, the interface is an important factor that influences the stress transfer between the fiber and matrix. The present study is an effort to characterize and model the fiber-matrix interface in polymer matrix composites. Finite element models were developed to study the interfacial behavior during pull-out of a single fiber in continuous fiber-reinforced polymer composites. A three-dimensional (3D) unit-cell cohesive damage model (CDM) for the fiber/matrix interface debonding was employed to investigate the effect of interface/sizing coverage on the fiber. Furthermore, a two-dimensional (2D) axisymmetric model was used to (a) analyze the sensitivity of interface stiffness, interface strength, friction coefficient, and fiber length via a parametric study; and (b) study the shear stress distribution across the fiber-interface-matrix zone. It was determined that the force required to debond a single fiber from the matrix is three times higher if there is adequate distribution of the sizing on the fiber. The parametric study indicated that cohesive strength was the most influential factor in debonding. Moreover, the stress distribution model showed the debonding mechanism of the interface. It was observed that the interface debonded first from the matrix and remained in contact with the fiber even when the fiber was completely pulled out.

Keywords: fiber matrix interface; finite element analysis

1. Introduction

The interface plays a key role in strength translation and failure of a composite material [1,2]. It is well known that tailoring the interface for weak bonding improves energy absorption, while a stronger interface results in higher load bearing and environmental resistance of a composite. The sizing on glass and carbon fibers plays an influential role in the resulting fiber-matrix interface [2]. Typically, silane sizing on glass makes the surface compatible for bonding to epoxy, vinyl ester, and other thermoset, as well as thermoplastic, resins. Carbon fiber being non-polar is made reactive by applying a range of epoxy or urethane-compatible sizing. The sizing helps with the handling of the fiber and enhances the fiber-matrix interface [2].

A range of mechanical tests provide insight into the fiber-matrix combinations. It is known that many of the measured mechanical properties of composites are governed by the quality of adhesion between the fiber and the matrix. These tests include, but are not limited to, tensile, compression, shear, and fatigue testing, or a combination thereof. Without suitable interfacial interaction, optimal load sharing between the fibers does not take place, resulting in a weaker material [1].

There have been several techniques used by researchers to measure the fiber-matrix adhesion. These methods can be broadly classified into three categories: Direct methods, indirect methods, and composite lamina methods. The direct methods include the fiber pull-out method, single-fiber fragmentation method, embedded fiber compression method, and micro-indentation method [2]. The indirect methods for fiber matrix adhesion include the variable curvature method, slice compression test, ball compression test, dynamic mechanical analysis, and voltage contrast x-ray spectroscopy. The composite lamina methods include the 90° transverse flexural and tensile tests, three- and four-point shear, ±45° and edge delamination tests, short-beam shear test method, and mode I and mode II fracture tests [2]. Typically, the experimental setup for the direct test methods is very complex. Moreover, data reduction and interpretation are challenging because of several factors. These include experimental data scatter, the inability to always discern changes in the slopes of the recorded load-displacement plots, and the machine compliance from the recorded displacement.

Pithekethly et al. [3] devised a round robin test program to evaluate different techniques to evaluate the interfacial shear strength of a fiber/matrix bond in composite materials. The selected tests were the single-fiber pull-out test, micro-bond test, fragmentation test, and micro-indentation test. Twelve laboratories participated in this program, but the results were inconclusive and the scatter between laboratories for a given test was high. The researchers proposed a further investigation to devise a protocol.

The traditional pull-out and fragmentation tests suffer from a difficulty in specimen preparation [4]. The micro-bond test developed by Gaur and Miller [5] is one of the widely used single-fiber-matrix interfacial bond test methods to determine interfacial shear strength IFSS [4]. However, a standard procedure for the micro-bond is yet to be established with various researchers using different techniques to minimize the data scatter. This suggests the complexity of this technique and the need to devise a technique to measure interfacial properties.

Analytical and numerical models are preferred due to their time and cost-saving potential. Theoretical models have been very popular to understand the load-displacement behavior when a single fiber is pulled out from the matrix [6–8]. Stang and Shah [7] developed a closed-form solution to calculate the ultimate fiber tensile strength when fiber-matrix debonding occurred. Interfacial friction as a measure of debonding behavior was predicted using a simple shear lag model by Gao et al. [8]. They modeled the force-displacement behavior using interface toughness and friction as parameters. Residual clamping stresses and Poisson's contraction for the fiber were taken in consideration to analyze the stresses required to debond the interface by Hsueh [6].

Finite element models have also gained popularity due to technical advancements in commercial packages. Sun and Lin [9] analyzed the interfacial properties through parametric studies in which they varied the stiffness of the fiber and matrix coupled with irregular fiber cross-sections. The shear stress distribution across the interface and its effect on debonding was studied by Wei et al. [10]. Cohesive zone modeling (CZM) has emerged as a promising tool in formulating simulation models to study the interface. Dugdale et al. [11,12] were pioneers in implementing CZM, which relies on crack initiation and its propagation. Chandra [13] investigated interfacial fracture toughness and presented a detailed discussion on the cohesive damage model and its reliance on traction separation laws to simulate crack initiation.

This paper focuses on finite element modeling using Abaqus to study the interfacial behavior during pull-out of a single fiber in a continuous fiber-reinforced polymer composite. A 2D axisymmetric model was used to study the shear stress distribution across the fiber-interface-matrix zone. Studies on the sensitivity of interface stiffness, interface strength, friction coefficient, and fiber length were also conducted. A 3D unit cell cohesive damage model (CDM) was used to investigate the fiber/matrix interface debonding.

2. Cohesive Zone Modeling

The failure of the interface is conventionally studied using a linear elastic fracture mechanics approach. In this approach, the local crack tip field is characterized using parameters such as stress intensity factors (K_I, K_{II}, and K_{III}) and strain energy release rates (G_I, G_{II}, G_{III}). These parameters determine the initiation of the crack growth. However, traditional fracture mechanics approaches assume the existence of a sharp crack with stress levels locally approaching infinity. The crack tip is called the 'singular crack tip.' Moreover, the crack tip does not exist in the fiber matrix interface; hence, CZM is an alternative to traditional fracture mechanics approaches and this method is used for finite element analyses.

A bilinear cohesive law is implemented in this work (see Equation (1)), which reduces the artificial compliance inherent in CZM. Figure 1 illustrates the various stages of cohesive zone damage. τ_{is} is the average interfacial shear stress and δ is the relative tangential displacement. The traction across the interface increases and reaches a peak value, and then decreases and eventually vanishes, resulting in a complete decohesion, given by Equation (1).

$$\tau_{is} = K\delta \quad 0 \leq \delta \leq \delta_s \tag{1}$$

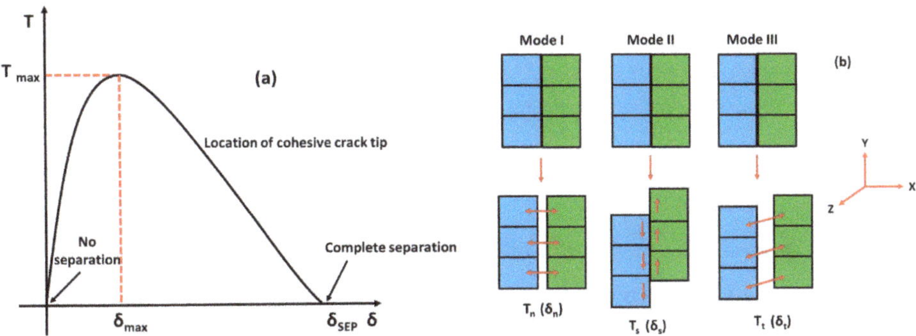

Figure 1. Cohesive zone damage; (**a**) typical traction separation behavior when a fiber is pulled out from the matrix [14], (**b**) traction separations for different fracture modes. Mode II fracture mode has been used in this analysis [15].

Commercial finite element code (Abaqus Version 6.13) was used to model the cohesive zone in the fiber/interface debonding and/or pull-out. The relation between traction stress and separation is given by Equation (2),

$$\left\{ \begin{array}{c} T_n \\ T_s \\ T_t \end{array} \right\} = \left[\begin{array}{ccc} K_{nn} & K_{ns} & K_{nt} \\ sym & K_{ss} & K_{st} \\ sym & sym & K_{tt} \end{array} \right] \left\{ \begin{array}{c} \delta_n \\ \delta_s \\ \delta_t \end{array} \right\} = K\delta \tag{2}$$

where T_n is the traction stress in the normal direction, T_s, T_t are traction stresses in the first shear and the second shear directions, respectively, K is the normal stiffness matrix, and δ_n, δ_s, and δ_t are separations in the normal, first, and second shear directions, respectively. The elastic stiffness and the cohesive strength would be obtained from experiments. The maximum stress criteria were used to predict damage initiation, given by Equation (3).

$$max\left\{\frac{T_n}{T_n^p}, \frac{T_s}{T_s^p}, \frac{T_t}{T_t^p}\right\} = 1 \tag{3}$$

T_n^p, T_s^p, T_t^p signify the peak values for traction stresses in the respective directions. Damage evolution law describes the rate at which the cohesive stiffness is degraded once the corresponding initiation criteria is reached. A scalar damage variable, D, represents the overall damage at the contact point, which is represented below. The value of D ranges from 0 to 1. Refer to Equation (4).

$$T_s = (1 - D)T_s \tag{4}$$

CZM was used to predict the initiation and evolution of damage at the interface of the unit cell comprising the fiber and matrix. For a composite unit cell that consists of multiple material systems, the number of potential failure mechanisms that must be accounted for exponentially increases the complexity of the analysis. Failure mechanisms include failure at the interface, fiber breakage, matrix cracking, and their interaction.

Different modeling approaches within CZM were employed, such as elements with 'zero thickness' and 'finite thickness'—(a) the interface was modeled as 'zero thickness' and adhesive properties were used to study the interface failure mechanism; and (b) the interface was modeled with a very small thickness for the 'finite thickness model,' respectively. Parametric studies were also conducted to study the most influential factors affecting the strength of the fiber matrix unit cell.

3. Finite Element Model Setup

A 3D finite element model of the fiber pull-out specimen was generated using the CZM approach. The finite element model of the unit cell and the boundary conditions are shown in Figure 2. The radius of the fiber was 7.5 µm and the fiber was encased in a square matrix that was 18.8 µm. The dimension of the matrix was based on a fiber volume fraction of 60%.

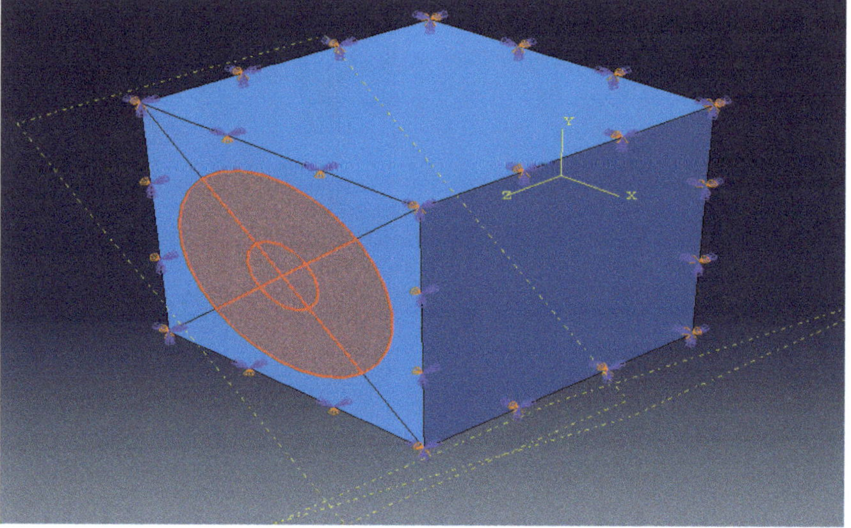

Figure 2. 3-D unit cell model where the purple circle indicates the fiber and the square represents the matrix within which the fiber is enclosed. The boundary conditions are also shown here.

Both the fiber and the matrix were modeled using 3D first-order (linear) hexahedron elements with incompatible modes (C3D8I) in Abaqus, which is an improved version of the C3D8 element. The boundary conditions that are constraints are also shown in Figure 2. More details about this type of element are available from [14].

The interface was modeled with zero thickness. In this model, the Young's modulus value for the interface was assumed to be equal to the matrix properties. However, the interfacial shear strength or the strength of the interface was taken from [16], which is 25 MPa. This value was measured by a single-fiber fragmentation test wherein a single fiber was embedded in an epoxy matrix by Kumar [16] who studied the effect of sizing on interfacial strength properties.

The material and input parameters are summarized in Table 1. The contact behavior of the fiber/matrix interface was modeled as discussed in Section 2 using surface-based cohesive behavior, which is similar to the cohesive element approach. This is a preferred approach when the interface or the adhesive layer is very thin [14]. A displacement-controlled load of 0.1 mm was applied at the free end of the fiber. The reason for imposing displacement on the fiber was that it results in a more gradual failure process than a similar loading using applied forces [17]. A similar approach was used by Bhemareddy et al. in Ref. [18] in their finite element model for the debonding of a silicon carbide fiber (SiC_f) embedded in a silicon carbide matrix (SiC).

Table 1. Material properties used in the 3D finite element model.

Material	Modulus (GPa)	Tensile Strength (MPa)	Coefficient of Friction (Static and Dynamic)	Poisson's Ratio
E-Glass Fiber	72	-	-	0.2
Epoxy Matrix	4.2	-	-	0.34
Interface (Cohesive Zone)	4.2	25	1 and 0.9	-

4. Results and Discussion

4.1. Effect of Continuous and Discontinuous Bonding between the Fiber and Matrix

Two cases were studied with this model to understand the effect of sizing coverage on the fiber. In one case, the complete surface of the fiber is bonded to the matrix while, in the second case, only half the surface of the fiber is bonded to the matrix. This simulates the situation where only half of the fiber has been coated with sizing while the other half is uncoated. This is very close to the real-life situation wherein a typical sizing applicator operates on at least two bundles of fibers. Each bundle of fibers travels from a bushing above this applicator down to a winder below [19]. Due to the nature of this process, uneven sizing is deposited on the fiber surface.

The stresses on the matrix region for discontinuous coating (shown in Figure 3b) are one magnitude lower than those for continuous coating (Figure 3a). This can be attributed to the fact that less force is required in the case of discontinuous coating. The force required to pull-out the fiber from the matrix for the continuous coating model was 0.06 N, while it was only 0.015 N for discontinuous coating, as shown in Figure 3c.

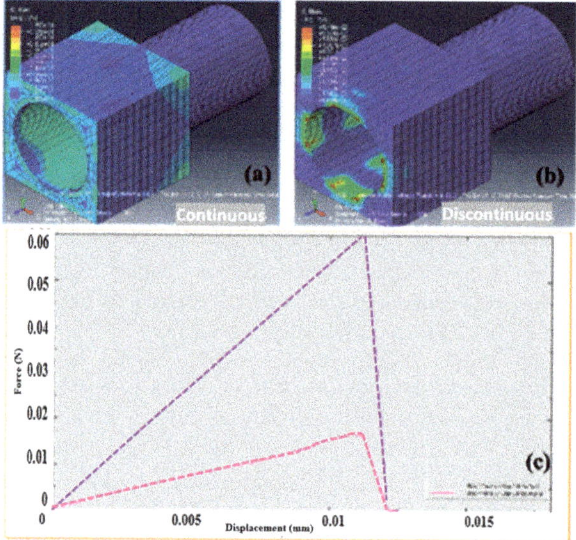

Figure 3. (**a**) Stress plot for continuous interface coating on the fiber, (**b**) stress plot for discontinuous interface coating on the fiber, and (**c**) force-displacement plot for the simulated fiber pull-out where force is measured in (N) and displacement in (mm).

The 3D finite element model simulates the situation when only half of the fiber is bonded to the matrix, and it has the potential of near-accurately predicting the load-displacement behavior when a single fiber is debonded from an encased matrix.

4.2. Parametric Study to Understand Influential Factors in Fiber Matrix Adhesion

A parametric study was undertaken on a 2D axisymmetric model to understand the influential factors affecting the fiber-matrix adhesion. Load-displacement behavior predicted by changes in these factors were recorded and compared to each other. The radius of the fiber was 7.5 µm and that of the matrix was 1.5 mm. Both the fiber and the matrix were modeled using four-node bilinear axisymmetric quadrilateral elements with reduced integration (CAX4R). CZM was used for this model. A displacement-controlled load of 0.1 mm is applied on the free end of the fiber. The factors studied are: (a) Coefficient of friction (static and dynamic), (b) cohesive stiffness of the interface, (c) cohesive strength of the interface, (d) fiber embedded length.

4.3. Effect of Coefficient of Friction

The parameter-coefficient of friction primarily comes into play only after complete debonding has taken place. Figure 4 shows the load-displacement plots for the finite element models with varying coefficients of friction. Four different sets of static and dynamic coefficients of friction were used in the parametric study. They were: (0.1 and 0.05), (0.4 and 0.3), (0.8 and 0.4), and (1 and 0.9).

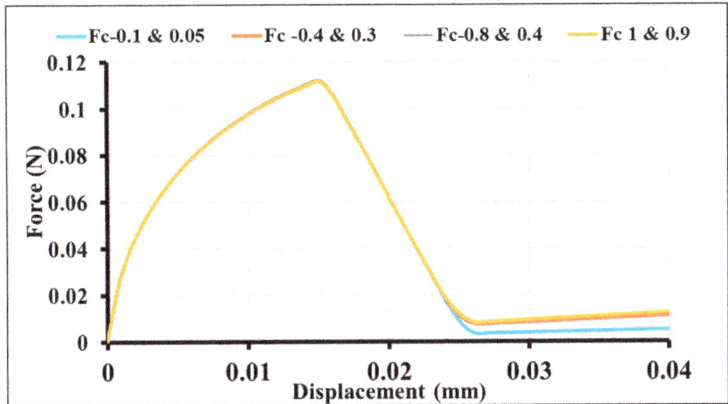

Figure 4. Force (or load)-displacement curve for varying coefficients of friction.

Figure 4 is a magnified version and, hence, the non-linearity of the force-displacement curve is exaggerated. However, the nature of the curve is non-linear to an extent and is not representative of an actual test. This could be due to the effects of coefficients of thermal expansion and residual stresses on the fiber, which have not been accounted for in this model. Nevertheless, finite element models generated by various researchers [4,20,21] have the same trends associated with the force-displacement curve.

4.4. Effect of Cohesive Stiffness of the Interface

The elastic modulus/stiffness of the interface was provided as an input property in the cohesive behavior for the interface in Abaqus. The basic assumption here was that the interface would behave similar to that of the matrix; hence, properties of the epoxy matrix were considered as the baseline. The interface stiffness was varied from 10% of the matrix stiffness to 1000% of the matrix stiffness in discrete values as 10%, 50%, 100%, and 1000% respectively.

As reported in Table 2, the elastic modulus for the epoxy matrix and the interface was taken to be 4200 MPa. Figure 5 shows the load-displacement behavior for varying moduli of the interface. It was noticed that the peak force required for debonding does not change even when the modulus of the interface is as low as 420 MPa. In addition, the higher the stiffness of the interface, the lower the displacement (complete separation). Furthermore, it was seen that for a very low interface modulus (420 MPa), the evolution of crack length was much higher when compared to other cases. This is along expected lines as the interface is too weak.

Table 2. Baseline material properties used in 2D axisymmetric model.

Material	Modulus (GPa)	Tensile Strength (MPa)	Coefficient of Friction (Static and Dynamic)	Poisson's Ratio
E-Glass Fiber	72	-	-	0.2
Epoxy Matrix	4.2	-	-	0.34
Interface (Cohesive Zone)	4.2	25	1 and 0.9	-

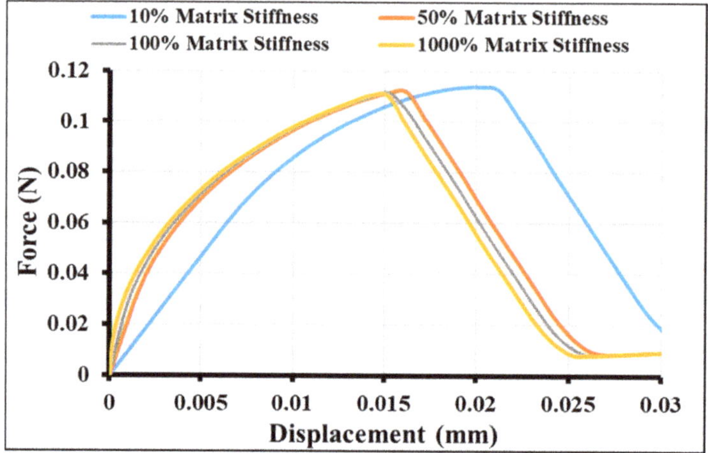

Figure 5. Force-displacement curve for varying interface stiffness.

4.5. Effect of Cohesive Strength of the Interface

The strength of the interface was varied starting from 1 to 10 MPa, keeping all the other variables at the baseline configurations. It can be clearly seen from Figure 6 that interfacial strength would directly affect the maximum load at which the interface fails. The peak load is almost proportional to the strength of the interface. It is also worth noting that at higher strength, ductile behavior of the interface is seen. In other words, the crack has initiated but, as the bond strength is too high, the crack evolution does not take place and debonding is delayed.

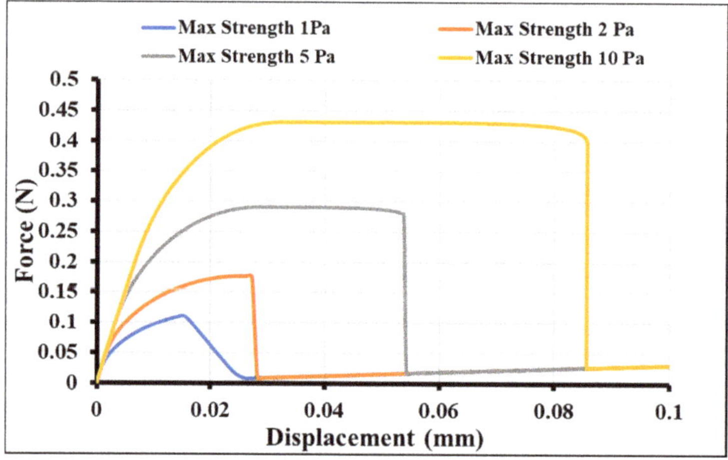

Figure 6. Force-displacement curve for varying cohesive strength.

4.6. Effect of Embedded Fiber Length

The effect of fiber-embedded length is more pronounced in terms of the maximum separation achieved. Three different fiber lengths (3, 6, and 9 mm) were used in the parametric study, and the respective load-displacement curves are illustrated in Figure 7. As the fiber length increased, it was observed that the fiber-matrix response became more compliant and delayed debonding was observed.

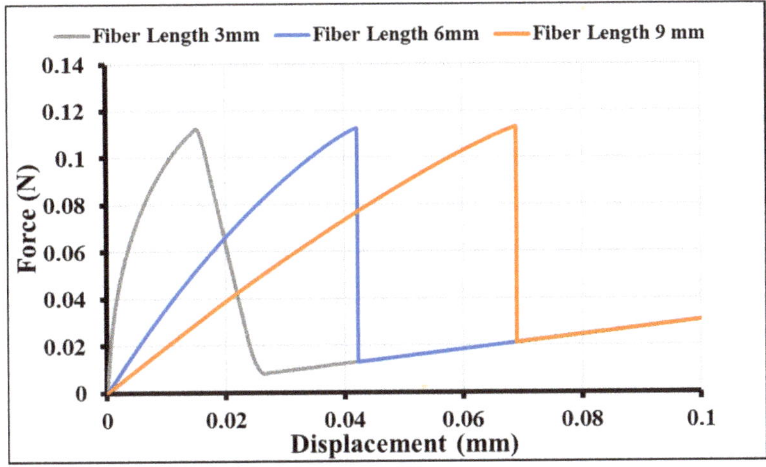

Figure 7. Force-displacement curve for varying fiber length.

A similar observation was observed by Sockalingam et al. [22] when they developed a finite element model of the micro-droplet test method. The micro-droplet test is one of the best techniques to study the interfacial properties between a fiber and matrix. This can be attributed to the fact that the fiber bears the load when it is pulled out of the matrix.

4.7. Stress Distribution at Debonding between the Fiber, Interface, and the Matrix

Shear stress distribution across the interface during the debonding stage is an important indicator of the interface and adhesion within the fiber/matrix. To investigate this effect, a 2-D axisymmetric model was created where the interface was given a thickness of 0.001 mm, the fiber diameter was 0.007 mm, and the matrix was 1.5 mm. This is shown in Figure 8. Furthermore, the interface was divided into three sections and each had its own isotropic material property assigned. The three sections were: (a) Interface close to the fiber, which was assigned fiber property; (b) interface in the middle, which was assigned the average constituent fiber and resin property; and (c) interface closer to the resin, which was matrix-dominated. The representation of the model is shown in Figure 8. Details of the input properties used in Abaqus are provided in Table 3. The assumption for the interface was that it would behave like a glass fiber when near the fiber and similar to the resin when in contact with the resin. CZM was employed here as well when considering the bond between the interface, fiber, and matrix.

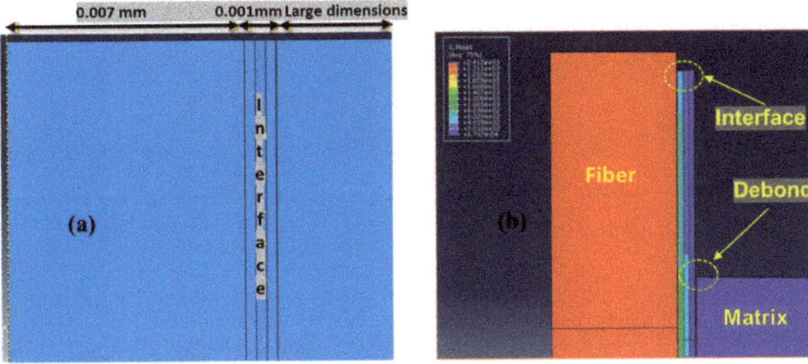

Figure 8. Shear stress distribution across the constituents—(**a**) dimension of the unit cell model, (**b**) debond stage. The stress contour plot is also seen here.

Table 3. Input properties for three-phase model where interface is modeled as a separate entity.

Material	Modulus (GPa)	Tensile Strength (MPa)	Poisson's Ratio
E-CR-glass	81	-	0.2
Epoxy Matrix	4.2	-	0.3
Interface (Fiber-Dominated)	4.2	-	-
Interface (Fiber/resin Average)	42.6	-	0.3
Interface (Resin-Dominated)	4.2	-	0.3
Fiber-Interface (Cohesive Zone)	50	40	-
Matrix-Interface (Cohesive Zone)	3.5	10	-

A pressure load was applied on the free end of the fiber, and the shear stress distribution across the fiber, interface, and the matrix was recorded. The boundary condition was applied to mimic a fiber pull-out where the bottom part of the matrix block is fixed and the fiber is pulled from the top end. Symmetry about the axis is also considered as it is an axisymmetric model. As higher strength was provided at the fiber-interface zone (Modulus 50 GPa), it was observed that debonding does not take place in this zone. However, the debonding occurs in the interface-matrix region (3.5 GPa). Figure 8 shows the shear stress distribution across the model. This was based on an assumption that the interface takes the property of the fiber in this zone, and the interface fails mostly in the matrix region, if the interface itself is not the weakest link.

The 2D axisymmetric model covers the entire range of intricacies involved in adhesion of the fiber and the interface and the matrix. It was observed that the interface, which was modeled as a thin film between the fiber and the matrix, continued to remain bonded with the fiber. Figure 9 shows a stress contour plot of all three sections of the model. As discussed above, the stress is mostly borne by the fiber and then then reduces gradually. The stresses calculated were 1980, 1870, 785, 39, and 11 MPa for the fiber, fiber-dominated interface, average interface, resin-dominated interface, and the matrix, respectively.

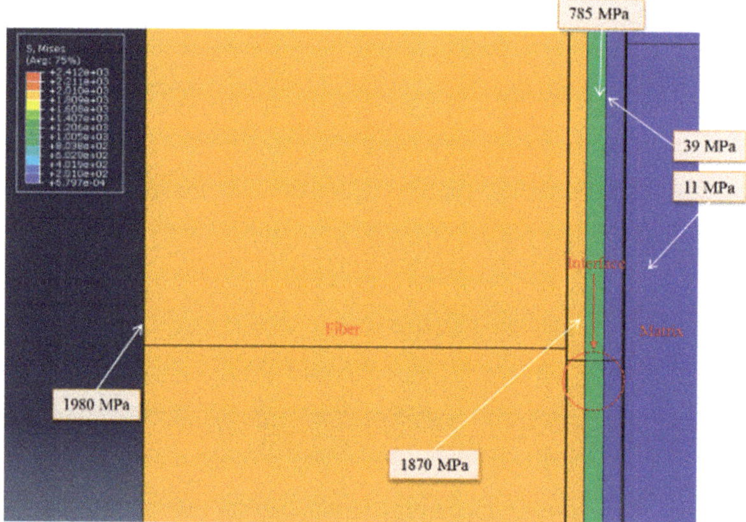

Figure 9. Stress values calculated during the simulation for each section (fiber, interface, and the matrix).

The CONTACT STATUS (CSTATUS) feature of Abaqus was used throughout the analysis between the bonded surfaces. It is divided into three parts—'stick, slip, and open.' The CSTATUS provides an indication (a) when the contact is closed and is intact; (b) when it has begun degrading; and (c) when it is completely open. Figure 10 illustrates the contact status at various stages of the analysis. At the initial stage, the contact is entirely intact between both the surfaces; in the middle stage of the analysis, progressive debonding takes place between the interface-matrix zones.

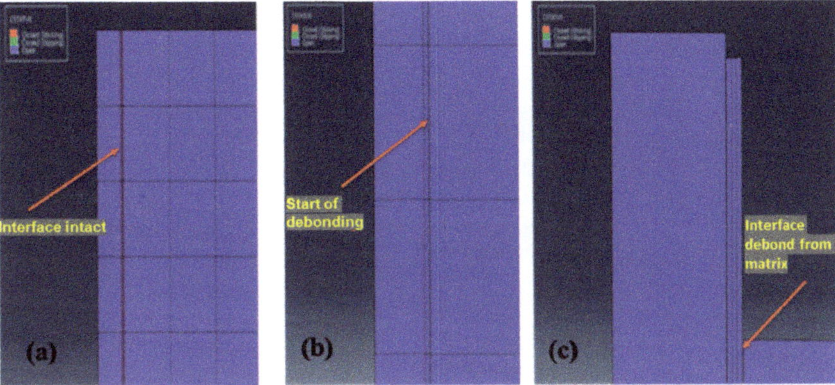

Figure 10. Finite element modeling of fiber pull-out and debonding: (**a**) Represents initial stage of pull-out where the red part indicates the interface is intact, (**b**) represents the middle stage of simulation where the absence of red spots indicates debonding, and (**c**) represents the debond between the matrix and interface.

5. Discussion

With the application of CZM, the finite element models demonstrate a single-fiber pull-out process very well. Even though the finite element models developed herein were for glass fibers, the same methodology can also be employed for carbon fibers. The parameters that would change are—radius of the fiber, dimension of the matrix block, friction coefficient, and interfacial crack initiation shear stress. More details on modeling parameters can be found in [23]. In Ref. [23], the authors have followed the CZM approach and simulated a single-carbon-fiber pull-out using the commercial finite element package Abaqus.

The finite element models in the current study have shown good potential to predict the load-displacement behavior. Availability of more data sets from exhaustive tests would make the model more robust and could be used for further investigation of adhesion between the fiber, interface, and matrix. The finite element models need to be validated by comparing to experimental test results, and they can be improved further to predict the load-displacement interfacial behavior during fiber-pull out.

Having analyzed the fiber/matrix interface using both 3-D models and 2-D models, it is clear that both the approaches have their advantages with respect to each other; however, the 2-D axisymmetric model with its relatively simple and user-friendly approach coupled with lower computational time was preferred. As both the models (3D and 2D) follow the same principles of CZM, the fundamentals remain the same.

6. Conclusions

This study addressed the interfacial characteristics in fiber-reinforced composites. A 2D and 3D finite element model of the fiber pull-out specimen was generated using the CZM approach. The key findings from the modeling were: (a) The effect of sizing coverage was found to be pronounced. While the force required to pull-out the fiber from the matrix for the continuous coating model was 0.06 N, it was only 0.015 N for discontinuous coating, a 300% increase with uniform sizing coverage; (b) The static and dynamic coefficient of friction had a moderate effect on the force-displacement past debonding of the interface. Friction coefficients greater than 0.3 resulted in about a 66% higher interface force magnitude over lower values of friction coefficients; (c) for varying degrees of interface stiffness ranging from 10% to 1000% of the matrix stiffness (modulus), it was noticed that the peak force required for debonding does not change even for 10% of the matrix stiffness. In addition, the higher the stiffness of the interface, the lower the displacement (complete separation); (d) varying the interface strength from 1 to 10 MPa directly affected the maximum load at which the interface fails. The peak load is proportional to the strength of the interface; (e) in terms of embedded fiber length, as the fiber length increased, it was observed that the fiber-matrix interface becomes more compliant and delays debonding; and (f) for the high-fiber-interface zone (for example, modulus 50 GPa), it was observed that the debond does not take place in this zone, but the debond takes place in the interface-matrix region (for example, modulus 3.5 GPa). With larger data sets available from experimental results in the future, these models can be used to capture details that are otherwise difficult to study. Furthermore, the findings from this study can be used in the composites industry to characterize and evaluate different fiber surfaces.

Author Contributions: Conceptualization, U.V., A.V.; methodology, D.K.S., V.T.; software, D.K.S.; validation, A.V., S.K., M.T.; formal analysis, D.K.S., S.K.; investigation, D.K.S.; resources, A.V., M.T.; data curation, D.K.S., S.K.; writing—original draft preparation, D.K.S., U.V.; writing—review and editing, U.V., A.V., V.T.; visualization, D.K.S.; supervision, U.V., A.V.; project administration, U.V.; funding acquisition, U.V., A.V. All authors have read and agreed to the published version of the manuscript.

Funding: This research was funded by US Department of Energy (DOE) Graduate Automotive Technology Education (GATE) Owens Corning (OC) and Institute for Advanced Composites Manufacturing Innovation (IACMI)-The Composites Institute. The information, data, and work presented herein were funded in part by the US Department of Energy (DOE), Office of Energy Efficiency and Renewable Energy, under Award Number DE-EE0006926 and in part by US DOE GATE DE-FG26-05NT42620.

Conflicts of Interest: The authors declare no conflict of interest.

References

1. Mallick, P.K. *Fiber-Reinforced Composites: Materials, Manufacturing, and Design*; CRC Press: Boca Raton, FL, USA, 2010.
2. Drzal, L. TheInterphase in Epoxy Composites. In *Epoxy Resins and Composites*; Dušek, K., II, Ed.; Springer: Berlin/Heidelberg, Germany, 1986; pp. 1–32.
3. Pitkethly, M.; Favre, J.; Gaur, U.; Jakubowski, J.; Mudrich, S.F.; Caldwell, D.L.; Lawrence, T. A Round-Robin Programme on Interfacial Test Methods. *Compos. Sci. Technol.* **1993**, *48*, 205–214. [CrossRef]
4. Sockalingam, S.; Nilakantan, G. Fiber-Matrix Interface Characterization Through the Microbond Test. *Int. J. Aeronaut. Space Sci.* **2012**, *13*, 282–295. [CrossRef]
5. Gaur, U.; Miller, B. Microbond Method for Determination of the Shear Strength of a Fiber/Resin Interface: Evaluation of Experimental Parameters. *Compos. Sci. Technol.* **1989**, *34*, 35–51. [CrossRef]
6. Hsueh, C.-H. Interfacial Debonding and Fiber Pull-out Stresses of Fiber-Reinforced Composites. *Mater. Sci. Eng.* **1990**, *123*, 1–11. [CrossRef]
7. Stang, H.; Shah, S.P. Failure of Fibre-Reinforced Composites by Pull-out Fracture. *J. Mater. Sci.* **1986**, *21*, 953–957. [CrossRef]
8. Gao, Y.-C.; Mai, Y.-W.; Cotterell, B. Fracture of Fiber-Reinforced Materials. *Z. für angewandte Mathematik und Physik ZAMP* **1988**, *39*, 550–572. [CrossRef]
9. Sun, W.; Lin, F. Computer Modeling and FEA Simulation for Composite Single Fiber pull-out. *J. Thermoplast. Compos. Mater.* **2001**, *14*, 327–343. [CrossRef]
10. Wei, G.F.; Liu, G.Y.; Xu, C.H.; Sun, X.Q. Finite Element Simulation of Perfect Bonding for Single Fiber pull-out Test. *Adv. Mater. Res.* **2012**, *418*, 509–512. [CrossRef]
11. Dugdale, D.S. Yielding of Steel Sheets Containing Slits. *J. Mech. Phys. Solids* **1960**, *8*, 100–104. [CrossRef]
12. Barenblatt, G.I. The Mathematical Theory of Equilibrium of Crack in Brittle Fracture. *Adv. Appl. Mech.* **1962**, *7*, 55–129.
13. Chandra, N. Evaluation of Interfacial Fracture Toughness Using Cohesive Zone Model. *Compo. Part A Appl. Sci. Manuf.* **2002**, *33*, 1433–1447. [CrossRef]
14. Hibbit, K.; Sorensen, A. *Abaqus Standard Analysis User's Manual*; Hibbit, Karlsson, Sorensen Inc.: Providence, RI, USA, 2007.

15. Cornec, A.; Scheider, I.; Schwalbe, K.-H. On the Practical Application of the Cohesive model. *Eng. Fract. Mech.* **2003**, *70*, 1963–1987. [CrossRef]
16. Kumar, G.K. *Studying the Influence of Glass Fiber Sizing Roughness and Thickness with the Single Fiber Fragmentation Test*; Degree Project; Kristianstad University: Kristianstad, Sweden, 2006.
17. Firehole Technologies. *Helius: MCT Tutorial 1*; Firehole Technologies: Laramie, WY, USA, 2011.
18. Venkata Bheemreddy, K. Chandrashekhara, Lokeswarappa R. Dharani, Greg Eugene Hilmas, Modeling of Fiber pull-out in Continuous Fiber Reinforced Ceramic Composites Using Finite Element Method and Artificial Neural Networks. *Comput. Mater. Sci.* **2013**, *79*, 663–673. [CrossRef]
19. Mason, K. Sizing up Fiber Sizings. 2006. Available online: http://www.compositesworld.com/articles/sizing-up-fiber-sizings (accessed on 6 September 2012).
20. Gao, X.; Jensen, R.; McKnight, S.; Gillespie, J., Jr. Effect of Colloidal Silica on the Strength and Energy Absorption of Glass Fiber/Epoxy Interphases. *Compos. Part A Appl. Sci. Manuf.* **2011**, *42*, 1738–1747. [CrossRef]
21. Scheer, R.; Nairn, J. A Comparison of Several Fracture Mechanics Methods for Measuring Interfacial Toughness with Microbond Tests. *J. Adhes.* **1995**, *53*, 45–68. [CrossRef]
22. Sockalingam, S.; Dey, M.; Gillespie, J.W., Jr.; Keefe, M. Finite Element Analysis of the Microdroplet Test Method Using Cohesive Zone Model of the Fiber/Matrix Interface. *Compos. Part A Appl. Sci. Manuf.* **2014**, *56*, 239–247. [CrossRef]
23. Jia, Y.; Yan, W.; Liu, H.-Y. Numerical study on carbon fibre pullout using a cohesive zone model. In Proceedings of the 18th International Conference on Composite Materials, Jeju Island, Korea, 21–26 August 2011.

Journal of
Composites Science

Article

Freeze-Thaw Performance Characterization and Leachability of Potassium-Based Geopolymer Concrete

Peiman Azarsa and Rishi Gupta *

Civil Engineering Department, University of Victoria, Victoria, BC V8P-5C2, Canada; azarsap@uvic.ca
* Correspondence: guptar@uvic.ca; Tel.: +1-250-721-7033

Received: 28 March 2020; Accepted: 23 April 2020; Published: 27 April 2020

Abstract: It is well known that concrete is one of the most widely used construction materials in the world, and cement as its key constituent is partly responsible for global Carbon Dioxide (CO_2) emission. Due to these reasons, high strength concrete with lower CO_2 emission, and concrete with lower reliance on natural resources is increasingly popular. Geopolymer Concrete (GPC), due to its capability to minimize the consumption of natural resources, has attracted the attention of researchers. In cold regions, frost action is one of the primary GPC deterioration mechanisms requiring huge expenditures for repair and maintenance. In this regard, two types of GPC (fly-ash based GPC and bottom-ash based GPC) were exposed to the harsh freeze-thaw conditions using a standard test method. The dynamic elastic modulus of both types of GPC was determined using a Non-Destructive Test (NDT) method called Resonant Frequency Test (RFT). The results of RFT after exposing to 300 freeze-thaw cycles showed that bottom-ash based GPC has better freeze-thaw resistance than fly-ash based GPC. Moreover, in this study, the leachability of bottom-ash based GPC was also investigated to trace the heavy metals (including Si, Al, Na, Cr, Cu, Hg) using Toxicity Characteristic Leaching Procedure (TCLP) test. The results of the TCLP test showed that all of the heavy metals could be effectively immobilized into the geopolymer paste.

Keywords: geopolymer concrete; fly-ash; bottom-ash; freeze-thaw; leachability; non-destructive test; TCLP; RFT

---

1. Introduction

Fly-ash and bottom-ash are by-products of the combustion of pulverized coal in thermal power plants. These by-products are pozzolans and typically consist of Silicon Dioxide (SiO_2), Aluminum Oxide (Al_2O_3) and Iron Oxide (Fe_2O_3) [1]. It is projected by Miller et al. [2] that coal production will increase up to 1000 million tons annually by 2040. So, due to the environmental regulations, utilization of by-products such as fly-ash and bottom-ash in different applications must be provided to protect natural resources and avoid landfill disposal of ashes. Although, broadly speaking, the use of by-products in various industries leads to a cleaner environment, the Environmental Protection Agency [3] reported that by-products can pollute groundwater and can increase a person's health risk to incurable diseases. So, besides creating a durable concrete made using by-products, it is worthwhile to assess the environmental impact of concrete.

Ordinary Portland Concrete (OPC) is one of the most used construction materials all over the world due to its high durability, high mechanical properties and long service life. However, the production of cement as a main constituent of OPC requires energy that leads to the generation of Carbon Dioxide (CO_2). According to the Portland Cement Association (PCA), 1000 kg of Portland Cement Production releases 927 kg of CO_2 into the atmosphere [4]. Hence, to alleviate this impact on the atmosphere, it is critical to reduce the use of Portland cement in the concrete by replacing it with other pozzolanic materials such as fly-ash, slag and bottom-ash.

Due to the issues mentioned above, scientists have been trying to use waste materials such as fly-ash to make durable concrete. However, the use of bottom-ash to produce concrete is not widely reported in the literature. French scientist Joseph Davidovits initially found one such concept, that of the "Geopolymer". Then, slag-based geopolymer cement was made in the 1980s [5]. Subsequently, in 1997, based on the obtained results for slag-based geopolymer cement, Silverstrim et al. [6], Van Jaarsveld and Van Deventer et al. [7] created an inorganic polymer concrete called Geopolymer Concrete (GPC). GPC is produced by reacting aluminate and silicate bearing materials with a caustic activator [8]. Since then, numerous works have reported on the development of Sodium-based (Na-based) GPC durability [9–12]. However, there are only limited studies conducted on the development of mix design, durability and service life of Potassium-based (K-based) GPC made by the combination of bottom-ash and fly-ash. Habert et al. [13] compared the environmental impact of GPC made by two types of by-products (fly-ash and slag), metakaolin and Ordinary Portland Concrete (OPC) using the Life Cycle Assessment (LCA) methodology. The results confirmed that GPC has a lower environmental impact in terms of emission of CO_2.

Freeze-thaw damage is a potentially critical deterioration mechanism that occurs not only in cement-based concrete but also in GPC structures. The reason for the deterioration of GPC in a cold condition is the water in cracks/capillary pores. The water in the cracks/capillary pores converts into ice when the material was exposed to the freeze-thaw cycles, where this issue leads to the internal expansion stress in the paste. This expansion stress causes internal micro-cracks. Continuously, an increase of freeze-thaw cycles leads to an extension of micro-cracks. Fu et al. [14] studied the freeze-thaw resistance of Na-based alkali-activated slag concrete. Six specimens of five mix proportions were cast and tested to measure the freeze-thaw resistance of alkali-activated slag concrete. Mass and Relative Dynamic Modulus of Elasticity (RDME) of each sample were measured after 25 freeze-thaw cycles. The result of the freeze-thaw test showed that RDME and mass of all mix proportions decrease about 10% and less than 1% after 300 cycles of freeze-thaw respectively. The authors also developed damage mechanics-based models using RDME. The authors reported that attenuation and power function models are more suited to accumulative and exponential damage models respectively.

This paper deals with the leachability of heavy metals of GPC. Basically, the potential leaching of heavy metals into the groundwater is a crucial concern with the use of by-products as a constituent of GPC, especially once the by-products are mixed with chemical materials such as Potassium Hydroxide (KOH) and Potassium Silicate (K_2O_3Si). Thang et al. [15] investigated the leachability of hazardous metals including Copper, Cadmium, Lead, Iron and Chromium of fly-ash based and red-mud-based geopolymer. Nine mix proportions were made to use Inductively Coupled Plasma- Atomic Emission Spectrometry (ICP-AES) at a pH of 7 to characterize the heavy metals in the geopolymer. The authors reported that raw materials leached a high amount of Lead, Palladium, Chromium and other hazardous metals. However, at a pH of 7, the level of leached metals from geopolymer materials was within the range specified by the European Standard. It is also found that geopolymer materials had lower concentrations than their constituent raw materials (fly-ash and red-mud). Arioz et al. [16] studied the leachability of fly-ash based geopolymer paste cured at different temperatures, including 40 °C, 80 °C and 120 °C for 6, 15 and 24 h. The hazardous metals including Arsenic, Lead, Chromium, Cadmium and Mercury were characterized using Toxicity Characteristic Leaching Procedure (TCLP) test for samples and ICP-AES for solutions extracted from GPC and fly-ash. The results of the TCLP test showed that fly-ash has higher heavy metals concentration than geopolymer paste. Moreover, the higher concentration of Arsenic and Mercury were measured in the solution of the samples cured at 120 °C for 15 and 24 h.

2. Research Significance

First of all, ample studies on the durability and service life of Na-based GPC have been reported. However, limited studies have been reported on the mechanical properties of K-based GPC. Moreover, it is reported that bottom-ash has similar physical, chemical and mechanical properties to fly-ash [17]. Since the use of bottom ash is relatively limited, dealing with this industrial material is now posing to be one of the most significant challenges in recent years. So, in the current study, attempts have been made to produce an appropriate mix proportion for K-based GPC made by a combination of 50% fly-ash and 50% bottom-ash (here noted as bottom-ash based GPC for further descriptions). K-based GPC synthesized only with fly-ash was also produced to compare its freeze-thaw resistance with bottom-ash based GPC.

Freeze-thaw delamination and heavy metals leaching are amongst the most persistent concerns of material in a frozen condition. Basically, aggressive environments damage the structure of GPC and consequently decrease their life span. Since there are limited studies that have reported on examination of the durability of K-based GPC, it is important to investigate the freeze-thaw resistance and leachability of K-based GPC synthesized by both fly-ash and bottom-ash.

Regarding freeze-thaw resistance properties, numerous predictive models were proposed for Na-based GPC [9,18,19]. Hongfa et al. [20] proposed a damage model for cement-based concrete. While, authors could not find any empirical model for K-based GPC, this model was used for K-based GPC in the current study. Moreover, the damage variable of K-based GPC was calculated and compared with other research studies to find the applicability of this freeze-thaw resistance model for by-products-based GPC [21,22].

3. Experimental Work

3.1. Precursors

Both fly-ash and bottom-ash were procured from Lafarge Inc. in Vancouver, Canada. According to ASTM C618 [23], the term fly-ash is divided into three categories (N, F, C). Class F was selected for the present study as it is a pozzolanic material and is useful for developing a durable GPC. Bottom-ash used in the current study had coarser particles than fly-ash. So, bottom-ash was sieved (#1.18 mm) to remove bigger size particles to increase the surface area of particles to achieve reasonable strength. The chemical elements obtained from the X-Ray Diffraction (XRD) test measured by Lafarge Inc. (due to the cost prohibitive nature of this test) are indicated in Table 1. It should be mentioned that the values of Na_2O, K_2O, TiO_2, P_2O_3, Mn_2O_3 of fly-ash are not measured by are the main components of both fly-ash and bottom-ash. The specific gravity of fly-ash and bottom-ash was 2.5 and 2.3 respectively.

Table 1. Chemical Compositions of fly-ash and bottom-ash.

Properties	Fly-Ash (%)	Source: Lafarge Inc.'s Report Bottom-Ash (%)
SiO_2	47.1	60.11
Al_2O_3	17.4	14.35
Fe_2O_3	5.7	5.92
CaO	14	10.40
MgO	5.4	4.49
SO_3	0.8	0.10
LOI	0.19	0.00
Na_2O	N/A	2.232
K_2O	N/A	1.766
TiO_2	N/A	0.892
P_2O_5	N/A	0.200
Mn_2O_3	N/A	0.093

Basically, a combination of alkali solution and soluble silicate is needed to produce GPC with desire durability. Since the durability of K-based GPC has only been reported by a few researchers [24,25], the combination of KOH and K_2SiO_3 was used in the present study because at elevated temperature (higher than 30 °C, GPC made by K-based is more steady than Na-based GPC in terms of mechanical properties including compressive strength [26]. KOH flakes obtained from Sigma-Aldrich Private Ltd. (St. Louis, MO, USA), and K_2SiO_3 powder (AgSil 16) obtained from PQ Corporation (USA) were used in this study. Chemical elements of K_2SiO_3 obtained from the Material Safety Data Sheets (MSDS) of the product are shown in Table 2. The specific gravity of KOH and K_2SiO_3 was 1.45 and 1.26, respectively.

Table 2. Chemical composition of K_2SiO_3.

Compound	K_2O	SiO_2	H_2O
%W/W	32.4%	52.8%	14.8%

Microstructural Study of Fly-Ash and Bottom-Ash

The Scanning Electron Microscopy (SEM) of fly-ash and bottom-ash was investigated using Hitachi S-4800 (Chiyoda, Tokyo, Japan) at the Advance Microscopic Facility (AMF) of the University of Victoria. The SEM operated at an accelerating voltage of 15 kV. Both fly-ash and bottom-ash were analyzed under 16.0 mm × 900 magnification.

The SEM images of both fly-ash and bottom-ash are presented in Figure 1i,ii. Figure 1i shows that the fly-ash particles are spherical in shape and hence known as cenospheres (perfectly round smooth and intact) with the presence of a few irregular particles. Figure 1ii shows the bottom-ash particles that are larger in size and sub-angular to angular in shape. It also can be seen that bottom-ash particles are porous with tiny pores visible in Figure 1. This porous nature of bottom-ash causes bottom-ash to absorb more water than fly-ash [27]. This needs to be properly accounted for when using bottom-ash in GPC production since excess water might have a negative impact on the GPC properties.

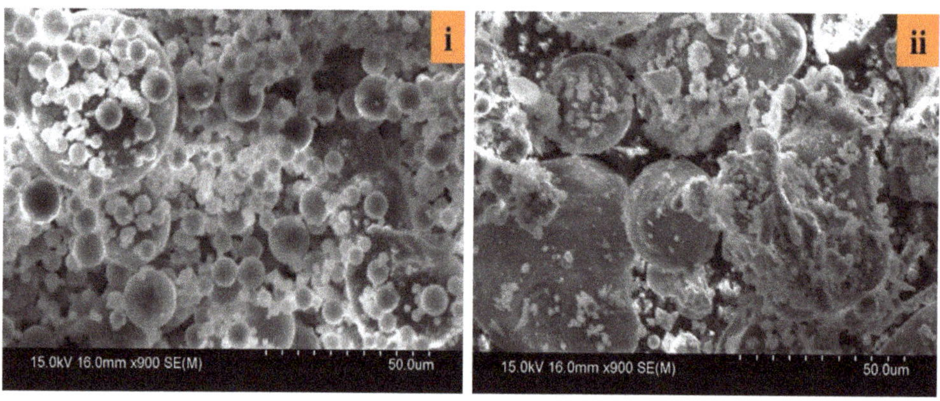

Figure 1. SEM images of fly-ash (**i**) and bottom-ash (**ii**).

3.2. Aggregates

Naturally available fine aggregate from Sechelt pit in B.C., Canada was used for the experimental work. Moisture content was measured by the change in weight of sample after keeping the sample for 24 h in an oven at 100 °C. The calculated moisture content was 4.10%. The fineness modulus, specific gravity and absorption of fine aggregate were 2.60, 2.65 and 0.79, respectively in accordance with ASTM C127 [28].

Coarse aggregate was also sourced from the Sechelt pit. The nominal size of the coarse aggregate was 12.5 mm. Calculated specific gravity, absorption and moisture content of coarse aggregate were 2.69, 0.69 and 1.39%, respectively in accordance with ASTM C127 [28]. The particle size distribution of coarse and fine aggregates was measured (shown in Figure 2) in accordance with ASTM C33 [29].

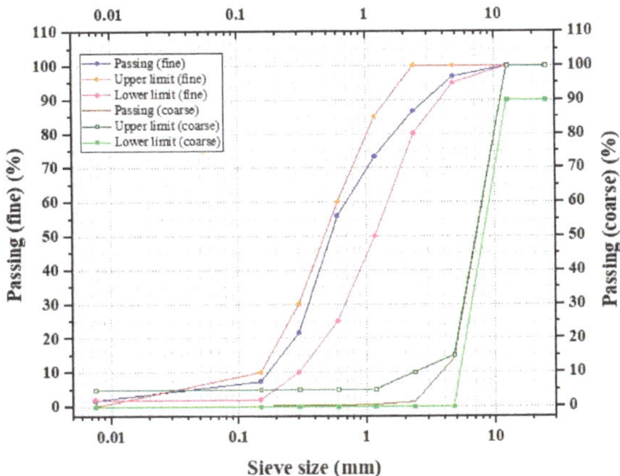

Figure 2. Aggregates size distribution.

3.3. Mix Design and Specimen Preparation of GPC

The mix proportion of GPC shown in Table 3 was derived after extensive initial trial experiments performed in the Facility for Innovative Materials and Infrastructure Monitoring (FIMIM) of the University of Victoria, Canada by the authors [12,30,31]. The KOH solution, with a concentration of 12 Molar (M), was prepared by dissolving KOH flakes in water. The KOH and K_2SiO_3 were mixed to prepare alkali activator solution in accordance with the mix design, provided in Table 3, 24 h before the casting day. The alkali activator ratio (K_2SiO_3/KOH) and a mass ratio of fly-ash to bottom-ash were considered to be 1.5 and 50:50 respectively to achieve a target strength of 35 MPa. This target strength was used as it is prescribed for numerous applications in practice.

Table 3. Mix Design.

Material	Bottom-Ash Based GPC (Kg/m^3)	Fly-Ash Based GPC (Kg/m^3)	Specific Gravity
Fly ash	194	388	2.5
Bottom ash	194	0	2.3
Coarse aggregates	1170	1170	2.69
Sand	630	630	2.60
KOH (12M)	85.16	85.16	1.45
K_2SiO_3	125.74	125.74	1.26
Extra Water	38.71	38.71	1
Air Entrained Admixture	1.5	1.5	1.03
Total	2439.11	2439.11	14.83

First, the alkali solution was prepared a day before casting day, as suggested by Davidovits [8]. Then GPC samples were produced in accordance with ASTM C192/C192M-15 [32] and ASTM C39/C39M-15 [33]. In order to produce GPC, dry materials were mixed in a concrete drum mixer for 1 min. Alkali solution was then added to the dry materials and was mixed for 3 min followed by a 3 min rest period. Finally, extra water (if the mixture was dry) was slowly added to the mixture for 2 min of final mixing. After mixing, GPC specimens were cast in molds (100 × 200 mm) for compressive test. Then, the Gilson vibrator table with a frequency of 60 Hz was used to discharge air bubbles to the surface, and for consolidating the specimens.

3.4. Curing of GPC Samples

Several efforts have been made for characterizing the influence of curing environments on different properties of Na-based GPC [34–36]. In previous studies performed by current authors [12,30,31] three methods of curing (ambient, steam and dry curing) were selected to accelerate the curing of K-based GPC and obtain higher compressive strength. So, in total, 54 cylindrical GPC samples (100 × 200 mm) were cured using the methods mentioned above. According to the results, steam-cured GPC samples showed greater compressive strength at temperature of 80 °C for 24 h. So, the steam curing method and temperature of 80 °C were used to cure fly-ash based and bottom-ash based GPC for this study.

Generally, for the steam curing method, after 24 h of ambient curing (approximate relative humidity range of 45% to 70% and approximate temperature range of 5 °C to 15 °C), the samples were left in a container surrounded by water. The container was packed tightly to prevent excessive evaporation during the curing process. The container was then put into the oven at a temperature of 80 °C. Lastly, the samples were removed from the oven, demolded, and was followed by 28 days of ambient curing. More details about the curing regime used are available in prior studies [12,30,31].

4. Methodology

Figure 3 shows the scope of work of this study. In this study, first of all, attempts have been made to produce bottom-ash based GPC and fly-ash based GPC using the steam curing method (at a temperature of 80 °C). After that, cylindrical bottom-ash based GPC and fly-ash based GPC specimens were made to measure their compressive strength. The freeze-thaw resistance, resonant frequency and leachability of bottom-ash based and fly-ash based beams were then tested using the Non-Destructive Test (NDT). Eventually, a comparative study was made between concrete parameter, compressive strength and number of cycles. It should be noted that the leachability of bottom-ash based GPC was separately studied using the TCLP test.

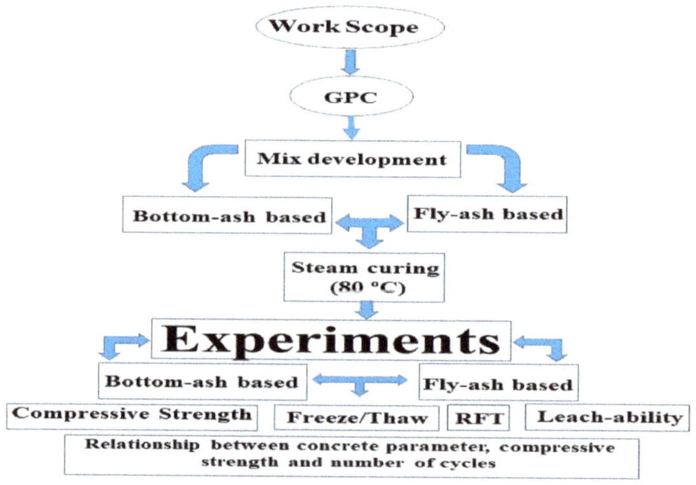

Figure 3. The work scope of the current study.

4.1. Compressive Strength

The steam-cured fly-ash based and bottom-ash based GPC samples (100 mm diameter and 200 mm height) were tested at the age of 28 days in accordance with ASTM C39/C39M-15 [33] using Forney compression testing machine model #AD 650.

4.2. Freeze-Thaw Test

In order to verify the durability of GPC under cold weather conditions, the freeze-thaw test was performed. Six GPC prisms ($76 \times 102 \times 406$ mm) were exposed to 300 freeze-thaw cycles, with temperature operating range from -17.8 °C to $+4.4$ °C and a humidity range from 10% to 95% in accordance with ASTM C666 [37]. Procedure "A" was selected in this study which arranges rapid freezing and thawing in water. After every 30 freeze-thaw cycles, samples were pulled off from the freeze-thaw cabinet to measure their mass loss, RDME and leaching.

4.3. Dynamic Elastic Modulus

One of the objectives of this research was to calculate the RDME using an NDT called Resonant Frequency Test (RFT)/Resonant Frequency Gauge (RTG). Figure 4 shows the components of the RFT/RTG device used for this study. Firstly, an accelerometer, with a frequency response measurement range of 20,000 Hz, was attached to the GPC surface using adhesive grease. After attaching the accelerometer and positioning GPC samples to the required mode of testing, a standard ball tip hammer weighing 110 ± 2 g with a tip diameter of 10 mm, is used to strike the surface at precise locations on the samples being tested. The achieved time domain signal was amplified and passed through BNC connection/cable. Finally, Olson instruments' RTG software records/shows the resonant frequency. ASTM C666 [37] suggested the following equation to calculate the RDME:

$$P_c = \left(\frac{n_1^2}{n^2}\right) \times 100 \tag{1}$$

where:

P_c = RDME, %.

n = fundamental transverse frequency at 0 freeze-thaw cycles.

n_1 = fundamental transverse frequency after 'n' freeze-thaw cycles.

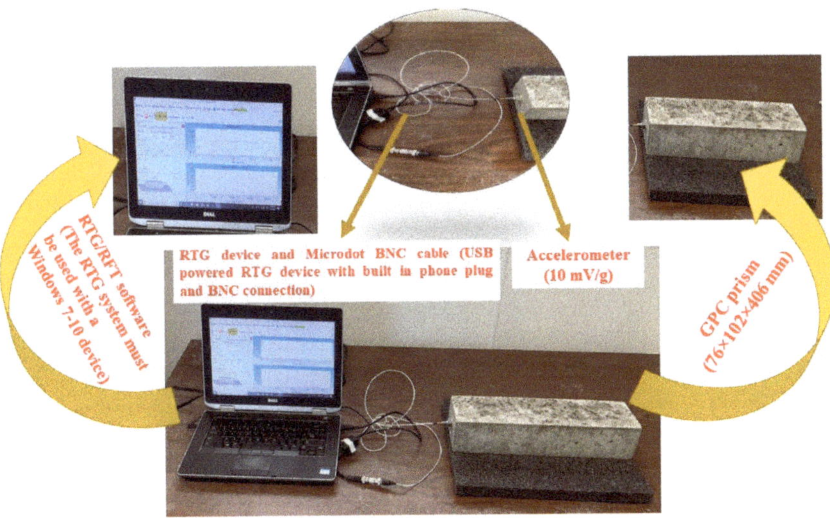

Figure 4. Resonant Frequency Test (RFT)/Resonant Frequency Gauge (RTG) Device.

5. Freeze-Thaw Damage Model

Hongfa et al. [20] used Aas-Jakobsen's [38] S-N equation (Equation (2)) and proposed a model (Equation (3)) for calculating the freeze-thaw damage of cement-based concrete in both the laboratory and real environmental conditions. According to their assumption, the freeze-thaw fatigue damage in concrete can be calculated as Equation (3):

$$S = \frac{\sigma'_{max}}{f_t} = 1 - \beta \, (1 - R) \, \log N \tag{2}$$

$$D_n = 1 - \frac{0.6 \, \log N}{\log N - 0.4 \log(N - n)} = 1 - \frac{E_n}{E_0} \tag{3}$$

where:

D_n = freeze-thaw damage variable expressed as the RDME.
N = fatigue life of concrete exposed to freeze-thaw cycles.
n = number of freeze-thaw cycles.
E_0 = initial dynamic modulus of elasticity.
E_n = dynamic modulus of elasticity after 'n' freeze-thaw cycles.
β = concrete material parameter

They also determined the value of β under freeze-thaw conditions using Equation (4) and established a relationship between β and compressive strength.

$$\beta = \frac{Dn}{\log N} \tag{4}$$

In this study, the value of β of fly-ash based GPC and bottom-ash based GPC and other by-products-based GPC performed by Zhao et al. [21] and Mengxuan et al. [22] was calculated to find the applicability of Equation4. Table 4 compares the mix proportions of other researchers [21,22] to the ones used in this study. Equations (3) and (4) were used to calculate the value of β of each mix proportion (shown in Table 4), and to establish the relationship between β, number of cycles and compressive strength of each mix proportion.

6. TCLP Test

Several materials are categorized as hazardous waste when they are dumped in landfills. GPC is considered as a toxic material due to the use of hazardous waste materials and chemical activators in its mixture. This is why it is vital to check its chemical metals leachability. In this study, the TCLP test was used to analyze the leaching of chemical metals from GPC. To study the environmental compatibility of K-based GPC made by 50% fly-ash and 50% bottom-ash, TCLP was performed as per USEPA 1311.

In this study, GPC was crushed to achieve a minimum of 100 g to characterize the chemical metals. Then, 5 g of GPC sample was mixed with 95 mL of distilled water to measure its pH level. After the determination of pH value, a proper extraction fluid was selected and was added to 100 g ± 0.1 g of GPC sample (< 9.5 mm). The solution was placed in the extraction vessel and was rotated at 30 ± 2 rpm for 18 ± 2 hrs at ambient temperature (23 ± 2 °C). At the end of the extraction period, the solution was transferred to the filter holder. The filtrate collected is called leachate.

Table 4. Comparison of mix proportions.

Sample	Mixture	Fly-ash	Slag	Cement	Sand	Gravel	Na_2SiO_3	NaOH	Water	Water reducer	Curing	Si/Al	L/S
1	OPCC [21]	56 (kg/m^3)	56 (kg/m^3)	448 (kg/m^3)	626 (kg/m^3)	1022 (kg/m^3)	-	-	174.7 (kg/m^3)	2.5 (kg/m^3)	Standard	-	-
2	GPC-10 [21]	346.7 (kg/m^3)	38.5 (kg/m^3)	-	601.7 (kg/m^3)	1203.5 (kg/m^3)	165.7 (kg/m^3)	66.2 (kg/m^3)	-	-	80 °C	-	-
3	GPC-30 [21]	269.6 (kg/m^3)	115.6 (kg/m^3)	-	601.7 (kg/m^3)	1203.5 (kg/m^3)	165.7 (kg/m^3)	66.2 (kg/m^3)	-	-	Standard	-	-
4	GPC-50 [21]	192.6 (kg/m^3)	192.6 (kg/m^3)	-	601.7 (kg/m^3)	1203.5 (kg/m^3)	165.7 (kg/m^3)	66.2 (kg/m^3)	-	-	Standard	-	-
5	RMSFFA-SA2.0-NA0.6-RT-28D [22]	-	-	-	-	-	-	-	25%	25%	Room temp. (23 °C)	2.00 mol	0.53
6	RMSFFA-SA2.0-NA0.6-80C-28D [22]	-	-	-	-	-	-	-	25%	25%	80 °C	2.00 mol	0.53
7	Fly-ash based GPC (current study)					Available in Table 3							
8	Bottom-ash based GPC (current study)					Available in Table 3							

7. Results and Discussion

7.1. Physical Characteristics

The slump test was performed for the analysis of viscosity behavior of fly-ash based GPC and bottom-ash based GPC to investigate the workability, according to ASTM C143 [39]. The average slump value of fly-ash based GPC and bottom-ash based GPC was measured as 245 mm and 215 mm, respectively. The bottom-ash based GPC specimens showed lower workability than fly-ash based GPC specimens because according to the microscopic study of GPC, the smooth surface and rounded-shape of the fly-ash particles improve ball-bearing effect, which increase the workability [40]. Moreover, it can be seen from Table 3 that bottom ash has a lower specific gravity when compared to fly-ash. Hence, when the same weight of fly-ash per cubic meter of the material is replaced with bottom ash, there is more dry volume of the material, which may also be partly responsible for reducing the workability of the mix.

The average dry density of fly-ash based GPC and bottom-ash based GPC at 7 days was 2415 kg/m^3 and 2422 kg/m^3, respectively. The dry density of these two types of GPC increased as the age of the GPCs increased. The average dry density of fly-ash based GPC increased from 2415 kg/m^3 to 2431 kg/m^3 when age of samples increased from 7 days to 28 days with an overall increase of 0.66%. In contrast, the average dry density of bottom-ash based GPC increased from 2422 kg/m^3 at 7 days to 2435 kg/m^3 at 28 days with an overall increase of 0.53%. The authors attribute this slight difference in density to in-batch test variability.

It is well-known that the specific gravity of fly-ash and bottom-ash is comparable because these two materials have similar chemical compositions. Table 3 indicates that fly-ash has higher specific gravity compared to its counterpart bottom-ash. It is reported that cenospheres and poor gradation of particles degrade specific gravity of bottom-ash [41].

7.2. Compressive Strength of GPC

It is well-known that elevated curing temperature and duration are beneficial toward the acceleration of the polycondensation process. In previous studies performed by current authors, three methods of curing (ambient, steam and dry curing) were used to achieve higher compressive strength. A minimum of six specimens (100×200 mm) were cured at ambient, 30, 45, 60, 80 °C for 24 h and then kept at room temperature for 28 days. These samples were tested at the age of 28 days by using Forney compressive testing machine model #AD 650. However, steam-cured GPC at a temperature of 80 °C for 24 h following by 28 days of room temperature curing achieved higher average compressive strength (35 MPa). Based on the other studies on the microstructure of GPC [42,43], the steam curing method improves the dissolution rate of chemical species, such as Silicon Dioxide (SiO_2) and Aluminum Oxide (Al_2O_3), from mixture where the rate of geopolymerization increases. This finding can be attributed to the full and uniform internal curing of specimens. This finding is also in good-agreement with Yewale et al. [44], where the optimum compressive strength of steam-cured GPC was achieved at 80 °C. So, in the present study, GPC samples were steam-cured at 80 °C for subsequent experiments. However, it should be mentioned that GPC samples can be steam-cured at 60 °C for some applications and it may not be necessary to cure at high temperature such as 80 °C.

The same method and temperature (80 °C) were then used to produce fly-ash based GPC. A minimum of three fly-ash GPC cylinders were cast to find their average compressive strength at the age of 28 days. The compressive strength of fly-ash based GPC specimens was 32 MPa, 30 MPa and 32 MPa with an average of 31 MPa. It can be seen from the results of the compression test that the compressive strength of bottom-ash based GPC steam-cured at 80 °C is higher than fly-ash based GPC with a similar curing regime. Although several studies have reported that bottom-ash based GPC has lower compressive strength compared to fly-ash based GPC [27,45,46], it can be attributed to the higher porosity of bottom-ash [47]. However, it should be noted that the ratio of SiO_2/Al_2O_3 also affects the compressive strength of GPC [48,49]. This means that GPC made by waste ashes with a higher of SiO_2/Al_2O_3 ratio tend to develop higher compressive strength. In this study, the bottom-ash had a higher SiO_2/Al_2O_3 ratio than fly-ash. This is why bottom-ash based GPC had higher compressive strength compared to fly-ash based GPC.

Generally, Na-based solution is mostly considered in various studies due to its low cost, availability, desired workability and durability. However, the K-based solution can also be used for high temperature applications [48,50–52]. According to Hounsi et al. [53], Na-based GPC gains lower compressive strength value than K-based GPC at the same alkali concentration of the current study (12 M). Hounsi et al. [53] attributed this phenomenon to the reduction of Si/Na ratio at a high concentration of sodium hydroxide (NaOH), where NaOH slows the polycondensation process and reduces the mechanical properties of Na-based GPC.

In previous studies performed by current authors, attempts have been made to cure the samples at ambient temperature. However, as aforementioned, the higher compressive strength is achieved at 80 °C. It is well known that only ambient temperature curing is a practical method in the construction field for GPC and to save energy. Hence, mixing calcium-based material such as cement with ashes is suggested to improve setting time, workability and durability of GPC cured at ambient temperature [54]. The microstructural investigation of fly-ash based GPC mixed with cement showed that the geopolymerization process is more likely as calcium alumino-silicate hydrate (C-A-S-H), which contributes to hardening and early strength gain of GPC mix with cement. Moreover, the enhanced strength of GPC mixed with cement is attributed to the generated heat during the geopolymerization process, where cement helps GPC to initiate condensation reaction at ambient temperature [35].

7.3. Freeze-Thaw Resistance

7.3.1. Mass Loss of Specimens during the Freeze-Thaw Process

The weight of GPC was taken every 30 cycles of freeze-thaw to calculate their mass loss. Figure 5 shows the average mass loss of six fly-ash based GPC and six bottom-ash based GPC up to 300 cycles of freeze-thaw. Fly-ash based GPC shows higher and rapid mass loss possibly because the specimens' structure failed at the early age of freeze-thaw cycles, and due to the poor bonding in the Interfacial Transition Zone (ITZ) which led to severe surface scaling. This finding is in good-agreement with the mass loss result of fly-ash based GPC-10 studied by Zhao et al. [21]. In contrast to fly-ash based GPC, the mass loss of bottom-ash based GPC was slow till 300 freeze-thaw cycles which indicates higher bonding strength of paste of bottom-ash based GPC.

7.3.2. RDME of Fly-Ash Based and Bottom-Ash Based GPC

Figure 6 indicates the average RDME reduction of six fly-ash based GPC and six bottom-ash based GPC over 300 cycles of freeze-thaw. As can be seen in Figure 6, bottom-ash based GPC exhibits higher RDME than fly-ash based GPC. The RDME of both types of GPCs dramatically dropped when cycles increased from 0 to 60. Authors attribute this RDME reduction to the existence of uncured by-product particles in the microstructure of GPC that caused GPC specimens to lose its microstructural strength at early freeze-thaw cycles [55].

It also can be seen that RDME loss of bottom-ash based GPC (13.4%) was higher than fly-ash based GPC (10.9%) until 60 cycles. However, RDME of fly-ash based GPC reduced intensely until 150 cycles (RDME $\approx$ 60%). In this study, the freeze-thaw test for fly-ash based GPC was continued up to 300 cycles, even though according to ASTM C666 [37], the freeze-thaw testing procedure of specimen must be stopped when its RDME reaches 60% of the initial modulus. Since concrete is a heterogeneous material, the issue mentioned above might be due to the various factors such as the density and RDME of the main constituents (such as fly-ash and bottom-ash) and the characteristic of the ITZ which affect the elastic behavior of the composite [56]. Moreover, in general, the RDME of the fly-ash based matrix and bottom-ash matrix are determined by their porosity. So, the parameters determining the porosity of the matrix, such as geopolymerization process, curing conditions, AEA amount, etc., could be the other reasons for rapid RDME reduction of fly-ash based GPC.

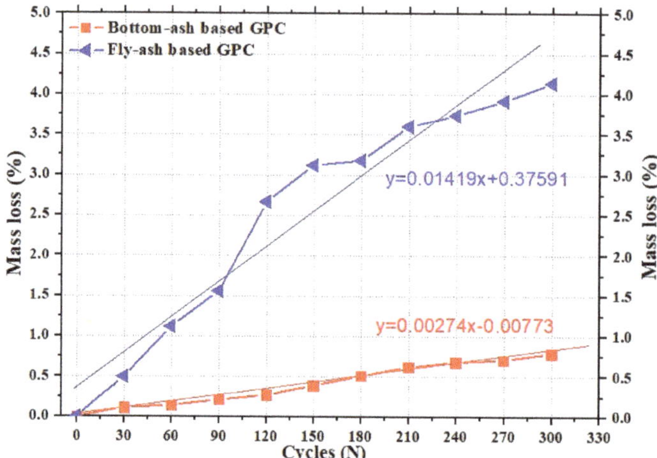

Figure 5. Average mass loss of fly-ash based Geopolymer Concrete (GPC) and bottom-ash based GPC over 300 cycles of freeze-thaw.

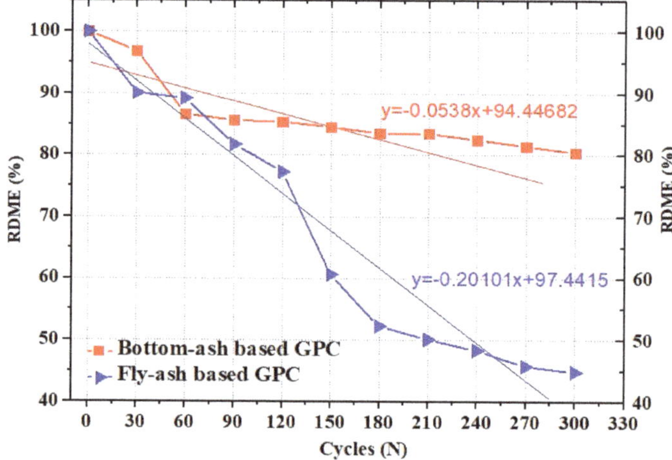

Figure 6. Average Relative Dynamic Modulus of Elasticity (RDME) reduction of fly-ash based GPC and bottom-ash based GPC vs. No. of cycles.

It is well-known that Air Entrained Admixture (AEA) provides more free spaces for the water to freeze. However, the pressure made by cryo (ice formation inside the air voids of GPC) can overcome the tensile strength of the matrix when these free spaces are overloaded with cryo. Consequently, this pressure creates micro-cracks in the ITZ and would reduce the ultimate strength of concrete. So, fly-ash based GPC can bear lower force when exposed to the freezing and thawing conditions since the initial compressive strength of fly-ash based GPC (~32 MPa) was lower than bottom-ash based GPC (~35 MPa).

7.4. Leachability of GPC

7.4.1. Laboratory Investigation of Bottom-Ash Based GPC

Identification of heavy metals and their leaching is one of the important factors that govern the utilization of GPC because of its influence on the environment. The toxicity of heavy metals depends on their concentration rate in the environment. With the increasing concentration rate of heavy metals in the environment, these toxic hazardous metals can be accumulated in living tissues and cause irreparable events. So, the TCLP test was conducted for obtaining the essential data about main chemical elements such as Si, Al, Na and other hazardous metals such as Cr, Cu, Hg, etc. It should be mentioned that the TCLP test was only possible on one sample due to the cost prohibitive nature of this test. Bottom-ash based GPC was selected in this study because of its higher freeze-thaw resistance than fly-ash based GPC. The TCLP test was performed by the Maxxam Analytics lab, Victoria, Canada. Table 5 shows the pH of bottom-ash based GPC at three levels of extractions. The initial pH of the sample was 11.6 due to the existence of alkali constituents (K-based) in the bottom-ash based GPC mixture. To prepare the leaching/extraction fluid, 5.7 mL glacial acetic acid (CH_3CH_2OOH) was added to 500 mL reagent/distilled water. When reasonably mixed, the pH of this fluid was 4.96. Then, to perform the TCLP test, an amount of the proper leaching fluid equivalent to 20 times the mass of the specimen (20:1 liquid to solid ratio) was added to the extraction vessel of TCLP equipment. Then, the measured pH of leachate was 6.36.

Table 5. pH values of bottom-ash based GPC.

pH	-
Initial pH of sample	11.6
Final pH of leachate	6.36
pH of leaching fluid	4.96

The results of the TCLP test showed that all the heavy metals could be effectively immobilized into the geopolymeric paste. Figure 7 indicates that Ba, B and Fe have the highest leaching concentration and the rest of heavy metals have a concentration of less than 0.10 mg/L. Moreover, the obtained results showed that the concentration of all the heavy metals is below the regulatory level in accordance with USEPA 1311 and USEPA CFR. This could be attributed to the cations of heavy metals (such as Cu^{2+}, Cd^{2+}, Fe^{3+}, Zn^{2+}, Pb^{2+}, and total Cr) that can participate in the balance of the negative charge of tetra-silicate ($[SiO_4]^{4-}$) and potassium tetra-aluminate ($[K^+AlO_4]^{4-}$) [57,58]. So, bottom-ash based GPC showed low porosity, which could help immobilize all the heavy metals [59,60].

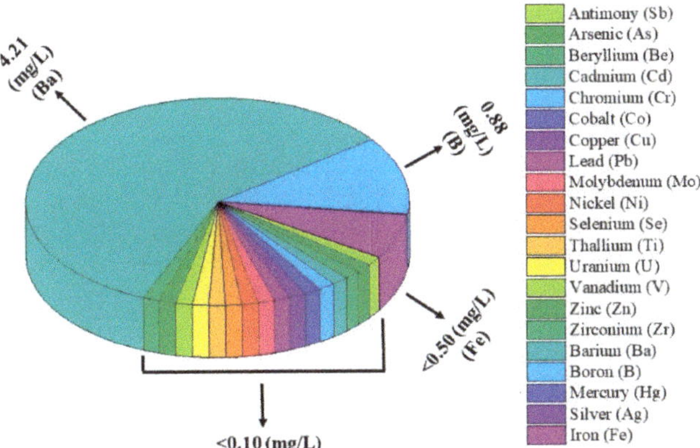

Figure 7. Toxicity Characteristic Leaching Procedure (TCLP) test results of bottom-ash based GPC.

7.4.2. Observational Study on Leachability of GPC during the Freeze-Thaw Process

In this study, the leachability of fly-ash-based GPC and bottom-ash based GPC was measured using a water testing method called HACH strips. The water sample was collected every 30 cycles from the freeze-thaw cabinet to measure the leachability of fly-ash based GPC and bottom-ash based GPC. Since the trend of data for every 30 cycles was quite similar only results obtained from the last cycle (300 cycles) were analyzed and plotted in Figure 8. The leachability of both fly-ash based GPC and bottom-ash based GPC was constant over 300 cycles of freeze-thaw (Figure 8). This phenomenon shows that the geopolymerization process was fully completed, and all toxic metals were trapped in the paste. Moreover, heat-treatment improved the microstructure of the GPC specimens, decreased the porosity and decreased the leachability of the paste [61].

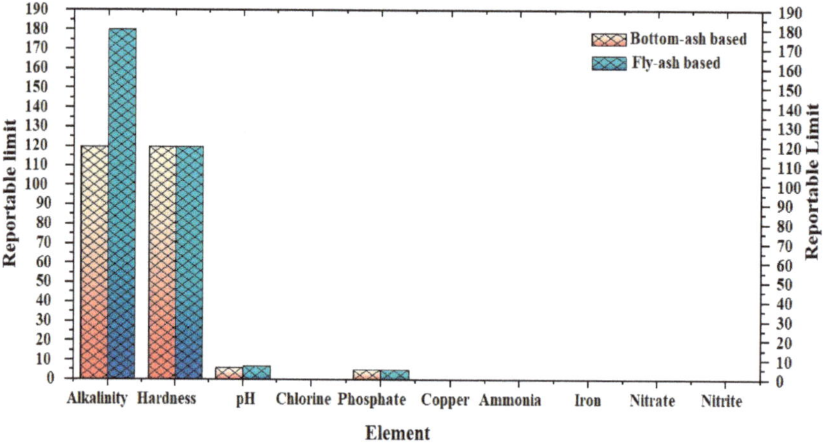

Figure 8. Leaching range of GPCs over 300 freeze-thaw cycles.

It also can be seen that the concentration of all the elements is almost in same range. However, fly-ash based GPC had more amount of alkalinity than bottom-ash based GPC, possibly due to poor geopolymerization of fly-ash based GPC compared to bottom-ash based GPC, which led to leaching of potassium from fly-ash based GPC.

7.5. The Relationship between β, Number of Cycles and Compressive Strength of GPC

Figure 9 shows calculated values of β for data obtained from Zhao et al. [21] and Mengxuan et al. [22] and compares with values of β of fly-ash based GPC and bottom-ash based GPC. The value of β for mix #1, 2, 3, 4, 5, 6, 7 and 8 is 0.19, 0.44, 0.22, 0.19, 0.27, 0.23, 0.18 and 0.16, respectively. According to Equation4, the value of β is inversely proportional to the logarithmic function of N, which means a higher number of freeze-thaw cycles give a lower value of β. The results of the current experiment, shown in Figure 9, confirm the finding mentioned above that bottom-ash based GPC (mix #8) with a lower value of β has higher freeze-thaw resistance than fly-ash based GPC (mix #7). This finding is also applicable to the rest of the GPC samples. Moreover, it can be seen that the value of β for GPC samples with compressive strength less than 50 MPa decreased abruptly. While, β value of GPC samples with compressive strength higher than 50 MPa reduced gradually. Although the results of the current study are in good agreement with Hongfa et al. [20], the authors suggest that compressive strength of GPC should be considered in Hongfa et al.'s [20] equation because compressive strength is one of the key properties of GPC revealing internal bonding of paste. Therefore, a greater number of freeze-thaw cycles were achieved when GPC had higher bonding strength. This finding can be seen in Figure 9 where the value of β was varied when compressive strength increased (the value of compressive strength of mix #1, 3, 4 and 8 was higher than their β, while the compressive strength of mix #2, 5, 6 and 7 was lower than their β).

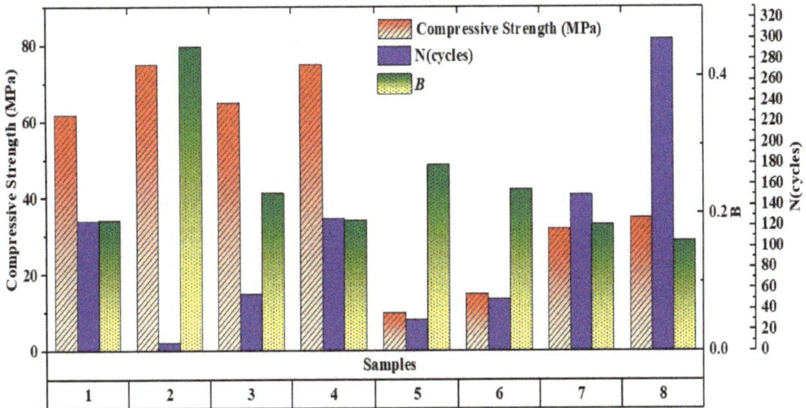

Figure 9. Value of β, compressive strength and number of cycles for different types of GPC.

8. Conclusions

In this study, the deterioration and mass loss produced in fly-ash based GPC and bottom-ash based GPC during 300 cycles of freeze-thaw were evaluated using RFT. Leachability of bottom-ash based GPC was also measured using the TCLP test to characterize the heavy metals. According to the obtained results:

- Bottom-ash based GPC indicated lower mass loss than fly-ash based GPC. Authors attributed the phenomenon mentioned above to the poor bonding of pastes in the ITZ in fly-ash based GPC.
- The resonant frequency of both types of GPC was measured after exposure to 300 freeze-thaw cycles with interval of 30 cycles. According to the results, bottom-ash based GPC showed better freeze-thaw resistance than fly-ash based GPC. It could be attributed to various parameters including geopolymerization process, curing conditions and the amount of AEA.
- Toxicity of heavy metals leaching from bottom-ash based GPC was measured using the TCLP test. The results showed that all the heavy metals including Si, Al, Na, Cr, Cu, Hg etc. were trapped and immobilized in the paste, and all of them were below the standard range of USEPA 1311 and USEPA CFR.
- A comparison between compressive strength, N and β of different types of by-products-based GPC was made. The experimental results showed that a higher number of freeze-thaw cycles give lower β. So, Bottom-ash based GPC with a higher number of cycles had lower β (0.1614) than fly-ash based GPC (0.1838). The authors also found that compressive strength should be accounted for in the proposed equation by Hongfa et al. [20].

Author Contributions: P.A. made different mix propositions and produced all the samples to find the higher compressive strength for these types of concrete. He also performed the compressive strength test, freeze-thaw test and resonant frequency test to investigate and estimate durability of fly-ash based and bottom-ash based GPC. R.G. provided the materials and shared knowledge on this subject. P.A. wrote the paper and R.G. revised it. All authors have read and agreed to the published version of the manuscript.

Funding: This research was funded by India Canada Research Centre of Excellence (IC-IMPACTS).

Acknowledgments: This research paper is made possible through the help and financial support from IC-IMPACTS. Authors also would like to thank all the staff in Civil Engineering department of University of Victoria.

Conflicts of Interest: The authors declare no conflict of interest.

References

1. Zimring, C.A.; Rathje, L. *Encyclopedia of Consumption and Waste: The Social Science of Garbage*, 2nd ed.; SAGE: New York, NY, USA, 2012.
2. Miller, B.G. The future role of coal. In *Clean Coal Engineering Technology*, 2nd ed.; Elsevier: Amsterdam, The Netherlands, 2017; ISBN 978-0-12-811365-3.
3. Environmental Protection Agency, Federal Registration, 17 April 2015. Available online: https://www.federalregister.gov/documents/2015/04/17/20150-0257/hazardous-and-solid-waste-management-system-disposal-of-coal-combustion-residuals-from-electric (accessed on 14 October 2015).
4. Marceau, M.; Nisbet, M.; Geem, M.V. *Life Cycle Inventory of Portland Cement Manufacture*; Portland Cement Association: Skokie, IL, USA, 2006.
5. Davidovits, J. Geopolymer cement: A review. *Geopolym. Sci. Tech.* **2013**, *21*, 1–11.
6. Silverstrim, T.; Rostami, H.; Larralde, J.; Samadi-Maybodi, A. Fly Ash Cementitious Material and Method of Making a Product. U.S. Patent 5,601,643, 11 February 1997.
7. Van Jaarsveld, J.G.S.; van Deventer, J.S.J.; Lorenzen, L. The potential use of geopolymeric materials to immobilize toxic metals: Part I. Theory and applications. *Miner. Eng.* **1997**, *7*, 659–669. [CrossRef]
8. Davidovits, J. *Geopolymer Chemistry and Application*, 4th ed.; Institut Geopolymere: Saint-Quentin, France, 2015.
9. Okoye, F.; Durgaprasad, J.; Singh, N.B. Fly ash/kaolin based geopolymer green concretes and their mechanical properties. *Data Brief.* **2015**, *5*, 739–744. [CrossRef] [PubMed]
10. Vora, P.R.; Dave, U. Parametric studies on compressive strength of geopolymer concrete. *Procedia Engineering* **2013**, *51*, 210–219. [CrossRef]
11. Shaikh, F.U.A. Mechanical and durability properties of fly ash geopolymer concrete containing recycled coarse aggregates. *Int. J. Sustain. Built Environ.* **2016**, *5*, 277–287. [CrossRef]
12. Gupta, R.; Rathod, H. Current state of K-based geopolymer cements cured at ambient temperature. *Emerg. Mater. Res.* **2015**, *4*, 125–129. [CrossRef]
13. Habert, G.; De Lacaillerie, J.B.D.E.; Roussel, N. An environmental evaluation of geopolymer based concrete production: Reviewing current research trends. *Clean. Prod.* **2011**, *11*, 1229–1238. [CrossRef]
14. Fu, Y.; Cai, L.; Wu, Y. Freeze–thaw cycle test and damage mechanics models of alkali-activated slag concrete. *Constr. Build. Mater.* **2011**, *25*, 3144–3148. [CrossRef]
15. Thang, N.H.; Ha, B.T.Q.; Ha, P.V.T.; Minh, D.Q.; Duc, H.M.; Quang, L.V. Leaching behavior and immobilization of heavy metals in geopolymer synthesized from red mud and fly ash. *Key Eng. Mater.* **2018**, *777*, 518–522.
16. Arioz, E.; Arioz, O.; Kockar, O.M. Leaching of F-type fly ash based geopolymers. *Procedia Eng.* **2012**, *42*, 1114–1120. [CrossRef]
17. Argiz, C.; Sanjuan, M.A.; Menedez, E. Coal bottom ash for Portland cement production. *Adv. Mater. Sci. Eng.* **2017**, *2017*, 7. [CrossRef]
18. Lee, W.-H.; Wang, J.-H.; Ding, Y.-C.; Cheng, T.-W. A study on the characteristics and microstructures of GGBS/FA based geopolymer paste and concrete. *Constr. Build. Mater.* **2019**, *211*, 807–819. [CrossRef]
19. Hamidi, M.R.; Man, Z.; Azizli, K.A. Concentration of NaOH and the effect on the properties of fly ash based geopolymer. *Procedia Eng.* **2016**, *148*, 189–193. [CrossRef]
20. Hongfa, Y.; Haoxia, M.; Kun, Y. An equation for determining freeze-thaw fatigue damage in concrete and a model for predicting the service life. *Constr. Build. Mater.* **2017**, *137*, 104–116.
21. Zhao, R.; Yuan, Y.; Cheng, Z.; Wen, T.; Li, J.; Li, F.; Ma, Z.J. Freeze-thaw resistance of class F fly ash-based geopolymer concrete. *Constr. Build. Mater.* **2019**, *222*, 474–483. [CrossRef]
22. Mengxuan, Z.; Zhang, G.; Htet, K.W.; Kwon, M.; Kwon, M.; Xu, Y.; Tao, M. Freeze-thaw durability of red mud slurry-class F fly ash-based geopolymer: Effect of curing conditions. *Constr. Build. Mater.* **2019**, *215*, 381–390.
23. ASTM C618. *Standard Specification for Coal Fly Ash and Raw or Calcined Natural Pozzolan for Use in Concrete*; ASTM: West Conshohocken, PA, USA, 2017.
24. Sakkas, K.; Panias, D.; Nomikos, P.P.; Sofiano, A.I. Potassium based geopolymer for passive fire protection of concrete tunnels linings. *Tunn. Undergr. Space Technol.* **2014**, *43*, 148–156. [CrossRef]
25. Hosan, A.; Haque, S.; Shaikh, F. Comparative study of sodium and potassium based fly-ash geopolymer at elevated temperature. In Proceedings of the International Conference on Performance-based and Life-cycle Structural Engineering, Brisbane, QLD, Australia, 9–11 December 2015.

26. Hosan, A.; Haque, S.; Shaikh, F. Compressive behaviour of sodium and potassium activators synthetized flyash geopolymer at elevated temperatures: A comparative study. *J. Build. Eng.* **2016**, *8*, 123–130. [CrossRef]
27. Haq, E.; Padmanabhan, S.K.; Licciulli, A. Synthesis and characteristics of fly ash and bottom ash based geopolymers–A comparative study. *Ceram. Int.* **2014**, *40*, 2965–2971. [CrossRef]
28. American Society for Testing and Materials. *Standard Test Method for Relative Density (Specific Gravity) and Absorption of Coarse Aggregate*; ASTM C127/C127M; American Society for Testing and Materials: West Conshohocken, PA, USA, 2015.
29. American Society for Testing and Materials. *Standard Specification for Concrete Aggregates*; ASTM C33/C33M; American Society for Testing and Materials: West Conshohocken, PA, USA, 2015.
30. Belforti, F.; Azarsa, P.; Gupta, R.; Dave, U. *Effect of Freeze-Thaw on K-Based Geopolymer Concrete and Portland Cement Concrete*, 6th ed.; Nirma University: Gujarat, India, 2017.
31. Yang, C.; Gupta, R. Prediction of the compressive strength from resonant frequency for low-calcium fly ash-based geopolymer concrete. *J. Mater. Civ. Eng.* **2018**, *19*, 435–533. [CrossRef]
32. American Society for Testing and Materials. *Standard Practice for Making and Curing Concrete Test Specimens in the Laboratory*; ASTM C192/C192M—15; ASTM: West Conshohocken, PA, USA, 2015.
33. American Society for Testing and Materials. *Standard Test Method for Compressive Strength of Cylindrical Concrete Specimens*; ASTM C39/C39M—15; ASTM: West Conshohocken, PA, USA, 2015.
34. Kumaravel, S. Development of various curing effect of nominal strength Geopolymer concrete. *J. Eng. Sci. Technol. Rev.* **2014**, *7*, 116–119. [CrossRef]
35. Nath, P.; Sarker, P. Effect of GGBFS on setting, workability and early strength properties of fly ash geopolymer concrete cured at ambient condition. *Constr. Build. Mater.* **2014**, *66*, 163–171. [CrossRef]
36. Hardjito, D.; Rangan, B. *Development and Properties of Low-Calcium Fly Ash-Based Geopolymer Concrete*; Curtin University of Technology: Perth, Australia, 2005.
37. American Society for Testing and Materials. *Standard Test Method for Resistance of Concrete to Rapid Freezing and Thawing*; ASTM C666; ASTM: West Conshohocken, PA, USA, 2015.
38. Aas-Jalobsen, K. *Fatigue of Concrete Beams and Columns*; Division of Concrete Structure, Norwegian Institute of Technology: Trondheim, Norway, 1970.
39. American Society for Testing and Materials. *Standard Test Method for Slump of Hydraulic-Cement Concrete*; ASTM C143/C143M; American Society for Testing and Materials: West Conshohocken, PA, USA, 2015.
40. Temuujin, J.; Williams, R.P.; van Riessen, A. Effect of mechanical activation of fly ash on the properties of geopolymer cured at ambient temperature. *J. Mater. Process. Technol.* **2009**, *209*, 5276–5280. [CrossRef]
41. Reddy, C.S.; Mohanty, S.; Shaik, R. Physical, chemical and geotechnical characterization of flyash, bottom ash and municipal solid waste from Telangana State in India. *Int. J. GeoEng.* **2018**, *9*, 1–23.
42. Pangdaeng, S.; Phoo-Ngernkham, T.; Sata, V.; Chindaprasirt, P. Influence of curing conditions on properties of high calcium fly ash geopolymer containing Portland cement as additive. *Mater. Des.* **2014**, *53*, 269–275. [CrossRef]
43. Li, X.; Wang, Z.; Jiao, Z. Influence of curing on the strength development of calcium-containing geopolymer mortar. *Materials* **2013**, *6*, 5069–5076. [CrossRef]
44. Yewale, V.V.; Shirsath, M.N.; Hake, S.L. Evaluation of efficient type of curing for geopolymer concrete. *Int. J. New Technol. Sci. Eng.* **2016**, *3*, 10–14.
45. Xie, T.; Ozbakkaloglu, T. Behavior of low-calcium fly and bottom ash-based geopolymer concrete. *Ceramic International* **2015**, *41*, 5945–5958. [CrossRef]
46. Chindaprasirt, P.; Jaturapitakkul, C.; Chalee, W.; Rattanasak, U. Comparative study on the characteristics of fly ash and bottom ash geopolymers. *Waste Manag.* **2009**, *29*, 539–543. [CrossRef]
47. Hardjito, D.; Fung, S.S. Fly ash-based geopolymer mortar incorporating bottom ash. *Mod. Appl. Sci.* **2010**, *4*, 44. [CrossRef]
48. Thokchom, S.; Mandal, K.K.; Ghosh, S. Effect of Si/Al ratio on performance of fly ash geopolymers at elevated temperature. *Arab J. Sci. Eng.* **2012**, *37*, 977–989. [CrossRef]
49. Khale, D.; Chaudhary, R. Mechanism of geopolymerization and factors influencing its development: A review. *J. Mater. Sci.* **2007**, *42*, 729–746. [CrossRef]
50. Barbosa, V.; MacKenzie, J. Synthesis and thermal behaviour of potassium sialate geopolymers. *Mater. Lett.* **2003**, *57*, 1477–1482. [CrossRef]

51. Kamseu, E.; Rizzuti, A.; Leonelli, C.; Perera, D. Enhanced thermal stability in K_2O-metakaolin-based geopolymer concretes by Al_2O_3 and SiO_2 fillers addition. *J. Mater. Sci.* **2010**, *45*, 1715–1724. [CrossRef]

52. Lizcano, M.; Kim, H.S.; Basu, S.; Radovic, M. Mechanical properties of sodium and potassium activated metakaolin-based geopolymers. *J. Mater. Sci.* **2012**, *47*, 2607–2616. [CrossRef]

53. Hounsi, A.D.; Lecomte-Nana, G.; Djétéli, G.; Blanchart, P.; Alowanou, D.; Kpelou, P.; Napo, K.; Tchangbédji, G.; Praisler, M. How does Na, K alkali metal concentration change the early age structural characteristic of kaolin-based geopolymers. *Ceram. Int.* **2014**, *40*, 8953–8962. [CrossRef]

54. Nath, P.; Sarker, P.K. Use of OPC to improve setting and early strength properties of low calcium fly ash geopolymer concrete cured at room temperature. *Cem. Concr. Compos.* **2015**, *55*, 205–214. [CrossRef]

55. Azarsa, P.; Gupta, R. Novel approach to microscopic characterization of cryo formation in air voids of concrete. *Micron* **2019**, *122*, 21–27. [CrossRef]

56. Mehta, P.K.; Monteiro, P.J. *Concrete Microstructure, Properties, and Materials*, 4th ed.; McGraw-Hill Education: New York, NY, USA, 2014.

57. El-Eswed, B.; Yousef, R.; Alshaaer, M.; Hamadneh, I.; Al-Gharabli, S.; Khalili, F. Stabilization/solidification of heavy metals in kaolin/zeolite based geopolymers. *Int. J. Miner. Process.* **2015**, *137*, 34–42. [CrossRef]

58. Qian, G.; Sun, D.; Tay, J. Immobilization of mercury and zinc in an alkali-activated slag matrix. *J. Hazard. Mater.* **2003**, *101*, 65–67. [CrossRef]

59. Ahmari, S.; Zhang, L. Durability and leaching behavior of mine tailings-based geopolymer bricks. *Constr. Build. Mater.* **2013**, *44*, 743–750. [CrossRef]

60. Nikolic, V.; Komljenovic, M.; Marjanovic, N.; Bascarevic, Z.; Petrovic, R. Lead immobilization by geopolymers based on mechanically activated fly ash. *Ceram. Int.* **2014**, *40*, 8479–8488. [CrossRef]

61. Izquierdo, M.; Querol, X.; Phillipart, C.; Antenucci, D.; Towler, M. The role of open and closed curing conditions on the leaching properties of fly ash-slag-based geopolymer. *J. Hazard. Mater.* **2010**, *176*, 623–628. [CrossRef]

Journal of
Composites Science

Article

Arbitrary-Reconsidered-Double-Inclusion (ARDI) Model to Describe the Anisotropic, Viscoelastic Stiffness and Damping of Short Fiber-Reinforced Thermoplastics

Alexander Kriwet [1,*] **and Markus Stommel** [2]

[1] Structural Dynamics Powertrain, NVH CAE, NVH Powertrain, Mercedes-Benz AG, Mercedesstr, 122, 70372 Stuttgart, Germany

[2] Institute Polymer Materials, Leibniz-Institute for Polymer Research Dresden, Hohe Straße 6, 01069 Dresden, Germany; stommel@ipfdd.de

* Correspondence: alexander.kriwet@daimler.com

Received: 25 February 2020; Accepted: 7 April 2020; Published: 9 April 2020

Abstract: Current state of the art, simulation methods to determine the frequency-, temperature- and humidity-depending stiffness and damping do not show an accurate prediction of the structural dynamics of short-fiber-reinforced thermoplastics. Thus, in the current work the new developed Arbitrary-Reconsidered-Double-Inclusion (ARDI) model has been used to describe the stiffness and damping. Thereby, a homogenization equation has been used to derive the transversal-isotropic stiffness and damping tensors. By rotating and weighting these tensors using orientation distribution functions (ODF), it is possible to create a material database. A validation of the developed ARDI model was performed on bending vibration specimens under variation of the fiber direction, temperature and humidity, to investigate the structural dynamics. In general, the comparison of the results of the simulation and experiments shows a good correlation of the eigenfrequencies and the amplitudes. The main differences in the simulation can be traced back to the used modelling of the damping behavior.

Keywords: structural dynamics; composite plastics; stiffness; damping; fiber orientation; ODF; viscoelasticity

---

1. Introduction

Engineering plastics have become an important component of modern industry as a construction material, due to their good mechanical properties at low weight. Under Noise-Vibration-Harshness (NVH) aspects, plastics show a reduced structure-borne sound transmission and sound radiation behavior compared to metallic components due to their viscoelastic damping properties [1]. The challenge is to describe the microstructural mechanics of the short glass fiber-reinforced plastics in a suitable material model for a reliable prediction of the structural dynamics behavior.

For the description of the microstructural mechanics of short fiber-reinforced plastics, different approaches were pursued, which are mainly based on the investigations of Eshelby [2]. Assuming that the fibers have an ellipsoidal shape, the local stress and strain fields of the individual phases in the composite were calculated with a localization equation. This describes the mean-field theory according to the Mori-Tanaka model [3] and builds the basis for extended models like Hill's self-consistent model [4], Benveniste´s reformulation of the Mori-Tanaka model [5] or the second-moment mean-field homogenization model of Dogri et al. [6].

A look at the microstructure of fiber-reinforced plastics shows that a high crystalline interphase is formed between the fibers and the matrix [7]. In this context, three-phase models such as the

Reconsidered-Double-Inclusion (RDI) model [7] were developed to characterize the material properties more precisely. An overview of currently available material models is given in [8,9].

For the prediction of the structural dynamics of plastic components, established simulation methods generate anisotropic, linear-elastic material models [1]. In this context, the integrative simulation method as a coupling between process and structure simulation has become an essential tool for the efficient design of short fiber-reinforced plastic components [10]. Here, the fiber orientation tensors were coupled to the mesh of the components to demonstrate the anisotropic material properties. This condition allows an adequate description of the anisotropic stiffness of short fiber-reinforced plastics in the elastic range.

For the description of the damping of plastics, reduced damping models such as Rayleigh or modal damping are still used today [1,11]. The main disadvantage is that the damping constants are usually iteratively optimized using a reverse engineering approach based on experimental data. Furthermore, the stiffness and damping of short fiber-reinforced plastics varies depending on the excitation frequency, temperature and humidity [12]. This effect is also neglected when defining a static material model for the stiffness and damping.

Under the usage of Prony-series it is possible to model the frequency-dependent viscoelastic material behavior of the plastics more accurate. Thereby, a linear interpolation of the material properties between a defined number of interpolation points is performed and weighted by a sum function. Park and Schapery [13] as well as Breuer et al. [14] have demonstrated a successfully application of Prony-series formation for the description of the frequency-dependent stiffness of fiber-reinforced plastics.

In this work, a novel anisotropic viscoelastic material model based on the RDI model and Prony-series formation is proposed, to describe the frequency-, temperature- and humidity-dependent stiffness and damping of short glass fiber-reinforced thermoplastics.

2. Methodology and Approach

In order to achieve the aim of this work, the first step is to define the material model. Therefore, the existing RDI model has been extended to the Arbitrary-Reconsidered-Double-Inclusion (ARDI) model, by using the complex modulus to consider the damping behavior. Afterwards, the orientation distribution function (ODF) has been reconstructed according to the maximum entropy method (MEM) [15] for a number of orientation tensors. This method allows the creation of a discrete ARDI material database.

In the next step, the execution and evaluation process of the experiments have been defined. For the execution of the bending vibration tests, a system for the uniaxial vibration excitation of test specimens [16] has been used. The positioning of this system in a climatic chamber allows the execution of the bending vibration tests under defined temperature and humidity conditions.

The evaluation of the bending vibration tests was carried out with the resonance-curve method [17] and the 3 dB method [18]. In this context, the evaluation of the bending vibration tests of pure thermoplastic material specimens provides the storage and loss modulus for the calibration of the ARDI model. The evaluation of the bending vibration tests of the glass fiber-reinforced thermoplastics provides the structural dynamics behavior in the form of an accelerance frequency response function (FRF) for a comparison to the simulation.

The present work focuses on the thermoplastics PA66 (polyamide 6.6) and PPA (polyphthalamide) as pure matrix materials and in the reinforced state with 30 (GF30) and 50 (GF50) weight percent glass fibers.

The bending vibration tests of the plastics were performed under conditioning of 23 °C with 0% relative humidity (rh), 130 °C with 0% rh and 23 °C with 50% rh. The selected temperatures and humidities were derived from the temperatures and humidities where the plastics of this work are typically used (23 °C, 0% rh) but also where the selected plastics can still be used in a technically useful way (maximum humidity 50% rh, maximum temperature 130 °C).

The micro-computed topography (μ-CT) measurements of the glass fiber-reinforced materials allow the determination of the experimental fiber orientation tensors. The μ-CT specimens have been prepared from the center of the plate material from the same batch as the bending specimens. Based on these μ-CT results, it is possible to adjust the setup of the filling simulation to improve the quality for predicting the fiber orientation. Based on the calculated velocity field of the flow, the Folgar-Tucker model has been used to determine the fiber orientation tensor [19].

In the next step, a nearest neighbor mapping process has been used to couple the calculated fiber orientation tensors from the computational fluid dynamic (CFD) mesh of the filling simulation to the finite-element (FE) model mesh. Then, the material properties from the ARDI material database have been assigned for each element of the component according to the smallest error deviation of the Euclidean standard of the orientation tensors.

In the last step the calculation of the structural dynamics for each continuum element under update of the stiffness and damping tensor has been performed. For this, a linear-viscoelastic material law based on a Prony-series formulation via subroutine (UMAT) has been used [14].

The comparison of the results of the existing RDI to the ARDI model allows the assessment of the significant extension to the state of the art. For this purpose, the existing RDI model has been combined with a global modal damping (GMD) behavior approach. The GMD behavior has been modeled as ratio between stiffness and damping tensor and thus corresponds to the dissipation factor.

The calculation of the FRF of the bending specimens allows a comparison to the experiments and an assessment of the quality of the developed ARDI model for the selected temperatures and humidities. Thus, it is possible to assess whether the ARDI model is able to simulate a wide range of humidity and temperature conditions under the usage of the defined calibration method. The concept of this work is shown in Figure 1.

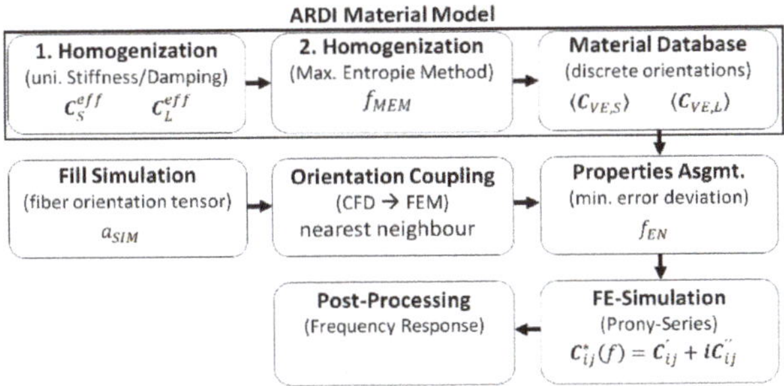

Figure 1. Concept for embedding the Arbitrary-Reconsidered-Double-Inclusion (ARDI) material model in a simulation routine.

3. Arbitrary-Reconsidered-Double-Inclusion (ARDI) Material Model

This section presents the equations of the ARDI model and corresponding, the implementation of the model in a simulation routine. The ARDI model consists of four significant steps: the first and second homogenization step, the generation of a material database and the property assignment.

3.1. First Homogenization Step

The ARDI model is based on the RDI model according to literature [7]. The RDI model is a two-step homogenization method to characterize three-phase composite structures with respect to a unidirectional, transversal-isotropic effective composite stiffness. In semi-crystalline thermoplastics, the three phases are the fibers, the matrix and a highly crystalline interphase between matrix and fiber. The effective stiffness tensor has been calculated by Equation (1):

$$C_S^{eff} = C_S^M + f^I \Delta C_S^{I-M} A^I + f^C \Delta C_S^{C-M} A^C, \tag{1}$$

with

$$\Delta C_S^{I-M} = C_S^I - C_S^M,\tag{2}$$

and

$$\Delta C_S^{C-M} = C_S^C - C_S^M.\tag{3}$$

Here, C_S represents a fourth rank stiffness tensor, f the fiber volume fraction and A a strain concentration tensor. The index ^{M} stands for the matrix, ^{I} for the inclusion (e.g., glass fibers) and ^{C} for the coating or interphase. A detailed derivation of Equation (1) is given in [7]. Assuming that the main damping contribution in the composite is introduced by the matrix [20], Equation (1) has been modified to Equation (4):

$$C_L^{eff} = C_L^M - f^I C_L^M A^I - f^C C_L^M A^C.\tag{4}$$

Here, the fibers and interphases have a negligible damping behavior. Therefore, the determined total damping tensor of the matrix material C_L^M is reduced by the volume fraction of the fibers and the interphases. Furthermore, C_L represents a fourth rank damping tensor.

The tensors C^M, C^I and C^C have been assumed as transversal-isotropic stiffness C_S or damping C_L matrices and have been determined by the inverse of the matrix C^{-1} according to Equation (5):

$$\frac{1}{C_k^\alpha} = \begin{pmatrix} \frac{1}{E_k^\alpha} & -\frac{\nu^\alpha}{E_k^\alpha} & -\frac{\nu^\alpha}{E_k^\alpha} & & 0 & 0 & 0 \\ -\frac{\nu}{E_k^\alpha} & \frac{1}{E_k^\alpha} & -\frac{\nu^\alpha}{E_k^\alpha} & & 0 & 0 & 0 \\ -\frac{\nu^\alpha}{E_k^\alpha} & -\frac{\nu^\alpha}{E_k^\alpha} & \frac{1}{E_k^\alpha} & & 0 & 0 & 0 \\ & 0 \quad 0 \quad 0 & & \frac{2(1+\nu^\alpha)}{E_k^\alpha} & 0 & 0 \\ & 0 \quad 0 \quad 0 & & 0 & \frac{2(1+\nu^\alpha)}{E_k^\alpha} & 0 \\ & 0 \quad 0 \quad 0 & & 0 & 0 & \frac{2(1+\nu^\alpha)}{E_k^\alpha} \end{pmatrix},\tag{5}$$

with E_k^α as unidirectional module and ν^α as Poisson´s ratio of the respective phase $\alpha : M; I; C$ and depending on whether the storage or loss module $k : S; L$ have been used. The values of E_k^α correspond to experimental determined storage E_S^M or loss module E_L^M.

In the next step, the effective tensors have been rotated in discrete directions. For this, the spatial construction of an icosphere has been used. A detailed description can be found in [15]. In the next step, the effective tensors are rotated into a discrete number of directions of the icosphere, according to Equation (6):

$$C_{R,k}^{eff} = K C_k^{eff} K^T.\tag{6}$$

The rotation matrix K is defined by:

$$K = \begin{bmatrix} K^{(1)} & 2K^{(2)} \\ K^{(3)} & K^{(4)} \end{bmatrix},\tag{7}$$

with

$$K_{ij}^{(1)} = \Omega_{ij}^2,\tag{8}$$

$$K_{ij}^{(2)} = \Omega_{imod(j+1,3)}\Omega_{imod(j+2,3)},\tag{9}$$

$$K_{ij}^{(3)} = \Omega_{mod(i+1,3)j}\Omega_{mod(i+2,3)j},\tag{10}$$

$$\begin{aligned} K_{ij}^{(4)} &= \Omega_{mod(i+1,3)mod(j+1,3)}\Omega_{mod(i+2,3)mod(j+2,3)} \\ &+ \Omega_{mod(i+1,3)mod(j+2,3)}\Omega_{mod(i+2,3)mod(j+1,3)}, \end{aligned}\tag{11}$$

and $i, j = 1 \dots 3$. Furthermore:

$$mod(i,3) = \left\{ \begin{array}{ll} i & if\ i \leq 3 \\ i-3 & if\ i > 3 \end{array} \right\}. \tag{12}$$

Ω corresponds to an orthogonal transformation matrix from the basis vectors $e = \{e_1, e_2, e_3\}$ and $m = \{m_1, m_2, m_3\}$, whereby:

$$\Omega_{ij} = m_i \cdot e_j, \tag{13}$$

and

$$\Omega\Omega^T = \Omega^T\Omega = I, \tag{14}$$

with I as unit tensor. A detailed derivation of the rotation matrix K can be found in [21,22].

3.2. Second Homogenization Step

In the second homogenization step, the ODF is reconstructed for a discrete number of eigenvalue combinations according to the MEM. As result, it is possible to calculate the effective composite stiffness and damping over a sum function for all of these discrete eigenvalue combinations.

First, a discrete number of eigenvalue combinations has been generated. The eigenvalues represent the reduced second order orientation tensor and can accept values between the following equations:

$$f_1(x_1) = -x_1 + 1, \tag{15}$$

$$f_2(x_2) = -2x_2 + 1, \tag{16}$$

$$f_3(x_3) = x_3, \tag{17}$$

with $\{x_1 \in \mathbb{R} \mid 0 \leq x_1 \leq 0.5\}$, $\{x_2 \in \mathbb{R} \mid 0 \leq x_2 \leq 1/3\}$ and $\{x_3 \in \mathbb{R} \mid 1/3 \leq x_3 \leq 0.5\}$. The intersection points of these equations represent a two-dimensional triangle. A meshing of the inner area of this triangle has been implemented on basis of a Delaunay method [23]. As a result, the coordinates of the mesh nodes correspond to a discrete number of two dimensional eigenvalue combinations. With regard to the following two boundary conditions:

$$a_{xx} + a_{yy} + a_{zz} = 1, \tag{18}$$

$$a_{xx} \geq a_{yy} \geq a_{zz}, \tag{19}$$

the complete second order orientation tensor:

$$a = \left\{ a_{xx}, a_{yy}, a_{zz} \right\}, \tag{20}$$

can be derived for all of these mesh nodes. Figure 2 shows the meshed two-dimensional triangle with the corresponding eigenvalue combinations.

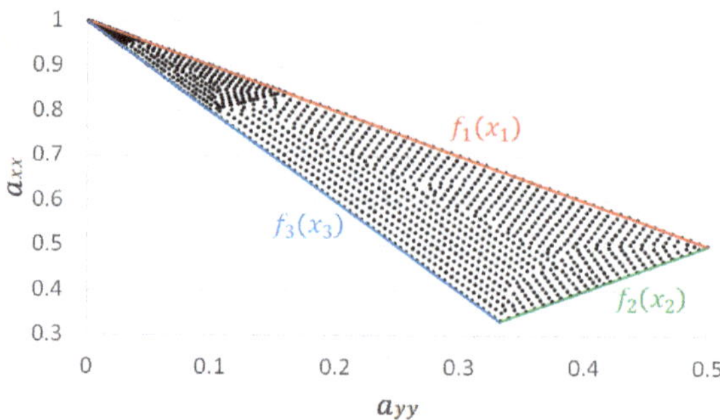

Figure 2. Delaunay-based triangle mesh with the corresponding eigenvalue combinations.

Next, the MEM is applied to reconstruct the ODF for the discrete number of eigenvalue combinations of the previous step. This leads to a discrete number of ODFs. The MEM uses the entropy to assess the coherence to an existing information content. In case of ODF reconstruction, the entropy represents the deviation of the calculated probability. The MEM has been used due to the following advantage: The method is based on a physical approach without the necessary of a closure formulation and can reconstruct a large area of fiber orientation distribution, with respectively small error deviations compared to other reconstruction methods [15].

A Bingham distribution function has been used for reconstructing the ODF. This implementation corresponds to a normal distribution on the unit sphere. The general bivariate MEM has been reformulated to a minimization problem. Thereby the deviation (entropy) between a calculated and a given second order fiber orientation tensor is minimized. The minimization problem can be defined with the MEM as:

$$f_{MEM} = (d_{ix}w_i - a_{xx})^2 + \left(d_{iy}w_i - a_{yy}\right)^2 + (d_{iz}w_i - a_{zz})^2, \tag{21}$$

whereby:

$$d_{ij} = \overline{p}_{ij}^2, \tag{22}$$

and $\overline{p}_{ij}$ correspond to the surface points of the icosphere. a_{xx}, a_{yy} and a_{zz} correspond to the entries of the major axis of a given second order fiber orientation tensor of the previous step. w_i is the probability which the fibers in a volume element point in discrete direction $\overline{p}_i$. A detailed derivation of the corresponding equations of the MEM is given in [15].

3.3. Material Database

For the determination of the representative composite stiffness, the orientation averaging approach after Advani and Tucker [24] has been used. Thereby, the individual stiffness in discrete directions multiplied by the corresponding value of a given ODF results in the effective composite stiffness, according to:

$$\langle C_{VE,S} \rangle \approx \sum_{i=1}^{N} C_{R,S,i}^{eff} \cdot w_i. \tag{23}$$

and transferred to the composite damping applies:

$$\langle C_{VE,L} \rangle \approx \sum_{i=1}^{N} C_{R,L,i}^{eff} \cdot w_i. \tag{24}$$

This approach of a sum function for describing the effective composite stiffness has already been successfully used in existing studies [7,25–28]. The approach of a sum function to describe the effective damping of a composite material represents a new methodology. On the use of a discrete numbers of

ODFs, it is possible to create a material database consisting of a defined number of composite stiffness $\langle C_{VE,S} \rangle$ and damping tensors $\langle C_{VE,L} \rangle$.

In the next step, an assignment of the properties from the material database to the volume elements of the components to be calculated has been performed. It is based on the principle of the smallest error deviation of the Euclidean standard between the fiber orientation tensor of the elements of the components and the stored orientation tensors of the database. Here:

$$f_{EN} = (a_{xx,G} - a_{xx,C})^2 + (a_{yy,G} - a_{yy,C})^2 + (a_{zz,G} - a_{zz,C})^2, \tag{25}$$

has been minimized, where:

$$a_G = \{a_{xx,G}, a_{yy,G}, a_{zz,G}\}, \tag{26}$$

corresponds to the orientation tensors of the volume elements of the component and:

$$a_C = \{a_{xx,C}, a_{yy,C}, a_{zz,C}\}, \tag{27}$$

corresponds to an orientation tensor of the material database.

3.4. User-Defined Material Subroutine

In the present work, the frequency-dependent, linear-viscoelastic material behavior of the plastics has been calculated using a Prony-series into a user subroutine (UMAT) in the FE software Abaqus version 2017. Thereby, the frequency dependent stiffness tensor $C'_{ij}(f)$ has been determined by a Prony-series:

$$C'_{ij}(f) = C'_{ij,0} + \sum_{r=1}^{n} C'_{ij,r} \cdot e^{\frac{(f'-f)}{\tau_r}}. \tag{28}$$

$C'_{ij,0}$ corresponds to the stiffness tensor at start frequency f_0 and is determined by inserting the experimental determined $E_S^M(f_0)$ in Formula (5) and inversion. The Prony-series elements of the stiffness matrix result from the evaluation of $E_S^M(f)$ at the selected support point r. For the Prony support points result:

$$E_S^M\left(f_0 + \frac{(f_{max} - f_0)}{n} \cdot r\right), \tag{29}$$

where n is the total number of selected interpolation support points. The Prony support point E_S^M has been inserted in Equation (5) and inverted to the Prony stiffness tensor $C'_{ij,r}$. Following, the stress of a volume element with viscoelastic material behavior can be defined as:

$$\sigma_i = \int_{f_0}^{f} C'_{ij}(f) \cdot \dot{\varepsilon}_j \, df, \tag{30}$$

whereby $\bar{\varepsilon}_{j,r}$ is determined as Prony strain. A detailed description of the used Prony series formulation can be found in [14].

In the present work, an extended approach is followed to consider the frequency-dependent damping behavior. The frequency-dependent stiffness defined in Equation (30) has been extended by the damping tensor to:

$$\sigma_i = \int_{f_0}^{f} \left(C'_{ij}(f) + i C''_{ij}(f) \right) \cdot \dot{\varepsilon}_j \, df, \tag{31}$$

where $C_{ij}''(f)$ corresponds to the frequency-dependent damping tensor and i to the complex number. Thus:

$$C_{ij}^*(f) = C_{ij}'(f) + iC_{ij}''(f), \tag{32}$$

can be treated as a complex module. The frequency-dependent damping tensor $C_{ij}''(f)$ is also determined by a Prony-series with

$$C_{ij}''(f) = C_{ij,0}'' + \sum_{r=1}^{n} C_{ij,r}'' \cdot e^{\frac{(f'-f)}{\tau_r}}. \tag{33}$$

$C_{ij,0}''$ corresponds to the damping tensor at start frequency and $C_{ij,r}''$ corresponds to the Prony damping part, which are formulated with $E_L^M(f)$, analogous to the procedure described above.

4. Bending Vibration Tests

The determination of the dynamic material properties of the glass fiber-reinforced plastics has been carried out on bending specimens with the dimension $180 \times 15 \times 2$ mm which have been prepared from an injection molded plate material at 0° (Inflow - IF), 45° (Midflow - MF) and 90° (Crossflow - CF) to the flow direction, according to Figure 3. The injection molded plate material dimension is $180 \times 180 \times 2$ mm. With regard to Figure 3, the bending specimens of the pure matrix plate material have been prepared as 0° oriented respected to the injection flow direction. Preliminary investigations show that the dynamic material properties of the pure matrix material do not differ with variation of the direction of orientation.

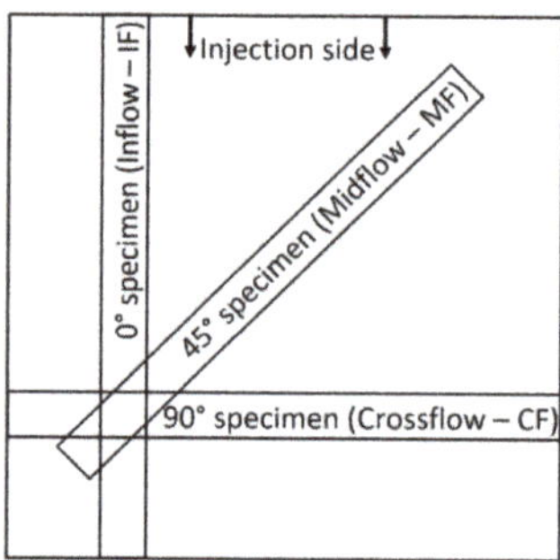

Figure 3. Positions of the 0° (IF), 45° (MF) and 90° (CF) orientated bending specimens.

The bending specimens have been applied to a connection structure via a clamp fit. A shaker with stinger connection to this structure allows a uniaxial vibration excitation of the test specimens. Furthermore, the testing structure has been placed in a climatic chamber and allows a vibration excitation of the specimens under defined temperature and humidity conditions.

To excite the specimens, the shaker was driven with a defined sine sweep in the range from 100 Hz to 8 kHz. The acceleration of the specimens under vibration excitation has been recorded by using the PSV-500-3D scanning laser Vibrometer of the company Polytec. The excitation force of the specimens has been recorded using a Type 8001 impedance sensor of the company Brüel & Kjaer. Thus, it is possible to determine the FRF of the bending specimens as ratio between the displacement of the lower end of the specimens and the applied excitation force. A detailed description of the method used to characterize the dynamic material properties is given in [16]. Figure 4 shows schematically the used test arrangement with resulting FRF.

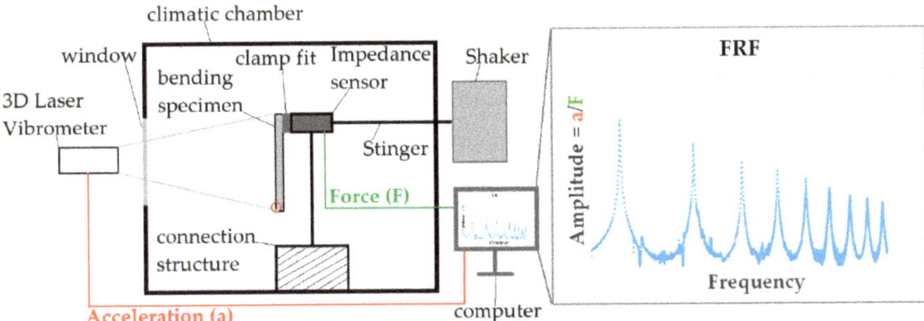

Figure 4. Principle sketch of the used bending vibration test arrangement.

Assuming a fixed-free connection and excitation of an Euler-Bernoulli beam, the *m*-th natural frequency $\omega_{0,m}$ has been obtained according to:

$$\omega_{0,m} = (\kappa_m l)^2 \cdot \sqrt{\frac{E'I}{\rho A_C l^4}}, \tag{34}$$

where E' is the storage modulus of the material, I is the moment of inertia of the bending specimen, ρ is the density of the material, A_C is the cross-sectional area of the bending specimen, l is the length of the specimen. $\kappa_m l$ is a characteristic factor for describing the shape of natural oscillation as a function of the bearing of the beam. This can be determined using:

$$\kappa_m l = (2m - 1)\frac{\pi}{2}, \tag{35}$$

for the *m*-th natural frequency and $m \in \mathbb{N}$. Switching Equation (34) to the storage module results with:

$$E' = \frac{\omega_{0,m}^2}{(\kappa_m l)^4} \cdot \frac{\rho A_C l^4}{I}. \tag{36}$$

Equation (36) allows to evaluate the FRF of the bending vibration tests at the occurring resonance frequencies to the storage module. Using linear inter- and extrapolation, the storage module as function of the frequency has been determined. Furthermore, under conditioning and testing the plastics in the climatic chamber, it is possible to determine the frequency-, temperature- and humidity-dependent storage modulus $E_S^M(f, T, rh)$.

Following, the dissipation factor $\tan(\delta)$ at *m*-th natural frequency can be determined using the 3 dB method [18] according to Figure 5.

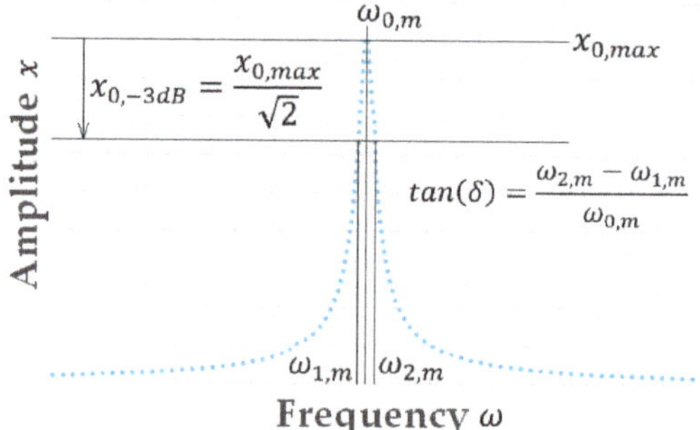

Figure 5. Estimation of the dissipation factor $tan(\delta)$ at m -th natural frequency of a FRF.

Equation (37) describes the relationship between dissipation factor, storage and loss module with:

$$tan(\delta) = \frac{E''}{E'}.$$
(37)

Thus, it is possible to convert Equation (37) to the loss modulus E'' and determine the loss modulus as function of the frequency, temperature and humidity $E_L^M(f, T, rh)$ by linear inter- and extrapolation over the frequency. As result, the values of E_S^M and E_L^M allow calibrating the ARDI model.

5. Simulation

In the present work, the developed ARDI material model has been applied to investigate the structural dynamics of a bending specimen. Starting point is the geometry of the bending specimen with the dimension 180x15x2mm. Using Abaqus CAE version 2017, this geometry has been discretized into 27,000 C3D8 tetrahedral elements. This corresponds to an element size of 1x1x0.2 mm and thus 10 elements across the thickness. A fixed-free boundary condition has been applied. As result, in the simulation, the upper area of the bending specimen has been excited unidirectional with a force amplitude, whereas all other directions have been blocked against movement. Figure 6 shows the discretized bending specimen with applied boundary conditions.

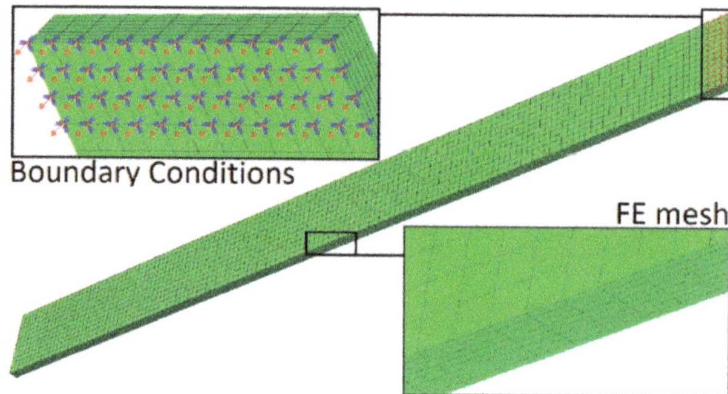

Figure 6. Finite-element (FE) mesh of the discretized bending specimen with position of applied boundary conditions.

6. Results

6.1. Stiffness and Damping Behaviour

The unidirectional, transversal-isotropic, fourth ranked stiffness C_S^{eff} and damping C_L^{eff} tensors were rotated by the rotation matrix K into different planar spatial directions of the icosphere and thus resulting fiber directions in the plain, corresponding to Figure 7.

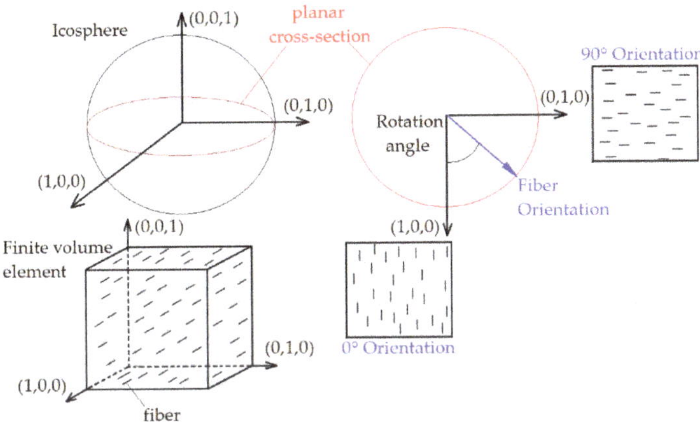

Figure 7. Schematic explanation of the rotating procedure of the unidirectional, transversal-isotropic tensors in different fiber directions.

$C_{k,1111}^{eff}$ describes the modulus orientated to the main global $(\mathbf{1,0,0})$–direction and $k : S; L$ depending on whether the experimental determined storage S or loss modulus L has been used to calibrate the tensor. Following, $C_{k,2222}^{eff}$ describes the modulus in the $(0,1,0)$–direction. $C_{k,3333}^{eff}$ is equal to $C_{k,2222}^{eff}$ due to the used transversal-isotropic behavior. and $C_{k,2323}^{eff}$ describes the main transverse modulus components.

Figure 8 shows the main matrix entries of the fourth ranked stiffness C_S^{eff} and damping tensor C_L^{eff} of PA66-GF30 at 23 °C and 0% rh. Figure 8a shows that the stiffness in the main axis direction $C_{S,1111}^{eff}$ is highest, when all fibers pointing in the main $(1,0,0)$–direction (0°). As the angle deviation of the fibers from the main axis increases, the stiffness $C_{S,1111}^{eff}$ decreases, whereas the stiffness perpendicular to the main axis $C_{S,2222}^{eff}$ remains relatively constant. This applies up to a rotation angle of the fibers of about 45°. The transverse modulus $C_{S,1212}^{eff}$ increases continuously with a rotation from 0 to 45° and reaches its maximum under 45° fiber orientation. At this point, the trend of the modulus $C_{S,1111}^{eff}$ and $C_{S,2222}^{eff}$ reverses for a further rotation from 45 to 90°. In this case, the transverse modulus $C_{S,2323}^{eff}$ is constant due to the planar rotation.

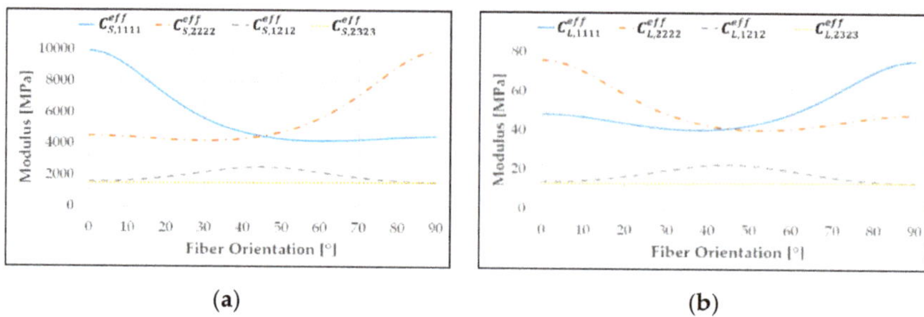

Figure 8. Components of the fourth ranked (**a**) stiffness and (**b**) damping tensors of PA66-GF30 at 23 °C and 0% rh.

Figure 8b shows that the damping $C^{eff}_{L,1111}$ is lowest, when all fibers pointing in the main $(1,0,0)$–direction (0°), while the damping perpendicular $C^{eff}_{L,2222}$ is highest. This is due to the used definition of the damping tensor. Here, the strain concentration tensor, which includes the geometry tensor of the ellipsoid inclusions, increase the more fibers point in the main axis direction. Thus, the effective composite damping decrease.

As the angle deviation of the fibers from the main axis increases, the damping perpendicular $C^{eff}_{L,2222}$ decreases, whereas the damping in the main axis $C^{eff}_{L,1111}$ remains relatively constant up to a rotation angle of the fibers of about 45°. Following, the trend of the modulus $C^{eff}_{L,1111}$ and $C^{eff}_{L,2222}$ reverses for a further rotation from 45 to 90°, comparable to the stiffness tensor.

Figure 9 shows the matrix entries of the fourth ranked stiffness C^{eff}_S and damping tensor C^{eff}_L of PPA-GF50 at 23 °C and 0% rh. Due to the same definition of the stiffness and damping tensors, the trend of the curves of the tensor components of PPA-GF50 are comparable to the PA66-GF30, according to Figure 8. Due to the increased fiber volume fraction, the stiffness tensor of the PPA-GF50 is increased. The damping tensors of the PPA-GF50 are comparable to the PA66-GF30. This is due to an increased damping tensor of the pure PPA matrix material, but stronger reduction of this damping tensor due to an increased fiber volume fraction.

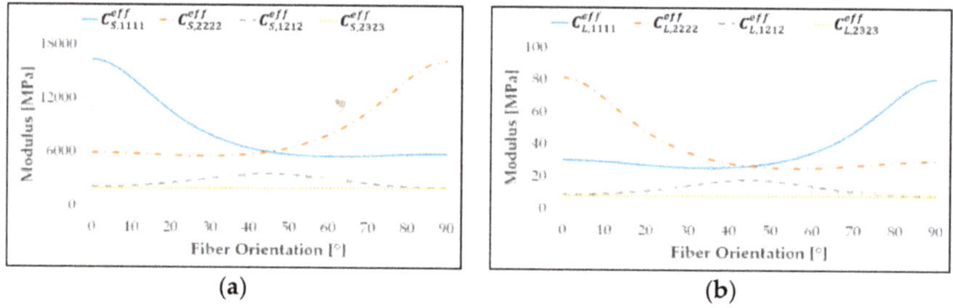

Figure 9. Components of the fourth ranked (**a**) stiffness and (**b**) damping tensors of PPA-GF50 at 23 °C and 0% rh.

Figure 10 shows the main matrix entry of the stiffness $C^{eff}_{S,1111}$ and damping tensor $C^{eff}_{L,1111}$ of PA66-GF30 at 23 °C and 0% rh, 23 °C and 50% rh, 130 °C and 0% rh. Figure 10a shows that the stiffness decreases with increasing humidity and temperature. Figure 10b shows that damping increases sharply with an increase in humidity and temperature. The influence of temperature increase is more pronounced than the influence of humidity increase. Both, the decrease in stiffness and the increase in damping is generally due to the softening behavior of the matrix material with an increase in humidity or temperature [29,30].

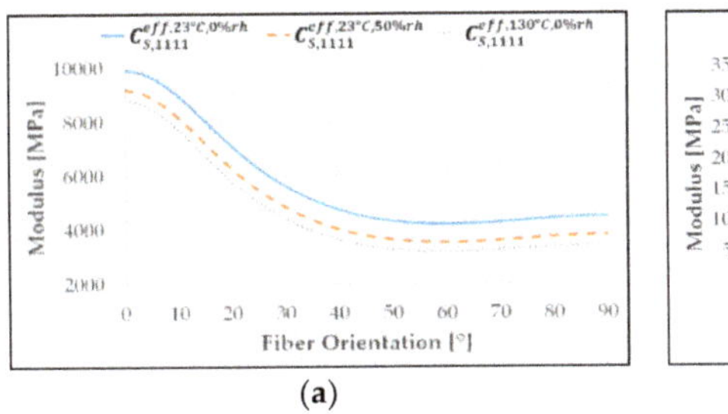
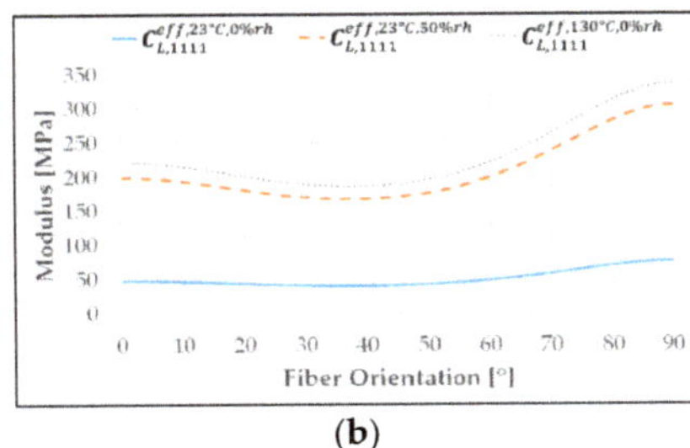

(a) **(b)**

Figure 10. Components of the fourth ranked (**a**) stiffness and (**b**) damping tensors of PA66-GF30 at 23 °C and 0% rh, 23 °C and 50% rh, 130 °C and 0% rh.

Figure 11 shows the main matrix entry of the stiffness $C^{eff}_{S,1111}$ and damping tensor $C^{eff}_{L,1111}$ of PPA-GF50 at 23 °C and 0% rh, 23 °C and 50% rh, 130 °C and 0% rh. In comparison to Figure 10, Figure 11 also shows that the stiffness of the PPA-GF50 decreases with an increase in humidity and temperature, while damping increases. In comparison to PA66-GF30, the influence of the increase in humidity on the stiffness and damping of the PPA-GF50 is less pronounced, as the PPA matrix material has a reduced water absorption capacity [31].

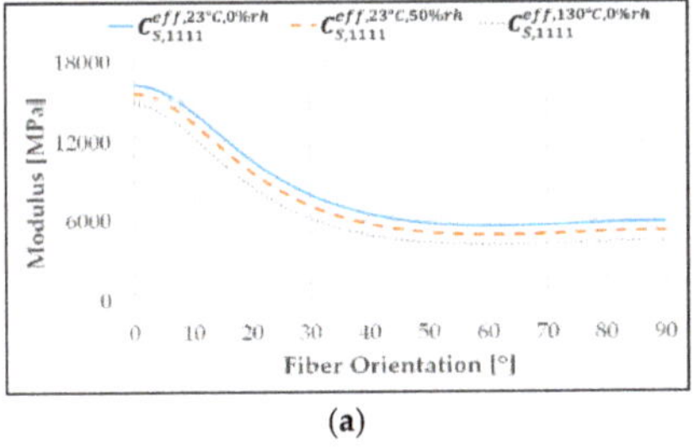
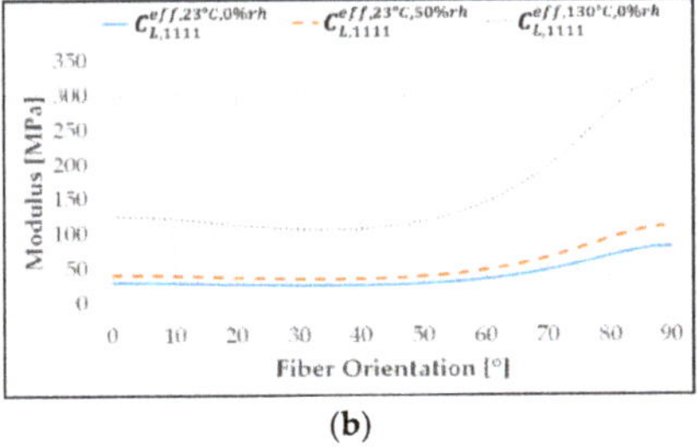

(a) **(b)**

Figure 11. Components of the fourth ranked (**a**) stiffness and (**b**) damping tensors of PPA-GF50 at 23 °C and 0% rh, 23 °C and 50% rh, 130 °C and 0% rh.

6.2. μ-CT Measurement and Filling Simulation

The convention of the axis system of the μ-CT specimens is shown schematically in Figure 12. In this case, the z-direction corresponds to the injection flow direction. Using the filling simulation of the three-dimensional geometry of the plate material, it is possible to derive the second order fiber orientation tensor a_{SIM}. The filling simulation in this work was performed with the software CADMould by the company PART Engineering.

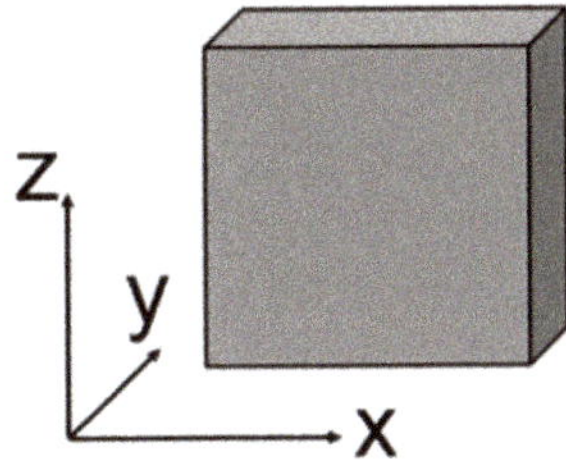

Figure 12. μ-CT specimen with convention of the associated axis system.

In the first step of the filling simulation, a CFD mesh of the plate material has been generated. Then the process parameters for the production of the material plate have been set. The process parameters of the PA66-GF30 originates from the material database of the CADMould software and the parameters of the PPA-GF50 have been taken from a specification of the manufacturer. In the next step, the filling simulation has been performed. The fiber orientation tensors of the filling simulation a_{SIM} of the plate material have been evaluated at the same center positions of the plate material comparable to the μ-CT investigations. The measured a_{CT} and calculated a_{SIM} fiber orientation tensors of the PA66-GF30 plate material are shown in Figure 13.

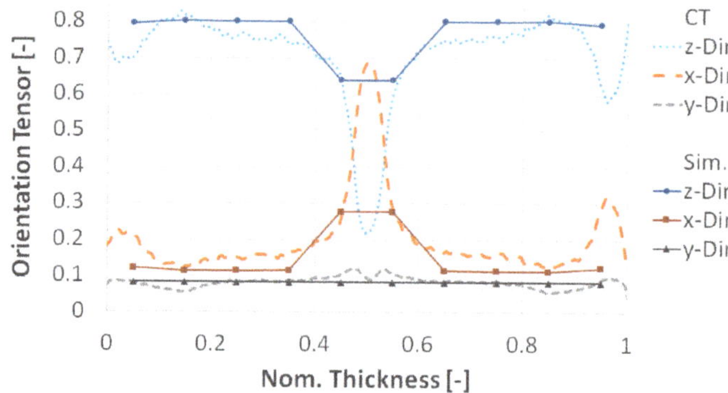

Figure 13. Fiber orientation tensors of the measured (CT) and calculated (Sim.) PA66-GF30 specimens.

Figure 13 shows that the PA66-GF30 has a narrow, but very pronounced edge and middle layer. Furthermore, Figure 13 shows a good correlation of the trend of the measured fiber orientation tensors of the PA66-GF30 in comparison to the filling simulation. Only the pronounced orientation gradient in the middle layer is not covered due to the absence of an evaluation point in the middle of the plate. This is associated with loss of information. As result, the orientation tensor in flow direction is slightly increased in the middle of the flow channel.

The measured a_{CT} and calculated a_{SIM} fiber orientation tensors of the PPA-GF50 plate material are shown in Figure 14. Figure 14 shows that the PPA-GF50 has a continuous transition from edge to middle layer. As result, there is a broad and strong pronounced middle layer. Figure 14 also shows that the continuous transition and thus the change in the orientation gradient of the PPA-GF50 material does not show a good correspondence to the filling simulation. The result is a loss of information in the middle of the flow channel comparable to the PA66-GF30, which also leads to an increase of the orientation tensor in the flow direction.

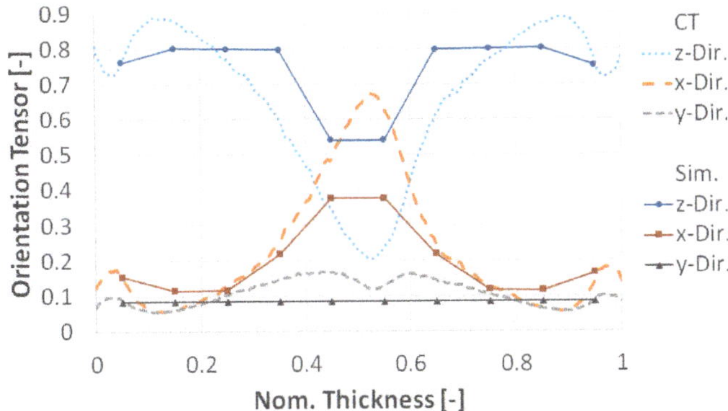

Figure 14. Fiber orientation tensors of the measured (CT) and calculated (Sim.) PPA-GF50 specimens.

The comparison shows that the calculated fiber orientation tensors of the filling simulation are sufficient to be used for the calculation of the direction-dependent material properties. In this case, an adjustment of the settings of the filling simulation was purposeful.

6.3. Orientation Tensor Mapping Procedure

In order use the calculated second order fiber orientation tensor a_{SIM} for the FE simulation of the structural dynamics, the data of the filling simulation (CFD) have been transferred. This has been done using the nearest neighbor principle.

The vector distance between the nodes of the CFD and FE mesh has been evaluated. Any node of the FE mesh with the smallest distance to another node of the CFD mesh has been assigned to each other. As result, the eigenvalues have been transferred from the CFD to the FE mesh. The assignment of the eigenvectors of the elements has been performed to the same principle. The eigenvectors describe the position of the elements (local coordinate system) relative to a base (global coordinate system).

In order to keep the loss of information low, a fine discretization of the FE mesh is necessary. For the mapping process the software Converse has been used. In addition, [10,32] describes the principle of the nearest neighbor in more detail. Figure 15 shows the fiber orientation tensor of the filling simulation of the plate material, as well as the transferred fiber orientation tensor to the geometry of an IF bending specimen.

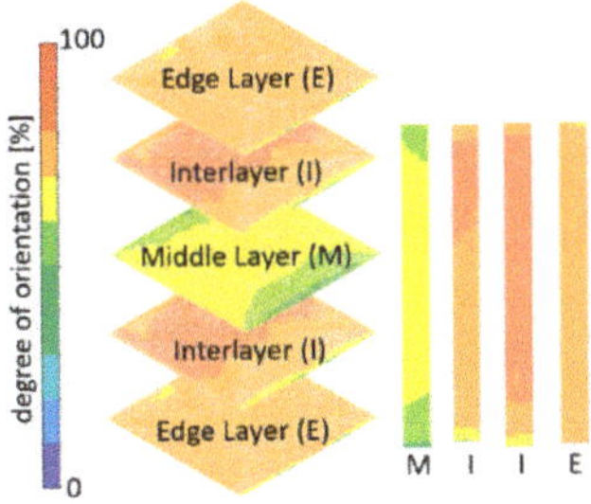

Figure 15. Fiber orientation of the fill simulation of the plate material and mapped bending specimen.

6.4. Comparison between RDI and the ARDI Model

Figure 16a shows the FRF of the RDI model combined with the approach of a GMD behavior.

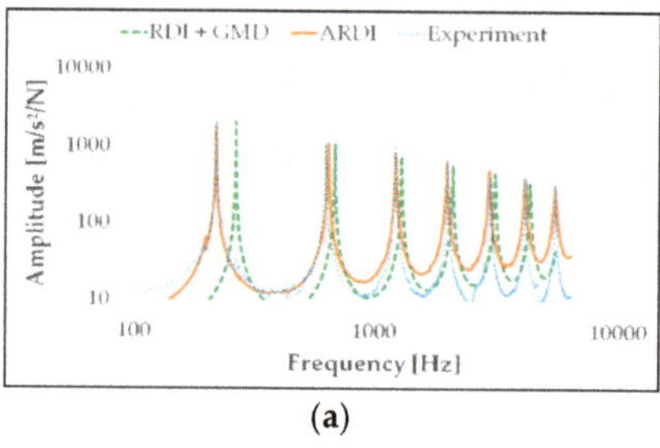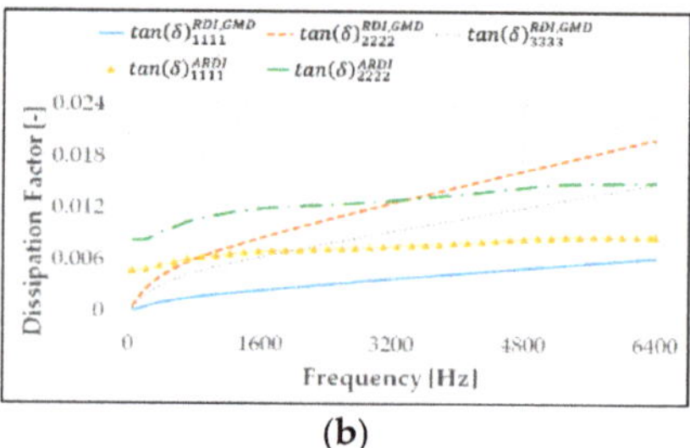

(a)   (b)

Figure 16. (**a**) Frequency response function (FRF) of the Reconsidered-Double-Inclusion (RDI) model with global modal damping (GMD), the ARDI model and the experiment of PA66-GF30 of the IF orientation at 23 °C and 0% rh and (**b**) dissipation factors of the RDI and ARDI model.

The RDI model is based on the definition of a homogenized, unidirectional, transversal-isotropic stiffness tensor and does not provide a reconstruction of ODFs for different fiber orientation tensors. As result, the stiffness tensor can only be rotated in discrete spatial directions.

Following, the comparison of the FRF of the RDI model with the experiment shows that the natural frequencies are shifted towards higher frequencies, corresponding to Figure 16a. This is due to a stiffer behavior of the modelled stiffness tensor of the RDI model compared to the experiments. In structural dynamics, an increase in the stiffness is shown by a shift of the natural frequencies to higher frequencies. This is derived from the resonance-curve method and Equations (34) to (36).

Furthermore, Figure 16a shows the FRF of the extended ARDI model. Using the MEM to reconstruct discrete ODFs, the effective composite stiffness can be reconstructed very well by the ARDI model. The result is that the ARDI model shows a good agreement of the natural frequencies of the FRF compared to the experiment.

Corresponding to the 3 dB method, the amplitude of the FRF is a parameter to assess the damping of a system. In the present case, both the RDI model with GMD behavior and the ARDI model shows a good agreement of the amplitude of the FRF in comparison with the experiment. The optimization of the GMD behavior has been performed numerically in a reverse engineering approach using Altair HyperStudy software and the FRF of the experiment.

Figure 16b shows the dissipation factors $tan(\delta)$ as ratios of the main entries of the stiffness and damping tensors of the RDI and ARDI model as a function of frequency. The dissipation factors of the RDI model $tan(\delta)^{RDI,GMD}$ start uniformly at 0 Hz with a value of 0. After a curved increase a linear course follows with different slopes up to the maximum frequency. The dissipation factor pointing in the main axis direction $tan(\delta)^{RDI,GMD}_{1111}$ is lowest and the dissipation factor perpendicular to the main axis direction $tan(\delta)^{RDI,GMD}_{2222}$ is highest. The dissipation factor in thickness direction $tan(\delta)^{RDI,GMD}_{3333}$ is slightly reduced compared to $tan(\delta)^{RDI,GMD}_{2222}$.

In comparison, the dissipation factors of the ARDI model $tan(\delta)^{ARDI}$ do not start with a value of 0 at 0 Hz, whereas the further course is approximately linear to maximum frequency. Furthermore, the dissipation factor in the main axis direction $tan(\delta)^{ARDI}_{1111}$ is lowest and the dissipation factor perpendicular to the main axis direction $tan(\delta)^{ARDI}_{2222}$ is highest. The dissipation factor in thickness direction $tan(\delta)^{ARDI}_{3333}$ is identical to the dissipation factor $tan(\delta)^{ARDI}_{2222}$, due to the used transversal-isotropic behavior.

On the one hand, the comparison shows that the reverse engineering approach is a good method to determine the dissipation factors and thus the components of the damping tensor, if there exists no knowledge about the damping of an entire system. This is shown by a good agreement of the amplitudes of the FRF of the RDI model with GMD behavior compared to the experiment.

On the other hand, the stiffness tensor of the RDI model does not represent a reliable input for the simulation due to a too stiff behavior. This shows the main disadvantage of the reverse engineering approach, it is a pure numerical optimization of the damping behavior based on the defined input of the simulation.

In comparison, the ARDI model provides a material-specific description of the stiffness and damping tensors based on calibration data of bending vibration tests. This is shown by a good agreement of the natural frequencies and the amplitudes of the FRF of the ARDI model compared to the experiment.

6.5. FRFs of ARDI Model

Figure 17 shows the FRF of the IF, MF and CF bending specimens of the PA66-GF30 at 23 °C and 0% rh. Figure 17 shows that the IF-FRF has the highest stiffness (high natural frequencies). This is due to the high orientation of the fibers in the flow direction of the IF specimens, which correlates with the resulting stiffness. As result, the stiffness of the PA66-GF30 decreases (reduced natural frequencies) with increasing deviation to the IF direction of the material.

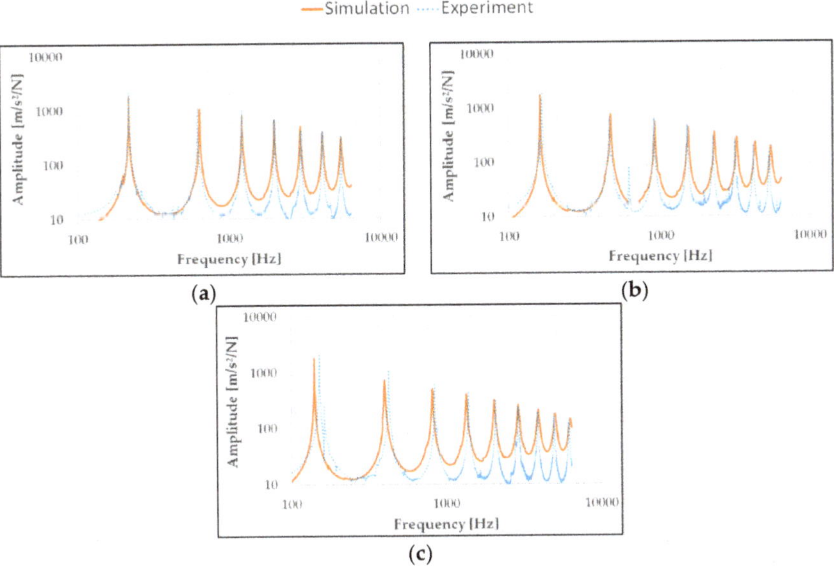

Figure 17. FRF of the simulation and the experiments of PA66-GF30 of the IF (**a**), MF (**b**) and CF (**c**) orientation at 23 °C and 0% rh.

Furthermore, the damping increases with increasing deviation of the fibers from the direction of propagation of the vibrations. Energy dissipation in the matrix and at the fibers can be considered as potential physical reason. The more the fibers are transverse to the direction of loading, the more energy is dissipated at the fibers and the interphases. Vice versa, the more fibers in the direction of the vibrations, the lower the damping, since the fibers and interphases act like a vibration conductor. This effect correlates to a number of existing investigations such as [33,34]. Corresponding to Figure 17, the damping of the PA66-GF30 increases (reduced amplitudes) with increasing deviation to the IF direction. The FRF of the PPA-GF50 according to Figure 18 shows a similar behavior.

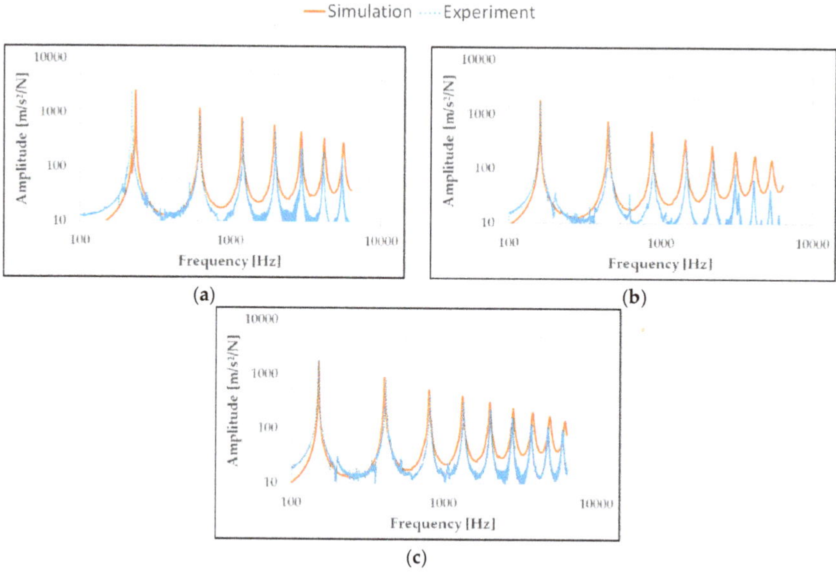

Figure 18. FRF of the simulation and the experiments of PPA-GF50 of the IF (**a**), MF (**b**) and CF (**c**) orientation at 23 °C and 0% rh.

The MF orientation of the PPA-GF50 according to Figure 18b shows a significant deviation of the amplitude towards higher frequencies and can be traced back to an error during the execution of the bending vibration tests.

Furthermore, all FRFs of Figures 17 and 18 show that the oscillation grounds in the simulation are shifted towards higher amplitudes with increasing frequency. This is mainly due to the defined modelling of the damping using the ARDI model with inter- and extrapolation via Prony-series formulation. The evaluation of the damping behavior of the glass fiber-reinforced PA66-GF30, as well as the PPA-GF50, generally tends to a non-linear course of the damping, especially in the direction of higher frequencies.

Figure 19 shows the IF-FRF of the PA66-GF30 bending specimens at 23 °C, 50% rh and 130 °C, 0% rh. It shows that an increase in temperature and/or humidity leads to a strong reduction in stiffness (decrease of natural frequencies) and a strong increase in damping (decrease of amplitude). This corresponds to the results of chapter 6.1. In the present case with selected test conditions, the influence of the temperature increase on stiffness and damping is more pronounced than the increase in humidity. The PPA-GF50 shows a similar behavior with an increase in temperature and humidity, according to Figure 20.

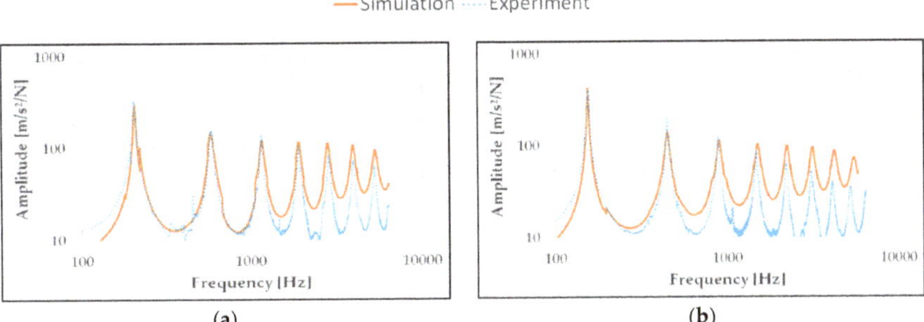

Figure 19. FRF of the simulation and the experiments of PA66-GF30 of the IF orientation at 23 °C and 50% rh (**a**) and 130 °C and 0% rh (**b**).

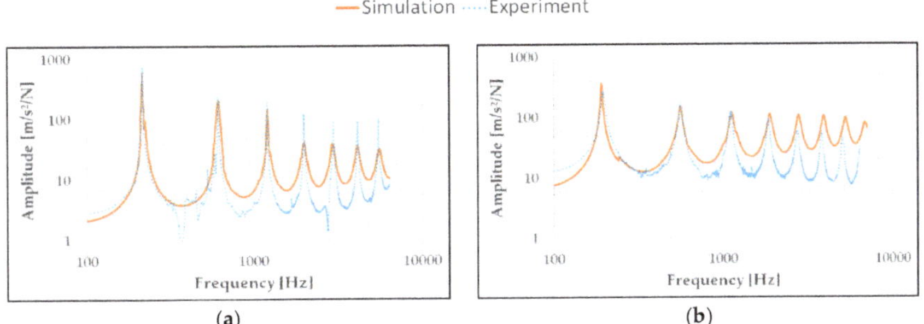

Figure 20. FRF of the simulation and the experiments of PPA-GF50 of the IF orientation at 23 °C and 50% rh (**a**) and 130 °C and 0% rh (**b**).

The comparison between the FRFs of the simulation with the ARDI model and the experiments generally shows a good correlation of the natural frequencies up to 6 kHz and a good correlation of the amplitude up to 1 kHz.

The ARDI model is not sufficient above 1 kHz to describe the damping behavior. This applies in particular to an increase in temperature or humidity. This is shown by a strong shift of the oscillation grounds in the simulation with increasing frequency in the direction of higher amplitudes and tends to the assumption, that the damping behavior is nonlinear.

On the other hand, this effect is enhanced by the used calibration method. The ARDI model have been calibrated with the stiffness and damping of the pure matrix material. However, due to the increased humidity and temperature, only a small number of natural frequencies can be evaluated with the usage of the resonance-curve method and the 3 dB method below 1 kHz. Higher frequency ranges are too noisy and do not provide acceptable values for the stiffness and damping. Consequently, the linear extrapolated stiffness and damping tensors are not sufficient to describe the structural dynamics behavior on the use of the ARDI model.

7. Conclusions

In the present work, the novel ARDI model for the description of the anisotropic, viscoelastic stiffness and damping of short glass fiber-reinforced thermoplastics has been presented.

The comparison between simulation and experiment of the FRF of an IF-, MF- and CF-bending specimen of the PA66-GF30 and PPA-GF50 shows in general a sufficient correlation of the natural frequencies and the amplitudes. Thus, the ARDI model is in general able to predict the frequency-, temperature- and humidity-depending stiffness and damping of short glass fiber-reinforced thermoplastics.

At higher frequencies, the linear inter- and extrapolation of the experimentally determined damping results in a shift of the oscillation grounds to higher amplitudes and thus in a significant deviation from the experiments. This effect is more pronounced with an increase in humidity and temperature and tends to the assumption, that the damping behavior is nonlinear. Due to the small number of investigated conditioning states, a nonlinear course of the damping can only be assumed as possible reason for the deviation of the structural dynamics behavior.

Taking into account that the properties of the composite are homogenized from the properties of the matrix, the fibers and the interphase and the resulting composite properties are linearly inter- and extrapolated, the ARDI model provides a good prognosis for the structural dynamics of short glass fiber-reinforced thermoplastics.

Further investigations concentrate on the execution and evaluation of bending vibration tests with a larger variation of humidity and temperature gradients. These tests form the basis for reconsidering the formulation of the stiffness and damping tensors of the ARDI model.

Author Contributions: A.K. and M.S. conceived the investigation. A.K. wrote the software. A.K. and M.S. analyzed the data. A.K. did the visualization. Writing the paper was done by all two authors. All authors have read and agreed to the published version of the manuscript.

Funding: This research received no external funding.

Conflicts of Interest: The authors declare no conflict of interest.

References

1. Raschke, K.; Korte, W. Faserverstärkte Motorbauteile besser berechnen. *Kunststoffe* **2019**, *109*, 184–189.
2. Eshelby, J.D. The determination of the elastic field of an ellipsoidal inclusion, and related problems. *Proc. R. Soc. Lon Ser. A* **1957**, 376–396.
3. Mori, T.; Tanaka, K. Average stress in matrix and average elastic energy of materials with misfitting inclusions. *ActaMetall* **1973**, *21*, 571–574. [CrossRef]
4. Hill, R. A self-consistent mechanics of composite materials. *J. Mech. Phys. Solids* **1965**, *13*, 213–222. [CrossRef]
5. Benveniste, Y. A new approach to the application of Mori–Tanaka's theory in composite materials. *Mech. Mater* **1987**, *6*, 147–157. [CrossRef]
6. Wu, L.; Adam, L.; Doghri, I.; Noels, L. An incremental-secant mean-field homogenization method with second statistical moments for elasto-visco-plastic composite materials. *Mech. Mater.* **2017**, *114*, 180–200. [CrossRef]
7. Schöneich, M.M. Charakterisierung und Modellierung viskoelastischer Eigenschaften von kurzglasfaserverstärkten Thermoplasten mit Faser-Matrix Interphase. Ph.D. Thesis, Saarland University, Saarbrücken, Germany, 2016.
8. Christensen, R.M. A critical evaluation for a class of micro-mechanics models. *J. Mech. Phys. Solids* **1990**, *38*, 379–404. [CrossRef]
9. Kaiser, J.-M. Beitrag zur mikromechanischen Berechnung kurzfaserverstärkter Kunststoffe–Deformation und Versagen. Ph.D. Thesis, Saarland University, Saarbrücken, Germany, 2013.
10. Stommel, M.; Stojek, M.; Korte, W. Prozess-Struktur-Kopplung. In *FEM zur Berechnung von Kunststoff- und Elastomerbauteilen*, 2nd ed.; Carl Hanser Verlag: München, Germany, 2018; pp. 289–292.
11. Wandinger, J. Dämpfung. In *Elastodynamik 2*; Hochschule für angewandte Wissenschaften München; 2007; Available online: http://wandinger.userweb.mwn.de/LA_Elastodynamik_2/v5.pdf (accessed on 21 February 2020).
12. Glißmann, M. Beanspruchungsgerechte Charakterisierung des mechanischen Werkstoffverhaltens thermoplastischer Kunststoffe in der Finite-Elemente-Methode (FEM). Ph.D. Thesis, RWTH Aachen University, Aachen, Germany, 2003.
13. Park, S.W.; Schapery, R. Methods of interconversion between linear viscoelastic material functions. Part I - A numerical method based on Prony series. *Int. J. Solids Struct.* **1998**, *36*, 1653–1675. [CrossRef]
14. Breuer, K.; Schöneich, M.; Stommel, M. Viscoelasticity of short fiber composites in the time domain: From three-phases micromechanics to finite element analyses. *Contin. Mech. Thermodyn.* **2019**, *31*, 363–372. [CrossRef]

15. Breuer, K.; Stommel, M.; Korte, W. Analysis and Evaluation of Fiber Orientation Reconstruction Methods. *J. Compos. Sci.* **2019**, *3*, 67. [CrossRef]
16. Pfeifer, F.; Armbruster, B. Vorrichtung zur Schwingungsanregung eines Probekörpers. Patent 10,2019,001,226,7, 14 August 2019.
17. DIN EN ISO 6721-3. *Determination of Dynamic Mechanical Properties–Part 3: Flexural Vibration, Resonance-Curve*; Beuth Verlag GmbH: Berlin, Germany, 1996.
18. Papagiannopoulos, G.A.; Hatzigeorgiou, G.D. On the use of the half-power bandwidth method to estimate damping in building structures. *Soil Dyn. Earthq. Eng.* **2011**, *31*, 1075–1079. [CrossRef]
19. Folgar, F.; Tucker, C.L. Orientation behavior of fibres in concentrated suspensions. *J. Rheol.* **1982**, *26*, 98–119.
20. Lenk, P.; Coult, G. Damping of Glass Structures and Components. In Proceedings of the Challenging Glass 2–Conference on Architectural and Structural Applications of Glass, Delft University of Technology, Delft, The Netherlands, 20–21 May 2010.
21. Bower, A.F. Basis change formulas for anisotropic elastic constants. In *Applied Mechanics of Solids*; CRC Press, Taylor & Francis Group: Boca Raton, FL, USA, 2010.
22. Ting, T.C.T. Transformation of C and s. In *Anisotropic Elasticity–Theory and Applications*; Oxford University Press Inc.: New York, NY, USA, 1996; pp. 53–56.
23. Engwirda, D. Locally optimal Delaunay-refinement and optimisation-based mesh generation. Ph.D. Thesis, University of Sydney, Sydney, Australia, 2014.
24. Advani, S.G.; Tucker, C.L. The use of tensors to describe and predict fiber orientation in short fiber composites. *J. Rheol.* **1987**, *31*, 751–784. [CrossRef]
25. Doghri, I.; Tinel, L. Micromechanics of inelastic composites with misaligned inclusions: Numerical treatment of orientation. *Comput. Methods Appl. Mech. Eng.* **2006**, *195*, 1387–1406. [CrossRef]
26. Camacho, C.W.; Tucker, C.L.; Yalvac, S.; McGee, R.L. Stiffness and thermal expansion predictions for hybrid short fiber composites. *Polym. Compos.* **1990**, *11*, 229–239. [CrossRef]
27. Lielens, G. Micro-macro modeling of structured materials. Ph.D. Thesis, Katholische Universität Löwen, Leuven, Belgium, 1999.
28. Pierard, O.; Friebel, C.; Doghri, I. Mean-field homogenization of multi-phase thermo-elastic composites: A general framework and its validation. *Compos. Sci. Technol.* **2004**, *64*, 1587–1603. [CrossRef]
29. Jia, N.; Kagan, V.A. Mechanical Performance of Polyamides with Influence of Moisture and Temperature Accurate Evaluation and Better Understanding. *Plast. Fail. Anal. Prev.* **2001**, *2001*, 95–104.
30. Ultramid® A3W. CAMPUS®—A material information system for the plastics industry. CWFG GmbH:: Frankfurt, Germany, 2020. Available online: https://www.campusplastics.com/material/pdf/113709/UltramidA3W?sLg=en (accessed on 30 March 2020).
31. Ultramid® T KR 4350. CAMPUS®—A material information system for the plastics industry. CWFG GmbH: Frankfurt, Germany, 2020. Available online: https://www.campusplastics.com/material/pdf/113792/UltramidTKR4350?sLg=en (accessed on 30 March 2020).
32. Samworth, R.J. Optimal weighted nearest neighbour classifiers. *Ann. Stat.* **2012**, *40*, 2733–2763. [CrossRef]
33. Bozkurt, Ö.; Bulut, M.; Özbek, Ö. Effect of Fibre Orientations on Damping and Vibration Characteristics of Basalt Epoxy Composite Laminates. In Proceedings of the World Congress on Civil, Structural, and Environmental Engineering, Prague, Czech Republic, 30–31 March 2016. Paper No. ICSENM 115.
34. Liang, Z.; Lee, G.C.; Dargush, L.G.F.; Song, J. *Structural Damping, Applications in Seismic Response Modification*; CRC Press, Taylor & Francis Group: Boca Raton, FL, USA, 2012.

MDPI
St. Alban-Anlage 66
4052 Basel
Switzerland
Tel. +41 61 683 77 34
Fax +41 61 302 89 18
www.mdpi.com

Journal of Composites Science Editorial Office
E-mail: jcs@mdpi.com
www.mdpi.com/journal/jcs